Monsoon Voyagers

THE CALIFORNIA WORLD HISTORY LIBRARY

Edited by Edmund Burke III, Kenneth Pomeranz, and Patricia Seed

Monsoon Voyagers

AN INDIAN OCEAN HISTORY

Fahad Ahmad Bishara

UNIVERSITY OF CALIFORNIA PRESS

University of California Press
Oakland, California

Library of Congress Cataloging-in-Publication Data

Names: Bishara, Fahad Ahmad, author.
Title: Monsoon voyagers : an Indian ocean history / Fahad Ahmad Bishara.
Other titles: California world history library ; 34.
Description: Oakland, California : University of California Press, [2025] | Series: California world history library ; 34 | Includes bibliographical references and index.
Identifiers: LCCN 2025004374 (print) | LCCN 2025004375 (ebook) | ISBN 9780520415911 (cloth) | ISBN 9780520415928 (paperback) | ISBN 9780520415935 (ebook)
Subjects: LCSH: Crooked (Dhow)—History. | Dhows—Indian Ocean—History. | Dhows—Persian Gulf Region—History. | Seafaring life—Persian Gulf—History. | Seafaring life—Indian Ocean—History. | Indian Ocean Region—Commerce—History. | Persian Gulf Region—Commerce—History.
Classification: LCC VM371 .B57 2025 (print) | LCC VM371 (ebook) | DDC 910.9165/309042--dc23/eng/20250604
LC record available at https://lccn.loc.gov/2025004374
LC ebook record available at https://lccn.loc.gov/2025004375

GPSR Authorized Representative: Easy Access System Europe, Mustamäe tee 50, 10621 Tallinn, Estonia, gpsr.requests@easproject.com

34 33 32 31 30 29 28 27 26 25
10 9 8 7 6 5 4 3 2 1

CONTENTS

Acknowledgments vii

Prologue: The Logbook 1

1 · Kuwait 21
Inscription: Debts 41

2 · The Shatt Al-ʿArab 47
Inscription: Freightage 67

3 · The Gulf 73
Inscription: Passage 93

4 · The Sea of Oman 98
Inscription: Guides 115

5 · Karachi to Kathiawar 122
Inscription: Letters 143

6 · Bombay 150
Inscription: Transfers 169

7 · Malabar 175
Inscription: Conversions 191

8 · Crossings 198
Inscription: Maps 214

9 · Muscat 222
Inscription: Poems 241

ACKNOWLEDGMENTS

This book has been a long time in the making. I had first thought I'd pursue it as a dissertation project but then chose to follow other routes across the Indian Ocean. In the meantime, I continued to read, ask questions, and collect material for what would eventually become this book. Throughout that process, my main guide—my nakhoda—was the inimitable Yacoub Al-Hijji, who at different times worked at the Center for Research and Studies on Kuwait. Without him, *Monsoon Voyagers* would have been impossible. His tireless work throughout the 1990s—in carefully documenting Kuwait's maritime past, in collecting and editing *ruznamahs*, in interviewing nakhodas, shipwrights, and sailors—forms the bedrock of this project; he already did it all, well before I ever thought to, and his work is much better. He is a national treasure, and I'm privileged to have benefited from his mentorship from early on.

As I began to pursue this book in earnest, I was lucky to have enjoyed access to research materials throughout Kuwait. Aslan Al-Matrook and Ghazi Fahad Muhammad Al-Sudairawi opened their homes to me from early on and have been deeply generous with their collections and family histories; I am deeply indebted to Mohammed Al-Habeeb for introducing me to them both. Ongoing publications by the Center for Research and Studies on Kuwait have made available documents relating to other families, thanks to the work of Drs. ʿAbdullah Al-Ghunaim and Faisal Al-Wazzan, both of whom deserve recognition. And ʿAbdulrahman Al-Failakawi, the son of the nakhoda ʿAbdulmajeed, has been unflinchingly supportive throughout this project. He waited as I photographed every last document he had, testing the boundaries of both his patience and generosity, and he answered every question I asked. Getting to know him and his family has been nothing short of joyful; I hope they find this book to be a worthwhile telling of their ancestors and their worlds.

My colleagues at the University of Virginia have provided me with as much moral and intellectual support as one could hope for. Paul Halliday and Josh White read parts of this manuscript early on and gave critical feedback and much-needed boosts of confidence, Brian Owensby seems to always know what questions to ask, and Chris Gratien has never failed to inspire me with his writing and his commitment to public humanities. Ira Bashkow, Emily Burrill, Indrani Chatterjee, John Handel, Kyrill Kunakhovich, Erik Linstrum, Sarah Milov, Oludamini Ogunnaike, Brian Owensby, Geeta Patel, Kristina Richardson, Joseph Seeley, David Singerman, Amir Syed—these and many, many more have made my intellectual and social life in Charlottesville truly rewarding. Special thanks go to Shankar Nair for his Persian tutorials and to Noah Salomon for being a partner in crime here at UVA in so many rewarding ways. The late Joseph Miller, whom I had the privilege of having so many conversations with during my first couple of years in Charlottesville, continues to inspire me. Murad Idris has been a friend and intellectual partner since the beginning of high school in Kuwait. T. C. A. Achintya, Madhumita Chatterjee, Soumya Johri, Robyn Morse, Ifadha Sifar, and Ayan Sharma have all inspired me with their commitment to the study of the Gulf, South Asia, and the Indian Ocean. My other students, especially those in my microhistory seminar, have helped me think through key aspects of this manuscript; Evan Richardson gave especially useful feedback. Joshua Morrison wrote the first truly enjoyable dissertation I ever read, and Scott Erich and Nick Roberts are masters of the craft.

The greatest surprise of the last eight years has been the Howell family. I would have never guessed that I'd end up in the same city and university as Nat Howell, the American ambassador to Kuwait during the Gulf War, and that he would play such an important role in my life. The Howells' support has made an enormous difference—not just for me, but for a much broader intellectual community of faculty and students at UVA. Nat, may he rest in peace, was a giant among us; he casts his long shadow over this project. His support, which lives on through Margie and Chip, has made the impossible possible. This book forms part of his legacy.

It turns out that one never loses mentors as they move through this career; they only gain more of them. My work here bears the imprint of Engseng Ho and Ed Balleisen's teachings, now more than a decade since I last studied with then. Abdul Sheriff, a fellow lover of the dhow and the Indian Ocean, has inspired me for nearly two decades now. Laleh Khalili has been a source of inspiration from the very beginning: her amazing work on shipping, her terrific public-facing writings, and her unyielding support continue to energize me. And anyone paying attention will see how Nandini Chatterjee's deep love of documents infuses my writing. I'm deeply grateful to them all, and to so many others: Rohit De, Tamara Fernando, Michael Gilsenan, Laura Goffman (who also read a version of this manuscript), Matthew Hopper, Mahmood Kooria, Mandana Limbert, Michael Christopher Low, Pedro Machado, Johan Mathew, Dodie McDow, Dilip Menon, Taylor Moore, Ghazi Al-Mulaifi, Michael O'Sullivan, Kalyani Ramnath, Emma Rothschild, Julia Stephens, Justin Stearns, Lindsey Stephenson, Lakshmi Subramanian, Itamar Toussia Cohen, Nancy Um, and many more. My good fortune is, truly, in the community I've found.

Many others have read and given feedback on various aspects and versions of this book. Emily Silk at Harvard University Press gave incredibly incisive comments early on; the book has much more of a sense of purpose because of her. Arash Khazeni allowed me to present very early thoughts on the project at Pomona College and also introduced me to Niels Hooper at UC Press, who very enthusiastically took the project on. Scott Reese and Eric Tagliacozzo both served as reviewers for the manuscript; their feedback was both encouraging and enormously helpful. And I've benefited from feedback from participants at the Borderlands Colloquium at UVA, the International and Global History Workshop at Columbia University, the Yale Global and International History Workshop (special thanks to Laurie Benton and Alan Mikhail), and from colleagues at NYU Abu Dhabi. And at UC Press, Niels Hooper and (especially!) Nora Becker have been fabulous.

As for Ahmed Al-Maazmi: what is there to say? There aren't enough languages in the world for me to communicate how much he has meant to me. He has read every word I've written here (but for this paragraph) and has pushed me to clarify, connect, and contextualize in ways that only the most astute and engaged reader could. To say that I'm grateful is an understatement. I look forward to many years of collaboration with him.

Throughout this project, there have been people who have sustained other aspects of my life, who have been alongside me during my journey. My mother has put up with me being too far away, and for far too long. My sisters Shahad, Reem, and Farah, my brother-in-law Khaled, and my nephew Ahmad all keep my Kuwaiti home alive (sometimes a little too alive). My friends in Kuwait (really, brothers) have saved my spot in the *diwaniya* over the many years I've been away. And in Charlottesville, John Boyd, Joe Young, Zane Havens, Stacey Carter, and Jeremy Layell all filled out my musical world. The Islamic Society of Central Virginia families, the Livingston family, and the Cempre family—especially Chris—have given me the gift of community. And Carl Briggs always made things just a little easier; he also always asked about "the book."

My wife, Rose Buckelew, is the greatest teacher I know. She has sharpened my thinking on so many things and has moved me to be more active around the university and within our community. Her commitment to pedagogy and to justice—and she never thinks of them as being separate—is a constant source of awe and inspiration. She's also a constant source of support, to me and to many others. I am lucky to have her in my life. I dedicate this book to our two children, Jossem and Tala, both of whom have anchored me whenever I've felt lost. Jossem has kept me on my toes: in addition to thrusting me toward new adventures, he asks really great questions (sometimes too many of them!). Tala never gets tired of my dad jokes, and I never get tired of her hugs. They are both a source of unalloyed joy in my life.

Speaking of dads, I also dedicate this book to two other people who are no longer around: my father and my grandfather, both of whom loved the sea. They both, in their own way, instilled in me an appreciation for a world beyond the borders of Kuwait. My grandfather, the nakhoda ʿIsa Bishara, has inspired me in ways that are so obvious I hardly need to list them. My father listened to me talk about this project and fed me ideas from early on; he would have been eager to see it committed to print. His light continues to guide me through this world—through my scholarship and fatherhood. Everything I've done is a tribute to him.

Prologue

THE LOGBOOK

THE PAGES OF THE LOGBOOK belonging to the dhow captain (*nakhoda*) ʿAbdulmajeed Al-Mulla Ahmad Al-Failakawi are yellowed and dusty with age; its marbled covers are slightly brittle, asking to be handled with care. It was nearly a century old by the time I saw it; and for much of that time, it sat wrapped in a linen cloth inside a Malabar teak chest tucked away in a house in a quiet suburb of Kuwait.

It was not the only one in the chest. There were three logbooks—called *ruznamahs*—each covering a stretch of time. They logged twenty-six voyages in total, running between 1920 and 1944. The one in front of me was the first of the three *ruznamahs* and announced its origins on its first page. "Bought in Calicut for Rs. 1, with the *baghla* of ʿAli bin Muhammad," it declared, continuing: "In the name of Allah, the gracious, the merciful: it entered into the possession of ʿAbdulmajeed ibn Mullah Ahmad, the son of Hajji Hassan Ka-Ibrahim, who lived on the island of Kharj [*Khārī*], an island between Bushehr and Basra, attached to Bandar Riq, [in] the Muhammadan-Prophetic year 1341, the Christian [*ʿIssawi*] year 1923, on the 23rd of Rajab [March 11]." A name, a date, and two places of origin—one for the *ruznamah* itself, and the second for its author, the nakhoda ʿAbdulmajeed. He tells us his grandfather Hassan lived on Kharj Island; however, he grew up on Failaka, a small island off the coast of Kuwait, from where the family gets its name. He was born on the island in 1901 and worked as a sailor, slowly moving up the ranks until his first voyage as a nakhoda in 1924.[1]

The *ruznamah* invites its reader into Al-Failakawi's world and travels but exacts a heavy price of admission. The nakhoda's handwriting is not particularly forgiving. At times, it is clear, evenly spaced, and elegant; at other times, sentences look more like a logjam of words that are difficult to distinguish

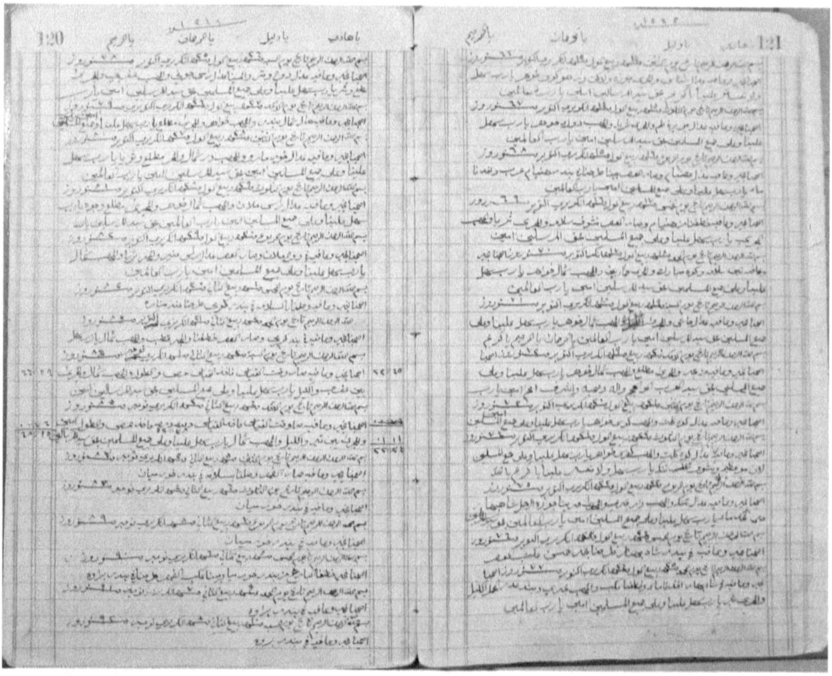

FIGURE 1. Pages from Al-Failakawi's *ruznamah*.

from one another. So many of the words were completely foreign to me, too: they were Arabic, but a nautical vernacular that had long since disappeared. Even for the patient reader, armed with a dictionary of nautical Arabic, the rewards are not altogether clear. Al-Failakawi's entries are terse, and he mostly limits himself to observations on wind, water conditions, and location. There is none of the excitement and adventure one imagines they'd find in a logbook—none of the daring action or thrill that we associate with life at sea. Put plainly, the *ruznamah* is boring.

And yet, in its heyday, the *ruznamah* had a remarkable career. It passed along the coasts of Arabia and Persia, India, and parts of East Africa. It weathered stormy seas, sat in the shade of bountiful date trees, and was carried through bustling marketplaces. It was taken on and off the deck of the dhow, in and out of different chests, and moved between different people in different places before it ended up in its final resting place, in the Malabar teak chest.

I hoped to read this world back into Al-Failakawi's *ruznamah*. But before I could begin, as I sat there in that suburb with the *ruznamah* and the contents of the Malabar chest spread out before me, I had to grapple with a more pressing question. How was I supposed to read this thing?

• • •

Al-Failakawi's *ruznamah* wasn't the first I had ever seen. I was home in Kuwait, back from college for the summer, when someone delivered a cardboard box full of books to our front door. The books were large, each roughly the size of a sheet of printer paper, and thick—around 650 pages long. Each had a glossy cover—a combination of fleshy pink and forest green I would come to know all too well years later—announcing that this was the *ruznamah* of the nakhoda ʿIsa Yacoub Bishara. ʿIsa Bishara was my grandfather, and though I had long known that he spent his early years as a dhow captain like his father and grandfather, and their fathers before them, I knew next to nothing about that world. Nor did the published *ruznamah*, carefully edited and clearly presented by the maritime historian Yacoub Al-Hijji, do much to help me understand it. At least not then.

Part of the problem was that I simply did not know enough about history, aside from what I learned in high school. But really, the main problem was I simply did not know enough about *that* history. The courses I had taken and books I had read on Middle Eastern history completely sidelined the Gulf. If it ever appeared—and it often did not—it was only in the context of the advent of the oil industry. The pre-oil Gulf was entirely absent from the picture. For the community of Middle Eastern historians writ large, it still is almost as if the region didn't exist before oil.

There was, of course, plenty of work by Gulf historians on the region's pre-oil past, but very little of it did anything to disabuse anyone of the notion that this was an entirely peripheral arena of a terrestrial Middle Eastern history. Studies that narrated the region's past from the vantage point of different empires—the Ottomans, the Portuguese, the Dutch, and above all, the British—did much to highlight how local rulers and tribes harnessed a changing imperial world to shape their political trajectories and how the region gradually emerged as a chessboard for the power politics of global empires.[2] And yet, this was a conversation that played out at the margins of Middle Eastern history; hardly any of it would register in the debates that shaped that field, much less make it into its textbooks. With so few historiographical rails to grasp, it is no wonder I felt so disoriented when reading the *ruznamah*, which wove together the Gulf, India, South Arabia, and East Africa.

To read the *ruznamah*, I start from the premise that the history of the Gulf only partly plays out in the region itself; it was a region whose "very boundaries reach out across the seas."[3] If I argue anything at all in this book, it is that the history of the Gulf can be found in the Middle East, but must equally be sought after in places like India, South Arabia, and East Africa. To thread the needle through the Gulf and its Indian Ocean world, I chart out a transregional society of merchants and mariners operating out of different ports around the ocean's littoral, from Basra to Zanzibar and beyond, between the late nineteenth and mid-twentieth centuries. I stitch their story into a longer history of states and empires around the Indian Ocean world, overlaying it onto the circuits of scholars and statesmen who moved along the coasts of the Arabian Peninsula and across the waters of the Arabian Sea. Gulf history, I argue, is Indian Ocean history—and if historians of the Middle East want to take the Indian Ocean seriously, then they need to reposition the Gulf within their geohistorical imagination.

But what do I mean by Indian Ocean history, and what does it entail? As a field of study, oceanic history issued a direct challenge to the continental paradigms that had straitjacketed historians' thinking for so long. This wasn't immediate, of course; rather, it began as a somewhat muffled, overly cautious suggestion that historians might find more interesting histories if they looked across the water or along the shore than if they turned inland.[4] Then, as the global history craze swept the discipline, it received a shot in the arm. Historians joyfully abandoned the nation-state to turn seaward, charting out connections—Chinese pottery in Africa! African slave-governors in India! Indian scholars in Arabia!—that spoke to a dynamic world of mobility and exchange.

But Indian Ocean history is more than just connection for connection's sake. The promise of an Indian Ocean history lies in the space it gives the historian to think about world history from the vantage point of those who were long thought to be mere recipients of processes that originated elsewhere—namely, Europe. By writing world history from the Indian Ocean, the historian can anchor that history in an arena teeming with long-standing connections, institutions, and ideas; they can write that history "from the other boat."[5] And by drawing on different concepts, vocabularies, and ways of seeing, thinking, and being, Indian Ocean history offers a rich position from which the historian might put the Eurocentric world histories of old in their place. Rather than replicate older world historical narratives in which there was everything before European contact and then everything that came after it, the best Indian Ocean histories see European empires as one piece of a world that is already in motion.

And this is where we might begin to think of the history we are embarking upon here as being an oceanic history of the Gulf. By envisaging a Gulf history that is strewn across the littorals of India, South Arabia, and East Africa, I am making the claim that these communities were always shaped and reshaped by contact with groups from other parts of the Indian Ocean world, and that they were able to incorporate other people's languages, artifacts, practices, and signifiers into their cultural repertoires. Doing this requires that we first decouple society from the nation, and even more so from the state—and upon doing that, see how it spreads itself thinly across the vast expanses of the Western Indian Ocean without losing its internal density.[6] It then asks that we think of Gulf society as part of an open system rather than a closed one—a world in which merchants, mariners, nakhodas, scholars, and statesmen share dense business, social, political, and blood ties with individuals and groups in Persia, India, South Arabia, and East Africa, and in which distant markets for timber in Calicut, dates in Karachi and Aden, and mangrove poles in Zanzibar shape activity in places like Basra, Bahrain, and Kuwait.

"World" history matters here too. From their outset, the maritime communities that dotted the coasts from Basra to Ras Al-Hadd on one littoral and as far east as Gwadar on the other had to contend with and adapt to the presence of a series of empires, trading companies, and other quasi- and inter-imperial political projects, both European and non-European. The maneuverings of these empires through and along the Gulf littorals alternately opened up, closed off, expanded, or contracted—but always reshaped—the commercial and political arenas; political entrepreneurs along the coasts postured, positioned, and otherwise contorted themselves to seize the opportunities that the changing tides of trade and politics brought with them.

And so, what we are presented with is a history that is scattered, connected, and entangled, in which different communities, institutions, and histories bleed into one another, and in which it is difficult, if not wholly impossible, to capture an "original" moment. How might we tell a story like this one?

• • •

We might begin with a single entry from the *ruznamah* itself, to see this world in a few strokes of the pen. And we might as well begin with the first:

"In the name of Allah, and with our trust in Allah, in the name of Allah most gracious and merciful, on Monday the 10th of *Rabiʿ Al-Thani*, 19 December *Ingrīzī* [English], 101 *Nawruz*, we arose in good health and spirit, and boarded our *boom*."

It is hardly a colorful start to the logbook and exemplifies the sort of brevity that is typical of Al-Failakawi's entries in his *ruznamah*. And yet, mundane inscriptions like these tell us more than they let on. The ways in which the nakhodas marked time in their logbook entries are, on their own, revealing. Each entry corresponded with a day and invariably marked a position in one sort of space or another—a point on the coast or out on the open seas. On the face of it, the *ruznamah* appears to be the simple plotting of a voyage along linear time. But as nakhodas plotted their movements around the Indian Ocean, they used three different registers of time: the Hijri and Gregorian calendars, but also the *Nowruz* calendar, which is called the *Nairuz*.

There was a logic to it, of course, as each calendar fulfilled a particular function. The *Nowruz* calendar, which runs roughly 365 days a year, might be thought of as plotting the time of nature: the ascent and descent of different stars, the movement of the sun, and the different wind systems they had to contend with. This was the temporality of the voyage itself, which was simultaneously immediate and infinite: immediate in that it reflected the most pressing concerns the nakhoda had to deal with on a day-to-day basis, and infinite in that weather, wind, and water followed a regular pattern that the nakhoda could predict from one voyage to the next.

By contrast, the Hijri calendar, a lunar calendar that usually ran 354 days a year, mapped onto the different contracts, letters, rituals, and festivals that marked the dhow's voyage: the time of the oceanic marketplace. Business activity in this world was minutely choreographed: it required the careful attention of a range of different actors, and rested on a dense corpus of writings, agreements, accounts, bills, receipts, and other notes. Trade routes could connect places to one another; but for circulation to occur, there needed to be much more than just ships, merchants, and routes.[7] Trade necessitated the coordination of different actors through different forms of communication and recordkeeping, and through partnership and profit-sharing institutions. All of these were firmly plotted along the Hijri calendar, the calendar of the Indian Ocean marketplace.

Then what of the Gregorian calendar? Why would someone like Al-Failakawi mark it out when he already had two other calendars to rely on?

The answer may lie in the terminology he used to describe it: the *Ingrizi*, or English, calendar. Wherever the nakhodas sailed, they had to contend with the presence of British imperial actors: customs officials, political agents, naval officers, and more. They had to also contend with the artifacts of this imperial regime, manifested in bureaucratic documentation like receipts, permits, passes, and ship manifests, but also nautical manuals and other writings they actively engaged with. Many of these ended up in Al-Failakawi's chest, alongside his *ruznamah*. We might therefore imagine the *Ingrizi* calendar as signaling the time of empire—in Al-Failakawi's time, the British Empire, and in other times the Portuguese, French, Dutch, and other European empires in the Indian Ocean.

Voyage, marketplace, and empire: three temporal horizons, all plotted alongside one another in the *ruznamah*. Port by port, nakhodas and mariners felt the beating pulse of nature, of business, and of sailing, and saw the moments in which they converged with the syncopated rhythms of industrial capitalism, set by movements of steamships, telegraphs, and finance. All unfolded along different scales as well. At the microhistorical scale lies the voyage that forms the narrative core of this book, along with its attendant texts and artifacts. At the meso-level—the temporality of the marketplace—lies the world Al-Failakawi sailed through: the nakhodas, mariners, merchants, brokers, technologies, and institutions that produced the circulations that animated the Indian Ocean world during his time. And the macrohistorical scale maps onto the temporal horizons of empire—the British Empire, yes, but also past empires come and gone, and the local and regional political formations that emerged alongside them. Telling the story of Al-Failakawi and his world requires that we write across these different scales—all of which were present in the Malabar chest, if not in the *ruznamah* itself.

. . .

At the center of the voyage is Al-Failakawi himself, the nakhoda and author of the *ruznamah*. It is through Al-Failakawi's travels that we come to know the world of the dhow trade in the Indian Ocean, and it is through his writings that our understanding of that world is mediated. In the microhistory that lies at the heart of this book, Al-Failakawi is the protagonist.

In some ways, Al-Failakawi is not the ideal candidate for a good microhistory. Unlike the heretics, impostors, tricksters, and other idiosyncratic figures that have occupied center stage in the best microhistories, Al-Failakawi

FIGURE 2. The nakhoda ʿAbdulmajeed Al-Mulla Ahmad Al-Failakawi. Photo from ʿAbdulrahman Al-Failakawi.

was, in many ways, quite typical of nakhodas of his time.[8] His voyages, and the routes he traversed, were virtually identical to those of his contemporaries. I'm not suggesting that he's uninteresting; rather, he is about as ordinary of a nakhoda as they come.

And yet, it is because he's ordinary that Al-Failakawi is a perfect candidate for a microhistory. If microhistory, unlike biography, takes as its subject the more marginal figures of history, then a microhistory set in the Indian Ocean might as well feature a nakhoda.[9] Though there were thousands of other nakhodas and dhows that traversed the Indian Ocean—and hundreds in Al-Failakawi's lifetime alone—they almost never appear in histories of the region. They wash out into the background, acknowledged but not engaged with. Nakhodas and their crews are, at least in the English-language scholarship, "the lost peoples" of Indian Ocean history.[10] And even for the few historians who are familiar with the pantheon of Arab nakhodas, Al-Failakawi does not loom large in the imagination.[11]

But we might be misdirected in our focus on the individual as the subject of microhistory. We cannot confuse the point of microhistory with its scale; scale is the means to an end, not an end in itself.[12] The hope is that by reducing the scale of analysis and narrative, we might be better positioned to comment on broader phenomena—historical, historiographical, or analytical. From the vantage point of the micro, the historian can shed light on macro-level processes, intervene in debates, and ask their readers to think more closely about the categories they mobilize to understand the past and what they might have meant for their historical subjects.

To bring it back to nakhodas: while Al-Failakawi might be the main protagonist in this story, it might not be the nakhoda himself we're interested in—at least as historians. Compelling as he may be, he is not a particularly good candidate for a microhistory. But his *ruznamah*, and the expansive world it opens onto, might be.

. . .

Al-Failakawi's routes bore the weight of centuries past—of empires that vied for control of port cities and mastery over waterways. From at least the medieval period onward, states and empires around the Indian Ocean created built environments that sought to monitor and process the movement of ships—sometimes out of a desire to tax them, and other times out of a fear of attack by rivals. The arrival of European empires added another layer onto regional contests, as local chiefs, political entrepreneurs, and mobile communities of traders entangled the threads of global empire into their own projects. Around the Indian Ocean, global imperial contests anchored themselves in local contests over ports, forts, and waterways. Al-Failakawi's seascape accumulated the detritus of that history. Crumbling remnants of forts, dilapidated palaces, and abandoned customs houses littered the coastlines of Arabia, India, and East Africa, infusing the nakhoda's itineraries with the palimpsests of past struggles over the mastery of the seas.

The nakhoda knew well, though, that though the British Empire was only the latest in a long string of aspiring maritime hegemons, it arguably had the greatest role in reconfiguring the seascape he traversed. Over the nineteenth and twentieth centuries, Al-Failakawi and other inhabitants of the Indian Ocean world witnessed the emergence of an expansive transoceanic British imperial regime designed to track the maritime flows of goods and people. It emerged out of an imperative to combat "piracy" and slave trafficking but gradually came to be fueled by a desire to clear the way for a growing steamship-based transportation system. Nakhodas and their crews very quickly found themselves caught within British official crosshairs. Their comings and goings fell under British surveillance, and their cargoes, both human and nonhuman, were regulated in courtrooms and customs houses. By Al-Failakawi's time, dhow movements had for decades been punctuated by lighthouses, permits, quarantines, and reams of bureaucratic paper.

But Al-Failakawi's world of the twentieth century was no longer just imperial; it was colonial as well. Around the Gulf and Indian Ocean, officials from the Government of India staffed consulates and engaged in everyday administration. They wove themselves into regional diplomatic channels, crafted regulations, issued paperwork, regulated traffic into and out of the ports, and intervened in local disputes. These officials formed part of the local society, too: their wives and children often lived with them, shopped in the local marketplaces, and formed relationships with their neighbors, even if they mostly socialized with other Europeans. The British Empire, then, was not just something "out there" across the water and on the seascape; it was also very much a part of Al-Failakawi's everyday life at home.

British imperialism reshaped everyday life and movement in the Indian Ocean through other means too. Steamships and telegraph lines linked the Gulf and Indian Ocean world up to global circuits of information and commodity circulation. They also left their own imprint on the region's coastlines. As engineers dredged ports and harbors and built new facilities to accommodate a growing number of steamships, and as they laid submarine cables and linked them to overland telegraphs, they actively transformed the economic geography of the Indian Ocean world.[13] Alongside crumbling forts were modern port infrastructures that could breathe new life into old port cities like Aden, Basra, or Mombasa. Just as often, they created altogether new shipping hubs, like Karachi and Bombay, neither of which enjoyed much of a pedigree in the Indian Ocean world. Maritime mobilities in the Indian Ocean world had become channeled through new nodes.[14]

Other phenomena moved across Al-Failakawi's reconfigured maritime world, alongside the global networks of transport and communication. News of the world pulsed through the Indian Ocean, spilling out onto the pages of newspapers and journals that circulated between publishing centers in the Eastern Mediterranean, India, East Africa, and the Gulf. At the same time, printers around the Western Indian Ocean—in Basra, in Bombay, in Zanzibar—churned out new works of literature and scholarship and reprinted old ones, animating classrooms and coffeehouses around the region. As journalists and scholars rose to prominence, they also traveled the same circuits, giving lectures to rapt listeners and enjoying audiences with statesmen and patrons from Bombay to Zanzibar and beyond.

As we connect spaces and accumulate a sense of history, a kaleidoscopic picture of the world of the Gulf and the Indian Ocean begins to emerge, one of diverse local, regional, and transregional endeavors that were constantly changing over time. The voyages of the dhows, and the trading worlds they traveled through, were shaped by local and global events and rivalries originating in widely scattered parts of the Indian Ocean world. There was no overarching "global" context to their movements: only a series of intersecting local and regional contexts, many of which bore the imprint of a global past and present. By sailing through this world with the dhow, we link otherwise divergent interests and historical narratives together and entangle them into a textured transregional history.

• • •

Over the course of the *ruznamah* and the voyages it narrates emerges a thick account of Indian Ocean trade. Al-Failakawi did not drift aimlessly from one port to another; he sailed with purpose. To more fully grasp why he followed the routes that he did—and indeed, why the dhow would have been constructed in the first place—we had better appreciate the texture of the Indian Ocean marketplace, and the different actors, institutions, practices, and technologies that gave it its transregional shape. As Al-Failakawi sailed from one port to another, he engaged with the individuals who animated this particular Indian Ocean world of trade: landowners in Iraq, traders on the Persian coast, Arab and Indian merchants and brokers along the Western coast of India, Kuwaiti merchants in Aden, and Hadhrami middlemen on the coast of East Africa. We will come to know many of them—but especially Muhammad bin ʿAbdullah Al-Matrook, the Basra-based date trader, and Muhammad Salem Al-Sudairawi, the Bombay-based banker, both of whom actively corresponded with an enormous and far-flung world of commercial associates stretching out across the Western Indian Ocean and beyond. Between these actors circulated all manner of paper: letters, account ledgers, contracts, receipts, bills, and other ephemera. And though Arabic constituted the prevailing lingua franca, letters appeared in Persian, in Arabic written in an Urdu hand, and occasionally in Gujarati and Karnataka—and, once in a while, in English as well. Virtually everyone discussed the same sorts of goods, circulated the same kinds of notes and accounts, drew on the same stock phrases, and switched between different registers of commerce with surprising ease.

Dhows, nakhodas, letters, merchants, accounts, bills, brokers: these were all expressions of a much larger phenomenon in world history—the bazaar economy, a broad and deep marketplace that both ran alongside and intersected with the markets of European colonial capitalism.[15] Individuals like Al-Failakawi and his fellow nakhodas and merchants all maneuvered within the bazaar's capacious business culture: its weights and measures, its gradations and standards, and the financial practices that converted capital from ship to shop and back. This was a transregional economy that found expression in objects, technologies, money, labor, and nature—a world animated by a supple conceptual vocabulary and a shifting institutional landscape.[16] The bazaar was a site of capitalism around the Islamic world, a market like any other—but one that came with its own terminologies, ontologies, and genealogies.

Materials from this bazaar world, scattered throughout Al-Failakawi's writings, practically beg us to make sense of its place within the grand narratives of world economic history. For decades, economic historians have struggled to situate the trading worlds of non-Europeans within the history of global capitalism. Though they've come to see "the rise of Europe" as a more contingent phenomenon and have called into question the assumption that Europe clearly eclipsed the rest of the world—at least before 1800—they broadly agree that in the period following the Industrial Revolution, Asia had become yet another victim of the world-capitalist juggernaut, though the when and the how of it all is still up for debate.[17]

There is less doubt when it comes to the dhow trade. Gulf merchants and nakhodas are thought to have operated on the margins of the world market, multiple horizons away from more established global economic actors. By most observers' measure, theirs was a world that had died at least two deaths, if not more. For some, it was the coming of the Europeans, beginning with the Portuguese and continuing into the European trading company era and their encroachment upon Asian trading prerogatives. For others, a more decisive death blow came with the arrival of industrial capitalism by way of the steamship, which was thought to have completely displaced dhows and their merchants—if not by their technological superiority, then by the cartels, subsidies, and coercions that ultimately pushed dhows to the margins.[18]

Few could comprehend the persistence of nakhodas like Al-Failakawi and his community of merchants and mariners in the face of these global transformations. "If the Kwaiti booms and their Indian fellow-vessels continue to cross the Arabian Sea, they do so, it could be said, either as part of a backward economy or because their owners are not rational beings," wrote the pioneering Indian Ocean historian K. N. Chaudhuri.[19] Others were perhaps more charitable: one chronicler of the dhow framed its history as one of survival against adverse regulations, technological transformation, and "modernization in both material and mind"—a survival that owed much to the ingenuity and resourcefulness of its captains, who managed to find a way to eke out a living at the margins of a globalized economy.[20]

And yet it is the very position of actors like Al-Failakawi and the communities he was a part of that allows us to see just how the broader phenomenon of global capitalism emerged out of translocal business practices, stretched out across transregional spaces. The history of global capitalism, scholars have argued, cannot be reduced to the ways in which the logic of capital reproduces itself; it must account for the ways in which the diverse ways of being human interrupt the narrative.[21] Put differently, "capitalism" is not a universal, transhistorical phenomenon: it was always grounded in particular times and places, anchored in specific ways of doing and knowing. Around the world, different actors appropriated the technologies and artifacts of empire and industrial capitalism, incorporating them into their commercial and cultural repertoires and endowing them with new forms of meaning; they actively domesticated and vernacularized them. In place of a narrative in which global capital completely subsumed communities and individuals, we've come to appreciate that these moments of transition opened up new horizons and historical trajectories, and ultimately produced new "capitalisms" altogether.[22]

And so, rather than approach Al-Failakawi's world from the outside in—that is, rather than write a history in which a growing world of corporations, steamships, telegraphs, and banks constrain or suffocate the economic lives of Gulf merchants and mariners in the Indian Ocean—I instead want to flip the script. In this history, these communities recognized the possibilities offered by the infrastructures of empire and industry and used them to their advantage: they channeled them toward their own ends, through their own technologies and vocabularies. Their bazaar world was not a relic of the past, nor was it a localized peddler's market: it was broad, dynamic, and adaptive to the demands and opportunities of its present.

The dhow was deeply embedded in this world—in the weights, measures, contracts, and accounts that animated it, and in the circulations of people, commodities, and credit that pulsed life through it. Its movements inscribed and reinscribed those circulations into routes, forming alternative geographies and animating histories. A voyage on a dhow is a voyage through the world of the bazaar itself.

• • •

There is another *ruznamah*, virtually identical to the first. It also bears the name of ʿAbdulmajeed Al-Failakawi, but with an earlier date: 1339 AH, or 1920/21. Of the three *ruznamahs*, this one was by far the most interesting and invites the reader deep into the nakhoda's world. Unlike the one we just saw, this *ruznamah* offers little by way of itinerary: there is no colophon telling us when and where it entered into the nakhoda's possession, and the only voyage it details only picks up halfway through, on the return leg, beginning in Porbandar, India, and ending in Kuwait. It's a strange thing, to have the log of a journey start only on the return leg; almost as strange as the fact that Al-Failakawi would have been just nineteen years old at the time.

More curious, though, are the rest of its pages, which are full of notes, diagrams, charts, tables, and calculations. What comes across is less a logbook than a notebook; a place into which the nakhoda inscribed various observations, formulas, and helpful signposts—even stanzas of poetry. But why? Why would the nakhoda need to remind himself of things he already knew?

Al-Failakawi's isn't the only voice in that *ruznamah*. The handwriting in which the notes are inscribed is neater, more calligraphic, than the handwriting for the voyage entries. It doesn't take a paleographic expert to see that these are two distinct hands, belonging to different people.

I didn't realize this myself until I saw a yet another book, one published in 2006: *Al-Qawāʿid wal-Mīl wal-Natīja fī ʿIlm Al-Baḥar*, or *The Principles, Declinations, and Almanac in the Science of the Seas*. It was even more of a hodgepodge than the title suggested. There were principles, yes, but it was hardly a treatise; it would not even pass muster as an almanac. Instead, it contained a loosely connected series of observations, discussions, notes, and templates—and like Al-Failakawi's notebook, even poetry.

The notebook's author, its colophon declared, was Mansur bin Al-Hajj Ibrahim Al-Khalil, "inhabitant of the island of Failaka, and originally from the people of Khariy Island, of the coast of the Persians [the *ʿAjam*]"—the same island Al-Failakawi declared to be his ancestral home. In fact, Mansur came to be known as Mansur Al-Khariji, taking that placename as a surname, as so many often do. We know less about Al-Khariji's life: he was born in the late 1870s and moved to Failaka at age eight. He tells us he learned basic navigation from his older brother Ali, with whom he began sailing when he was in his late teenage years. Al-Khariji spent the next several decades on and off the dhow, compiling his notes piecemeal over roughly two decades. In one of his last voyages, Al-Khariji, who was by then almost completely blind, ferried the British traveler Wilfred Thesiger from Abu Dhabi

FIGURE 3. The nakhoda Mansur Al-Khariji aboard his dhow, as photographed by Wilfred Thesiger in 1945. Photo from Pitt Rivers Museum, Oxford, UK.

to Bahrain, traversing routes he could no longer see but which were etched deeply into his mind.[23]

The handwriting throughout Al-Khariji's *Al-Qawāʿid* bore a striking similarity to the notes in Al-Failakawi's notebook. In fact, one might reasonably guess they were all written by the same person.

When ʿAbdulrahman Al-Failakawi inherited the Malabar teak chest from his father, he was only seven years old and had little interest or understanding of what the chest contained; it was mostly full of old papers. His mother, however, knew to hold onto the chest; she knew the value of the logbooks, old maps, accounts, nautical instruments, and a host of other books, pamphlets, and ephemera her husband had accumulated over the course of his years at sea. Like many other wives and mothers around the Indian Ocean, she preserved the family's capital. And like many others, she joined men to one another too. Her own father was once a nakhoda; in fact, ʿAbdulrahman suggested, I may have heard of his grandfather as well. His name was Mansur bin Ibrahim Al-Khalil, better known as Mansur Al-Khariji, and he had trained the nakhoda ʿAbdulmajeed in navigation. Did the name ring a bell?

• • •

The presence of both Al-Failakawi and his teacher Mansur Al-Khariji's inscriptions in the *ruznamah* tells us the story of knowledge passed from one nakhoda to his protégé (and later son-in-law) and of the ways in which one learned to become a nakhoda. The *ruznamah* hints at what Al-Khariji thought Al-Failakawi might need to know in order to undertake his own voyages. Theirs was not a world of learning by doing alone; it also involved learning by reading, by studying, and by committing principles and formulas to memory. To that end, the *ruznamah* was more than just a book into which nakhodas inscribed their voyages. It was an artifact with a life of its own, passing from one person to another as the captains moved up the professional ladder and trained others.

The notes Al-Khariji passed down to his student open onto a world of texts and concepts that nakhodas engaged with over the course of their voyages. Some of these were texts written by nakhodas, for other nakhodas: treatises on navigation, manuals on trade, almanacs, and short pamphlets. Other notes involved forms of writing that nakhodas mobilized over the course of their voyages: debt contracts, safe conduct passes, money transfers, accounts, and other similar graphic artifacts. Together, they suggest the possibility of an intellectual history of nakhodas and dhow communities grounded in their own conceptual frameworks, emerging out of their own genres of writing. Theirs was a world of ideas, some internal and others drawn from a dynamic world of printing and publishing that swirled around them, one in which the nakhodas were both prolific producers and voracious consumers.

The notes in the *ruznamah* point to the interface between a sea of texts and concepts on the one hand and a world of maritime praxis on the other, joined together in the person of Al-Failakawi—and, peering over his shoulder, Al-Khariji. Through the notes, we can better appreciate the nakhodas as thinkers: as readers of books on nautical knowledge, and as theorizers of maritime and commercial practices. In the *ruznamah*'s notes, the oceanic marketplace comes across not as a disembodied arena of forces but as a phenomenon grounded in specific transactions, contracts, mathematical formulas, and genres of writing that enabled the connections, circulations, and conversions that propelled goods from one coast to another. Empire, too, becomes less a historical abstraction than a specific set of institutions, artifacts, and practices, all of which nakhodas and merchants were able to draw into their repertoires.

From the *ruznamah*, we peer onto the intellectual labor that made the Indian Ocean world possible: to better appreciate the process by which nakhodas came to know their oceanic worlds, and to more clearly see how nakhodas learned to "navigate" this world both physically and conceptually. From the micro-level, then, we come to better grasp the meso- and macro-level processes that conjured up the Indian Ocean marketplace as we know it. With the histories of commodities, trade, and empires in the Indian Ocean, we interweave a history of ideas.

• • •

"There has to be an argument here, Fahad." But does there?[24] Part of what I want to show, through Al-Failakawi's travels, is that an argument can be communicated through narrative, not just in direct terms—that we can appreciate the thickness of Al-Failakawi's world through his movement within it. Through Al-Failakawi's travels and his and Al-Khariji's writings, we get a grasp of the relationships, the technologies, and the intellectual toolkit necessary to forge connections and enable circulations around the Indian Ocean, year after year. What we end up with is a picture of how this world worked, occasionally colored by moments that show us how it broke down and picked itself back up.

But the narrative is the argument too. Throughout this book, I paint a picture of the nakhodas' life worlds: of the geographies they inhabited, the horizons they imagined, the worlds they moved through, in all of their texture—or as much of it as I can give here. Doing that is an argument in itself, a reminder that the people we write about lived textured lives that don't easily map onto the divisions by which we carve up the world today.

For those looking for arguments, I follow each chapter with an inscription in which I take a note by one of the nakhodas—usually Al-Failakawi—and use it to think about the broader world of ideas, concepts, and texts they inhabited, and their implications for how we understand bigger themes in world history: law, empire, and capitalism, to name a few salient ones. Readers who are uninterested in the more historiographical dimensions of Al-Failakawi's world can skip past the inscriptions. Because why get in the way of a good story?

• • •

Our voyage commences in August 1924, at the beginning of the sailing season.[25] We start with Al-Failakawi's dhow itself, which may have once been called *Fateh Al-Khair*, or the *Triumph of Good Fortune*, but was more colloquially known as *Al-'Away*, or *Al-A'waj*—the *Crooked*. We will look out from its deck, listening for the bustle of activity around the port town at the beginning of the sailing season and seeing the crowds of sailors standing by ready to push it out into the water. We will then hoist up its sails and let the winds fill them, allowing ourselves to sail out to sea, beyond the horizon, past the point when the coastline of Kuwait is but a faint glimmer in the distance.

Sailing out, we will gather histories that have been scattered. We move from port to port, traversing shifting political boundaries and crossing contested bodies of water as we trace new geographies. We connect date gardens in small towns along the Shatt Al-'Arab to markets in Gujarat and Bombay, timber traders in Calicut to brokers in Aden, and sultans and moneylenders in Muscat to Yemeni mangrove pole traders in East Africa, and meet a far-flung cast of captains, commanders, pirates, slaves, and mariners whose movements helped forge the contours of this world. This is a history in which small villages bridged their fortunes to distant cities, and in which the lives of individuals, families, and entire towns were inextricably intertwined with others across the sea. I unfurl these histories and themes slowly, over the course of an entire voyage, even if Al-Failakawi might have felt them all everywhere, all at once.

The process of gathering these histories involves picking up loose sheets of paper, stray letters, strewn manuals, and scattered reports, and stringing them together into a legible whole so that we might understand each part differently. But we won't stow these gathered materials in the belabored container of the nation, however tempting it might be to do so. Rather, we gather them onto the dhow's deck: we look out onto the land from the sea but resist the temptation to write a land-based history. We weave a narrative of histories formed across the water—histories forged in the act of crossing and animated by circulation. In a world of scattered histories, thinking with the route and writing along it helps gather them back together into a coherent whole.

Voyaging—what nakhodas called *al-safar*—constituted an enactment of history, threading together both time and space. Unpacking the voyage, then, involves sailing alongside the nakhodas around the ports of the Western Indian Ocean and stitching together different spaces. But it is also a voyage of discovery into the nakhodas' imaginaries—into how they and others like them actively thought about the world around them, in their daily entries, in the margins, and in other texts altogether. As we sail around the Indian Ocean, we will unravel the text of the *ruznamah* along the way. On this voyage, the past appeared in the present, and the old could run in parallel with the new without interfering in it.[26] In the act of the *safar*, routes of the present always bore the sediments of the past.

In this history of infinite pasts, among the hundred horizons, we have to begin somewhere. It might as well be Kuwait.

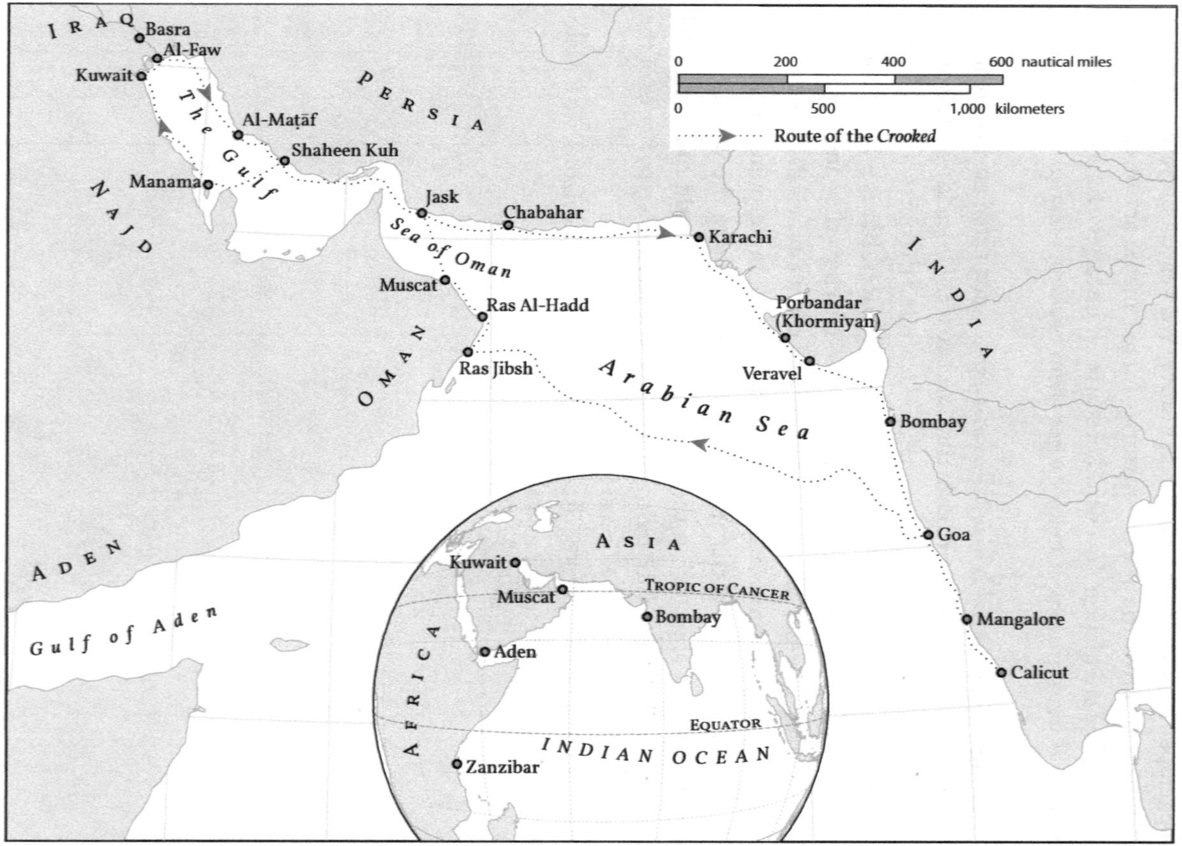

MAP 1. The Western Indian Ocean, including the route taken by the *Crooked*. Map produced by Nat Case, INCase LLC.

ONE

Kuwait

JULY AND AUGUST WERE ALWAYS busy in the port of Kuwait, and 1924 was no exception. Virtually everyone in the town was gearing up for the seasonal dhow voyages that would begin in August and set out, wave after wave, until the late autumn. The waterfront buzzed with activity, with mariners readying their dhows and nakhodas watching over the cargoes that were being loaded into their ships' holds.[1] Further away from the shore, all manner of merchants, brokers, and retailers gathered around reams of paperwork—contracts, receipts, ledger books, and letters—that the anticipation of the voyages generated. The joyful shrieks of children playing on the shore rang above the chorus of clanking hammers and the thuds of timber planks moving from ship to shore. This was the sound of a town getting ready to embrace yet another round of departures.

Among those who oversaw the work being done on their dhows was the nakhoda ʿAbdulmajeed bin Al-Mulla Ahmad Al-Failakawi. We don't know his dhow's formal name, though it was most likely the *Fateḥ Al-Khair*, or the *Triumph of Good Fortune*. Shipbuilders often tried to give their dhows auspicious names: *Al-Tayseer* (*Path of Ease*), *Muwafej* (*Fortunate*), and *Fateḥ Al-Raḥman* (*Triumph of the Merciful*); of these, *Fateḥ Al-Khair* was the most common. Nobody called the dhows by their formal names, though. Instead, they referred to the dhows by nicknames that reflected some aspect of the ship's character. Al-Failakawi's dhow was known as *Al-Aʿwaj*, the *Crooked*—hardly an auspicious name for any ship but one that captured one of its unique features. The dhow's builder, Jassim Al-Sbaghah, converted it from a short-distance trading vessel to one outfitted for longer distances and heavier cargoes. To enlarge its capacity to 3,000 *mann* of dates, or roughly 225 tons, Al-Sbaghah added more planks to its hull, and so the dhow lacked the straight, sleek lines that characterized most other vessels.[2] In a word, it looked crooked.

But it was a good dhow and had already proven itself seaworthy. Al-Failakawi had just returned from a six-month voyage that took him and his crew from Kuwait to Gujarat and then to Goa, where he spent two weeks before sailing back up the Arabian Sea and crawling along the Arab shores of the Gulf. There was no time to contemplate the voyage's success: preparation for the next season virtually always overlapped with the end of the one before. The dhows never arrived all at once, either; they came in waves, beginning in June and continuing into July and August. Those preparing to leave greeted those who just got back, and the moving procession of provisions making their way from the market to the shipyard passed cargoes moving in the other direction—bulging sacks of rice, sugar, flour, coffee, tea, and tobacco, among other goods. In and out, out and in.

For the dhow crews, the summer amounted to little more than a brief interlude. If nakhodas could spend the summer with their families while they cleared accounts and made arrangements for the next sailing season, the mariners enjoyed no respite. Many would have had to spend the summer months back out at sea, diving for pearls along with the thousands of other seasonal divers that the annual pearl dive attracted. They had debts to work off, and their earnings from the dhow voyage that took up the other nine months of the year were often not enough to repay debts and cover expenses at the same time. In any case, diving debts were owed to different nakhodas altogether.

For those who got to spend the summer in Kuwait, though, there was still a lot to do. The dhows, which had been propped up on stilts for several weeks, needed to be readied for another season's voyage. All along the shore, mariners scraped barnacles off the dhows' hulls and reapplied the foul-smelling fish grease (*ṣill*) that protected the timber from the relentless summer sun. Others sat cross-legged on the beach, carefully repairing the dhow's sails. The dhow's carpenter, the *gallāf*, examined its planks, repairing whatever damage he identified and caulking the gaps between planks with fresh raw cotton that had been soaked in oil. When the work was nearly done, the sailors used their bare hands to coat the hull with a mixture they called *shūnah*, after its Indian name—a combination of animal fat and powdered limestone that protected the dhow from coral and shipworms.[3] All took place under the nakhoda's watchful eye and to the throaty melodies of the mariners' singing, all amid the overwhelming stench of fat, fish, and seawater.

FIGURE 4. Dhows in a Kuwaiti shipyard, 1939. Photo from National Maritime Museum, Greenwich, UK.

Alongside the mariners, shipwrights put the final touches on new dhows that they hoped would be ready to sail out that season. Merchants placed orders for ships with master shipwrights (each called an *ustād*) months in advance, giving them enough time to assemble a team of carpenters to help with the work.[4] The timber came from Malabar, and those who were able to import it directly from there did; others bought it from among the stacks that piled up outside of the warehouses along the shipyards.

The marketplace also bustled with activity.[5] A cornucopia of different consumer goods from ports around the Western Indian Ocean—dates, rice, flour, spices, textiles, timber, perfumes and scented oils, and more—poured into the marketplace, as retailers arranged with the merchant-wholesalers to supply them with wares on credit. Daily supplies of perishables—fruits, vegetables, and fish—made their way into the harbor at dawn, where mariners unloaded them from their small boats into the eager hands of shopkeepers and hawkers looking to pick up their day's supplies before heading into their own corner of the marketplace. Meanwhile, merchants and shopkeepers scrambled to fill the orders that were coming in—orders from nakhodas to provision the dhows with food and water, and from mariners looking to ensure their families would have enough to eat during the eight to nine months they'd be at sea. The low hum of marketplace activity was punctuated by the ringing of hammers, as blacksmiths and coppersmiths rushed to forge the nails, pots, pans, utensils, and metal trunks that slowly streamed out of the marketplace into the dhow yard.[6] Around the marketplace, the prospect of the coming sailing season set off a flurry of activity. It was another summer in the small port town.

• • •

The inhabitants of Kuwait were all familiar to one another. The nakhodas, shipwrights, and carpenters were friends, relatives, neighbors, and colleagues. The shopkeepers, blacksmiths, and merchants, too, were members of their community: they drank tea together at the local coffeehouses, prayed together in the different mosques around the town, and celebrated ʿeid together in the public squares.[7] Their children chased one another up and down the maze of alleyways that spread throughout the town, at times huddling on the side of the street to play a game of marbles or jacks. They were all connected to one another through dense social and professional ties; if one framed the sailors out of the picture, it was a world that appeared to be deeply local in its contours.

Anyone who looked a little more closely, though, would notice how the outer edges of these networks spilled out into the sea and ran across it. The activities themselves might have been local, but nothing else was. The timber that the shipwrights used to build the dhows that lined the shore, the coir rope that made up a dhow's rigging, the linen cloth for its sails—to say nothing of the rice, flour, dates, sugar, tea, and other foods that made up its provisions—all came from across the sea and were all repackaged to make their way elsewhere.

Kuwait's success as an entrepôt meant that the bulk of the town's population also came from elsewhere. Over the course of its history, the town absorbed an astonishing number of migrants—often in steady streams, though sometimes in large waves. At the turn of the century, Kuwait's inhabitants included tribal Arabs from Eastern and Central Arabia, Africans, Persians, Baloch, Baharinah (the indigenous inhabitants of Bahrain), and a small number of Armenians and Jews. Of the town's thirty-five thousand inhabitants, the original settlers—the townspeople who made up Kuwait's social elite—comprised less than 1 percent. The bulk of the town's Arab inhabitants had migrated to it from elsewhere in the Arabian Peninsula:

Iraq, Najd, or somewhere on the coast. The largest minority groups were the Africans, the bulk of whom were brought over as slaves and who made up 11 percent; Persians made up roughly 3 percent of the town's inhabitants.[8] There were very few people who could claim a long-standing presence in Kuwait itself; the vast majority of its inhabitants were immigrants.

Much of the port's success lay in its ability to channel goods from abroad into its hinterland through caravan routes that stretched across the Arabian Peninsula.[9] In this world of the caravan trade, Kuwait's advantage was that it lay on one of the principal arteries running from Basra to Aleppo and Mecca, and that it could broker between that route and the markets around the Western Indian Ocean.[10] The town's role in bridging between the interior and overseas markets became clear to observers from early on. Not a single visitor failed to notice the presence of Bedouins from the interior in the town's market, where they would exchange horses and ghee in return for rice, tea, coffee, textiles, and more. By the beginning of the twentieth century, the place of interior traders in the market became institutionalized: a section of the town's market complex was set aside for Bedouin visitors and the animals they brought with them, and there emerged long-term credit arrangements between them and retailers in the marketplace.[11]

Kuwait's relations with the interior weren't always smooth. The Saudi ruler Ibn Saud grumbled that it was impossible to collect duties on the caravan traffic between Kuwait and Najd and pressed to establish a customs house of his own in the port town. Kuwait's rulers rejected the idea outright, largely out of fear of ceding their economic autonomy—and potentially their sovereignty—to the Saudi ruler. To retaliate, in 1923 Ibn Saud imposed a blockade of overland trade with Kuwait. A year later, as the nakhoda ʿAbdulmajeed readied the *Crooked* for another voyage, neither side showed any signs of willingness to reach a resolution; the blockade lasted another thirteen years.[12]

Although Ibn Saud's blockade slowed down the pace of Kuwaiti trade with the interior, it could not stop it altogether. Movement between Kuwait and Najd was difficult to monitor and police, and smuggling was rampant. A prospective trader could just as easily route his goods to Najd through Iraq rather than through the well-trodden paths that linked Basra to Kuwait and the interior. Those who wanted to move goods in bulk could also land them at any point on the stretch of coast between Kuwait and Dammam, including the harbor at ʿUjair. And in any case, the inland trade to Najd was just one outlet among many. Kuwaiti merchants found buyers for their wares up and down the Gulf.[13]

The hinterland may have served as the town's main market, but there was no doubt that Kuwait was, at its heart, a port town. Its layout reflected its maritime orientation; its buildings faced the sea, looking out across the water to the places with which it shared its fate. The town spread laterally, huddled around the bay that provided safe harbor for its dhows. Walking along the shore in the early twentieth century, observers would have gained a keen sense of appreciation for the central place that the sea occupied. Dotting the beach was a series of enclosed stretches of the shoreline—small wharfs called *niga‘* (sing. *nig‘ah*), encircled by jetties that thrust out into the water. These belonged to the town's merchants, and each *nig‘ah* became closely associated with the merchant family that maintained it. Dhows belonging to a merchant family and other shipowners moored in the family *nig‘ah* during the offseason, where they would be protected from the waves while the mariners and other workmen repaired them. On the shore, each *nig‘ah* was attached to a warehouse (called a *‘amārah*), where mariners would unload the ship's bulky cargo. Together, the *nig‘ah* and *‘amārah* marked the threshold where the sea and its wares met the land and its markets.

Peering out over the dhows unloading goods into the warehouses, on the waterfront, were the homes of the town's leading merchant families: large, yet inornate, buildings distinguished from others only by the fact that many had recently been rebuilt from coral rock and whitewashed with gypsum plaster—though some did feature elaborately carved teak doors.[14] Situated near their *nig‘ah* and *‘amarah,* merchants could quite literally have their vessels loaded and unloaded at their doors.

The family-owned wharf-warehouse-home complex ran uninterrupted for roughly two miles along an east–west axis—looking out northward onto the bay—punctuated only by the minaret of the occasional mosque and the flagstaff outside the residence of the British Political Agent. As it ran west, it gradually merged into the government wharf, called the *furḍah*—a complex that included the customs house, government-run storehouses, and warehouses belonging to the ruler and other members of the ruling family. It was a recent innovation to the port city's waterfront. Mubarak Al-Sabah, Kuwait's ruler at the turn of the century, had invested in expanding his customs administration to capture more of the wealth coming to Kuwait from the Indian Ocean; the *furḍah* was the physical manifestation of that goal. Its location reflected the ruler's desire to ensure compliance with the new administration; he placed it directly in front of his palace and residence.[15]

From the beach, the town spread inland for just around three-quarters of a mile, butting up against the walls that separated the town from its desert hinterland, built in 1920 to protect it from incursions by the *ikhwan*, the armed cavalrymen who had served Ibn Saud.[16] Though it seemed chaotic, the jumble of mud-brick houses connected by mazes of alleyways and thoroughfares settled into three quarters: Al-Jibla (literally, "the direction of prayer," or "the West") on the west, Al-Sharq ("the East") on the east, and Al-Wasat ("the middle") which, as the name suggests, was sandwiched in between. Each of these housed a distinct subsection of the town's inhabitants at the turn of the century. Al-Jibla was where many of the town's merchant elite connected with the overseas trade took up residence (distinct from the family homes that lined the waterfront), while Al-Sharq was home to the town's maritime community: its mariners, nakhodas, and shipwrights. By contrast, Al-Wasat was more of a commercial quarter, with only a few residential neighborhoods.[17]

To the town's three quarters, we might add another—the nakhoda ʿAbdulmajeed Al-Mulla Ahmad Al-Failakawi's own. Across the bay, at a distance of about ten miles from the easternmost part of the town, lay the island of Failaka—"Faylicha," as locals would have pronounced it—a wedge-shaped, low-lying flat of mud and sand roughly seven miles long and three miles wide. Despite its small size, the island of Failaka bore witness to a long history: it had been settled since at least 2000 BCE, when Mesopotamian traders established themselves on the island, and was famously colonized by Greek soldiers and settlers during Alexander the Great's march through the area in the fourth century BCE. Its name is thought to come from the Ancient Greek *fylakio*, or outpost. In the early twentieth century, visitors to the island could still very clearly see the remains of a Hellenistic fort and Greek temples.[18]

At the turn of the century, Failaka was home to only a small community. Estimates differ, but there were certainly no more than one thousand inhabitants in total.[19] Of these, the majority were engaged in maritime pursuits—principally fishing, but also the dhow trade; others pursued agriculture, farming wheat and barley, melons, and various vegetables.[20] The inhabitants of the island relied on the mainland for supplies of food, fuel, and news—and for its seafaring nakhodas, employment. They, too, had moved there from other ports: the island counted residents from the Persian coast, from Southern Iraq, and from as far away as Oman.[21] The nakhoda ʿAbdulmajeed himself moved to Failaka from Kharj, a nearby island off the coast of Persia, when he was a boy.

When Al-Failakawi was on the island during his brief returns from abroad, he spent his time visiting relatives and meeting friends at coffeehouses and would have taken meals with his wife and children at his waterfront home. Along with other mariners, he may have visited any number of the sixty to seventy tombs of the saints (*awliyāʾ*) of Failaka Island—but principally the shrine (*maqām*) of Al-Khiḍr, a pre-Islamic figure who was known as the patron saint of mariners—to make supplications for a safe and successful sailing season.[22] By the late summer, it was time for him to begin work. Every year, Al-Failakawi bid his family and friends goodbye and embarked on one of the ships that crossed the harbor to the mainland, where he would make preparations for the season's upcoming voyages, staying with his employer. Along the waterfront, in the coffeehouses, and among the *nigaʿ*, he saw his fellow nakhodas make the same preparations. All of them coalesced around a relatively small group of dhow firms, and all ended up on the same waterfront, humming the same tunes with the same level of anticipation that teemed along the shore. And as they looked out from the waterfront, they likely couldn't help but think about the long history that brought them all there.

• • •

In narrating the history of Kuwait, one is immediately confronted with the lore that surrounds its founding. As different historians tell it, the story of Kuwait begins with the migration of a group of families, the ʿUtub, led by the Al-Sabah and Al-Khalifah, from Central Arabia to the coast. The ʿUtub initially settled at the town of Zubara, on the western end of the Qatar peninsula, but then for various reasons decided to leave the settlement and migrate northward, ultimately settling in what is today Kuwait by the mid-1700s. The name Kuwait—the diminutive form of *kūt*, or "fort"—reflected the new settlement, which was near a fort held by the largest power in the area at the time, the Bani Khaled, who allowed the ʿUtub families to settle the area and remain under their protection, after which time they elected the Al-Sabah as the town's administrators. There are variations on this story, but the thrust of it remains the same.[23] Kuwait was a town established by migrants who, generations later, would tell the story to claim it as a place of origin.

It's unclear where the story originated, but it had already been circulating when the historian ʿAbdulaziz Al-Rushaid began writing his foundational history of the port town, *Tārīkh Al-Kuwayt,* in the early 1920s—just around the time Al-Failakawi began his voyages. In Al-Rushaid's telling, that foundational moment set in motion a history that lurched forward on dry land, one ruler after another, until it reached his own time—a long, unbroken chain of Al-Sabah shaikhs that culminated in his patron, Ahmad Al-Jaber Al-Sabah. It made for a compelling narrative at a time when Al-Sabah's claims to autonomy were being challenged by Ibn Saud.

The sea formed only a small part of the backdrop to Al-Rushaid's telling of Kuwait's past. The inhabitants of Kuwait, the historian was quick to point out, relied heavily on the summer pearl dive for their livelihood. Pearl diving formed a pillar of the port's economy; everyone in Kuwait, "young and old, poor and rich alike," could all attest to its importance. In more than a dozen pages, Al-Rushaid gave an encyclopedic description of the pearl dive: the variety of pearl dhows, the sorts of laborers they employed, its profit-sharing arrangements—down to the particular varieties of pearls divers fished and the ailments they suffered from.[24]

He had much less to say about the pearl *trade,* though—or really, any kind of trade. All he offered were generalizations. "Trade," he wrote, "was important—[Kuwait] could not do without it, as it was arid, with no agricultural wealth whatsoever, and no manufacturing sector that could produce what [Kuwaitis] needed." With few resources of their own to draw on aside from human capital, Kuwaitis turned a profit by building ships and carrying other people's products. His narrative of Kuwait's trade progressed in stages. In their early years, Kuwaitis traded across short distances, to nearby markets—Basra, which "they took to as a source for all of their urgent needs . . . rice, flour, wheat, dates, vegetables, fruit, clothing, tools, and the like," Baghdad, where they sold the pearls they fished, and the market towns of the Gulf's Arab littoral. And insofar as he understood why, it was fundamentally a question of maritime technology. "Their ships were not the sort that could sail past the countries that were on the Arab shores of the Gulf, because of their small size," he wrote; because of that, trade had been limited in scale—"provincial," he called it. Those who could build large oceangoing vessels that could reach the more distant shores of India and South Arabia had not yet migrated to Kuwait.[25]

Kuwait's traders also benefited from political developments at home. Throughout his history, Al-Rushaid asserted that Kuwait's growth was in part a result of the rule of the Al-Sabah family. The Al-Sabah, he declared, were savvy administrators and diplomats—especially Mubarak Al-Sabah, who ruled the town from 1896 to 1915. Mubarak, he wrote, kept a watchful eye over Kuwait's commerce and was keen to protect it from plunder—so much so that he attracted immigrants from all over the Gulf. To Al-Rushaid, Mubarak's commitment to protecting Kuwait's commerce found its most powerful expression in an exclusive agreement he signed in 1899 with the British government, in which he bound himself and his successors not to receive any foreign representatives in Kuwait and to not cede any part of the territory under his control without British consent. The agreement effectively established Kuwait as a protégé state of British India, granting Kuwaiti traders abroad access to British protection; the British government would extend Mubarak's writ beyond Kuwait's borders. This, Al-Rushaid asserted, "was the most apparent cause" for Kuwait's commercial efflorescence in the early twentieth century.[26]

There is something reassuring in the linearity of Al-Rushaid's narrative of Kuwaiti history. It's a history in which an essentially terrestrial people settled, gradually became amphibious, and under the guidance of far-sighted rulers, built a thriving country. It's a history on rails, one that takes its readers to their landed present with few diversions along the way. It's also a history in which it is exceedingly difficult to point to anything that might have happened outside of the town's borders.

We might read that history differently, though, and instead situate it within a much broader political landscape that was in flux in the two centuries leading up to Al-Rushaid's *Tārīkh*. The settlement of the ʿUtub in Kuwait in the mid-1700s reflected a moment of rupture in world history, in which the three major empires of the Islamic world all faced a series of political crises. In Persia, the halting collapse of the Safavid Empire in the 1720s opened the door to political aspirants on the coast who sought to shake off their allegiances to their imperial overlords. In India, the 1700s witnessed the gradual usurpation of Mughal dominions by a number of successor states and European companies. And although the Ottoman Empire was able to avoid the territorial losses that its neighbors experienced, it faced considerable military setbacks—first, from an expanding Saudi state in Central Arabia, and by the end of the 1700s with Napoleon's invasion of Egypt. To meet these challenges and address other tensions within the empire, the Ottomans adopted a policy of decentralization, gradually handing authority in more distant provinces over to their Mamluk slave governors.[27]

It is within this context that we might situate the migration of the ʿUtub and others from Central Arabia to the coast—and more importantly, their ability to settle in Kuwait under the nominal protection of the Bani Khaled, who dominated Eastern Arabia. In the late 1500s, when the Ottomans sought to extend their control over Eastern Arabia in order to establish a base from which they could battle the Portuguese and Safavids, the Bani Khaled were already clearly in control. Ottoman attempts to control the region were always tenuous, and the Ottoman government was largely unable to quell the restless Bani Khaled chiefs, even as it offered them different inducements. Over time, the Ottomans scaled back their investments in Hasa, and were ultimately expelled by the Bani Khaled in 1670.[28] By the time the ʿUtub arrived on the shores of the Gulf, the Ottoman presence was a distant memory; the Bani Khaled were the ones to whom they would approach for permission to settle.

But the overlords of Eastern Arabia would not last long. For all their prowess in throwing off the Ottoman yoke, the Bani Khaled were unable to defend themselves against repeated raids by the Wahhabis during the late eighteenth century. Although the Al-Hasa rulers were able to contain the Wahhabis in the mid-1700s, internal feuding within the tribal leadership in the later decades left them unprepared to face the Wahhabis' military expansion. The Bani Khaled were finally forced to capitulate in 1795.[29] Left to their own devices, the ʿUtub could manage their affairs however they deemed fit.

By the end of the eighteenth century, Kuwait had become one of a handful of independent port towns scattered across the Arab shores of the Gulf, from Basra down to Zubara. The bulk of the commerce from India passed through Basra, and despite their rulers' attempts to attract commerce by lowering customs duties or doing away with them altogether, the smaller ports of the Gulf could hardly hope to displace that port city, which constituted the main gateway into Ottoman Iraq and its markets. Kuwait and small ports like it were, for the most part, secondary market towns and alternative entry points for those who wanted to dodge the Ottoman customs administration.[30] Its proximity to Basra meant that Kuwait often benefited from that port's changing fortunes during the eighteenth century—endemic plague, recurring military conflicts with Safavid Persia, and tribal rebellion, all of which drove many of Iraq's merchants out of Basra and into nearby ports.[31] Kuwait's establishment, then, can only be understood as part of a changing fabric of politics and trade in the eighteenth-century Gulf.

The long-distance dhow trade fed Kuwait's growth. The town's inhabitants engaged in the India trade from at least the early 1800s, if not before. In 1829, one English naval officer wrote that the dhows from Kuwait would regularly "navigate the Gulf of Persia, Red Sea, Coasts of Sind, Guzerat, and Malabar, and to Bombay," and that they imported a wide variety of goods from those places.[32] Two years later, an English traveler described sailing from Bombay to Kuwait in a *baghlah* built in Malabar but "manned by forty or fifty natives of Grane, or Koete, on the western side of the Persian Gulph, and commanded by a handsome Nacquodah in the prime of manhood."[33] Local records bear witness to inhabitants of Kuwait shuttling goods between Basra, Bombay, and Malabar—and even to the East African coast—by the middle of the nineteenth century.[34] Rather than tell a story in which Kuwaiti dhow traders gradually developed the technology and capacity to travel greater distances, we might imagine those who settled in the port town as already having traversed those routes and built those networks, connecting all of it with the town's broad hinterland, and fusing its fortunes to places far away from it.

Understanding the history of Kuwait, then, necessitates that we read it as much from the sea as we do the land. Kuwait reaped the benefits of geopolitical reconfigurations around the Gulf and Indian Ocean, and the migrations that accompanied them. As the winds shifted in their favor, Kuwait's rulers aligned and realigned themselves with different regional polities as the opportunity presented itself—the Bani Khaled, whose protection allowed them to settle at the site of Kuwait; the Ottomans, to whom they lent military support for their campaigns in Basra and Eastern and Central Arabia; the Omanis, with whom they negotiated passage through the Straits of Hormuz and into India; and the English East India Company, whom the Al-Sabah rulers courted on and off throughout the nineteenth century. At least part of Kuwait's growth as a regional trading center, then, came from its rulers' ability to weave themselves into the broader political fabric of the Persian Gulf and Indian Ocean and establish a canopy under which Kuwaiti mercantile activity could flourish.

This, in short, was the world into which Al-Failakawi and Kuwait's fleet of dhows in the 1920s would sail: a world of political reconfigurations, contestations, and circulations of people, commodities, and capital. There is much more, of course—many more transformations around the region would shape Kuwait's relationship with the sea. The narrative we have so far feels incomplete, and perhaps necessarily so. Because the history of Kuwait did not only play out in that tiny stretch of land along a magnificent harbor in the upper Gulf. The history of Kuwait played out in other places, too—at times, far across the water. As we move along with the *Crooked*, we will come to grasp these.

• • •

As his sailors scraped, caulked, and greased away at the *Crooked*, Al-Failakawi's eyes scanned the *nigʿah* and its surroundings. He exchanged a quick glance with his employer, the merchant Muhammad Shaheen Al-Ghanim, who occasionally looked out onto the shipyard from his *ʿamarah* to watch over the upkeep of his family's capital. He'd nod at the other nakhodas in the yard and would hail those walking by—friends and colleagues with whom he spent much of the summer. They came from different families, neighborhoods, and backgrounds, but all ended up as part of the same tight-knit world of nakhodas and trading firms. This was his community—one he wove himself into over the course of many years of work. In this community of close ties, getting hired depended on who you knew.

Al-Failakawi's entry into the world of seafaring was not like many of the other nakhodas. He was not born into a seafaring family: his father was one of Failaka's *mullahs*, trained in Baghdad, and instructed the island's youth in reading, writing, and Quranic recitation. Al-Failakawi worked his way into the maritime trade by drawing on Failaka's small but reputable nakhoda community. He trained under a well-known Failaka nakhoda, Mansur Ibrahim Al-Khalil, known as Mansur Al-Khariji—who also happened to be from Kharj (hence the name) and who *did* come from a seafaring family; both Al-Khariji's brothers and his father served as nakhodas. Al-Khariji took on Al-Failakawi as a rank-and-file sailor and later taught him the principles of navigation.[35] He also introduced Al-Failakawi to his merchant employer, Muhammad Shaheen Al-Ghanim, who then hired him as one of his regular nakhodas. Al-Failakawi would later marry Al-Khariji's daughter, cementing the ties that helped establish him in the profession.

Others followed a more well-trodden path into captaining. The nakhoda Rashid bin ʿAli Al-ʿAsʿousi, whose voyages interlaced with Al-Failakawi's, enjoyed the social capital that came with having been born into a seafaring family. Like many other nakhodas in Kuwait, he was trained in the art of navigation by members of his own family. His uncle ʿAbdulrahman bin Hussain Al-ʿAsʿousi hired himself out as a nakhoda, ferrying dates from near Basra to the western coast of India on behalf of the ruler of the nearby emirate of Mohammerah—what is today Khuzestan, in modern Iran. The family history suggests that Rashid had learned how to navigate and captain a ship from ʿAbdulrahman. Others in the family had learned while on the job, too: during a voyage with his brother Hussain, ʿAbdulrahman's son ʿAbdulwahhab wrote his father that he was learning to use the sextant (*kamāl*) and had already learned the basics of nautical arithmetic.[36] Hussain had learned the basics of the trade from his uncle Khaled bin ʿAbdulaziz when he was just thirteen years old, and continued to develop his navigation skills as he sailed with Rashid for several years in the 1910s; he finally captained his own dhow in 1923, at the age of twenty-six.

The Al-ʿAsʿousi experience would have resonated with many other nakhodas in Kuwait. Nakhodas learned reading, writing, and basic arithmetic with a *mullah*, but the bulk of their training came on the job as they accompanied different family members on overseas voyages. ʿIsa Bishara learned to navigate from his own father, Yacoub (who in turned learned from sailing with his father, Yousef), whom he accompanied on voyages from the age of twelve. He sailed with his father for nine years, and then worked for two years as an expert navigator (*muʿallim*) on different dhows before being hired by a merchant to take the place of a retired nakhoda.[37] Those who could draw on family networks enjoyed a distinct advantage in learning the art of captaining a dhow but also in finding opportunities to work.

Having access to a family dhow helped as well. Members of the Al-ʿAsʿousi family had amassed enough capital to commission their own dhow in 1906—a *boom* they named *Hawalli*, after an area in Kuwait where wells of drinking water had recently been found (after all, one had to manifest good fortune). ʿAbdulrahman captained it for only a few years before retiring and transitioning into the mercantile side of the family business. By the time Rashid began working as a nakhoda in 1910, then, he already had access to a patron and a dhow from within the family.[38] Later, the *Hawalli* passed to his cousin. The dhow thus constituted a mobile form of capital that circulated within the family. It gave its owner a clear pecuniary advantage too. In the maritime economy, profits were divided between the owners of the capital stock (in all cases, the ship itself) and the owners of the labor (the crew members, or in some cases, their masters). As a general rule, on long-distance trading dhows, half the net profits went to the dhow's owner and the other half went to the crew, including the nakhoda. In this arrangement, a nakhoda who owned his dhow earned the lion's share of the season's profits.

Dhows also constituted a good form of security for those looking to borrow money: they enjoyed a clear pecuniary value that a lender could identify and later realize, should his debtor default. And around the Indian Ocean, nakhodas and merchants looking to raise business capital would pawn their dhows to their creditors. Oceangoing dhows, medium-sized coastal craft, or smaller longboats—all of these made their way into the credit market when money was tight.[39] Because they generated profit, dhows were a good investment no matter who owned them. Even someone with no interest in sailing the dhow could hire it out and collect a share of the profits at the season's end.

The nakhoda who had access to a family dhow was uncommon; rarer still was one who owned his own ship. The majority of Kuwait's nakhodas had neither; in the local parlance, they were the *ja'di* nakhodas—literally, the "sitting nakhodas." Nakhodas like Al-Failakawi, Bishara, and hundreds of others had the skill set but not the dhow. They were free agents, hiring themselves out to merchants who would supply them with a ship to captain. There were some benefits to being *ja'di*; a nakhoda earned money off his labor alone, and did not bear the risks and costs of owning a dhow. Writing from Kuwait in the late 1930s, the traveler Alan Villiers noted that in hiring nakhodas to run their ships, "the merchants did very well out of it, and the nakhodas made quite a good living too."[40]

Whether a nakhoda owned a dhow had little impact on his ability to find employment. Unless he captained the family ship, he had to find an employer for the season—someone who would hire him to ferry dates from Basra to ports around the Western Indian Ocean and to carry back goods for sale in the Gulf. Although they negotiated different terms, nakhodas did not frequently change employers. The ties between nakhodas and the merchants they worked for ran deep; the reciprocal exchanges of labor and capital that took place between them recurred every season, and each new iteration thickened the social bonds and deepened the accounts that ran between them. Al-Failakawi spent most of his career sailing for his first employer, Muhammad Shaheen Al-Ghanim.[41] Bishara worked for several nakhodas but enjoyed particularly deep ties with one in particular—the merchant Ahmad Al-Kharafi, for whom his father and grandfather had both captained at various times in their lives.[42]

To nakhodas, merchants were more than seasonal employers: they were gatekeepers to the marketplace and its wares. Through their merchant patron, nakhodas gained access to provisions and money that they needed to outfit their ships—supplies they would not have otherwise been able to procure in advance. As Al-Failakawi took goods from the merchant's stores, they both kept running accounts. We don't know what Al-Failakawi bought in the summer of 1924, but a handful of surviving account sheets from August 1939 give us a sense of what the nakhoda thought to purchase. Under the heading "An

FIGURE 5. Merchants and nakhodas in a Kuwaiti coffeehouse, 1939. Photo from National Maritime Museum, Greenwich, UK.

Account of Expenditures for Provisions [*Māchlah*] in Kuwait," the nakhoda listed multiple purchases of rice, sugar, salt, flour, dried fish, dates, tea, drinking water, and cooking oil. A week later, he drew up another account sheet under the heading "Expenditures on the *Boom*," in which he listed different supplies: coir rope, cordage, fishing equipment, gunny sacks, caulk, fish oil, lamps, kerosene, and a variety of containers.[43] It was important for Al-Failakawi to keep a list, as the expenses would be debited from the profits of the voyage; he and Al-Ghanim negotiated over who bore precisely which costs.

For his part, Al-Ghanim kept his accounts open with Al-Failakawi and the other nakhodas he employed, giving them access to their stores, much like he provisioned retailers in the marketplace itself. In most cases, he would push to subtract the cost of these provisions from the gross profits the crew earned by ferrying his goods around over the course of the nine-month sailing season. A savvy merchant was thus able to eke out extra margins on at least three ends: the provision of goods on credit at the beginning of the sailing season, the sale of his wares in overseas markets, and the share of the profits he took in return for the use of his dhow. It would be a mistake, though, to assume that merchants were after profits alone; to do so would be to strip society from the calculus altogether. Profit came with building a firm of skilled workers, not in squeezing them into penury.

But even then, after finding a teacher, a dhow, and a merchant patron, Al-Failakawi still needed to conjure up the most important element of a successful voyage: a crew.

. . .

The crew that huddled around the *Crooked* had been at it for weeks, singing their sea chanties and ribbing one another as they prepared the dhow for sail. As Al-Failakawi watched over these men, many of whom were in their late teens but some much older than him, he tried to remind himself of all the different places they came from—some from Kuwait itself, yes, but many from different towns on the Persian coast, or from further on down the Gulf, from Bahrain, Qatar, Dubai, Sharjah, and from as far away as Sur, Makran, and sometimes even Yemen. He remembered, too, how they had gathered in the shipyard weeks before, forming a single file in his direction. He asked their names and wrote them down in a notebook with a figure next to it—their advance for the season. Other nakhodas alluded to this with the brief note "*sallafnā*"—we loaned. Al-Failakawi knew some of his sailors from last year's voyage; others were recent additions to the crew, recruited from the transient mass of seafarers who milled about the waterfront at the beginning of the summer.

The work of assembling a crew began almost immediately after Al-Failakawi's return from the previous season's voyage, at the very beginning of the summer, right after paying his crew their season's earnings. Although merchants gave their dhows names that called for the intercession of good fortune, the nakhoda knew that it was pluck, rather than luck, that determined how much any one of them took home at the end of the year. Al-Failakawi's ability to realize gains from the voyage was highly dependent on there being a good crew.

His first step was to find a first mate, called a *mjaddimi*—"he who drives [the crew] forward"—whose immediate task was to set about putting together a body of reliable mariners. Al-Failakawi's *ruznamah* tells us nothing about the *mjaddimi* he employed, but it's likely that he had sailed with the nakhoda before. Crew lists from 1944 and 1945 suggest very little turnover from among Al-Failakawi's rank-and-file sailors, and he sailed with the same *mjaddimi* on both voyages. Other nakhodas also tried to hold onto their *mjaddimis*. The nakhoda ʿIsa Al-ʿUthman, who kept continuous crew lists for his voyages during the 1940s, relied principally on two *mjaddimis*: Hassan bin Yaʿqub, whom we know nothing about, and Jum'a bin Saleh, whom Al-'Uthman sometimes referred to as Jum'a bin Saleh Al-Suri—from Sur, in southeastern Oman—and who sailed with the nakhoda for six years in a row.[44] 'Isa Bishara cycled through a number of *mjaddimis* over his career, but his crew lists (also from the 1940s) indicate the consistent presence of at least one mariner—a 'Abdullah Al-Muhaini, whose name indicates that he, too, came from Sur, and who began as a helmsman in 1945 and had become the *mjaddimi* by 1949. One of Al-Muhaini's crewmates, Maskin bin Fairuz, also worked his way up, from rank-and-file sailor to assistant *mjaddimi* in just four sailing seasons.[45]

Depending on the size of the dhow, the size of the crew might be large: an average dhow of about 225 tons would require a crew of thirty sailors (roughly ten per 75 tons); a larger dhow would, of course, require an even bigger crew. Moreover, the crew had to include members with specialized skills: in addition to able-bodied and versatile seamen, a long-distance trading dhow often needed a shipboard carpenter to help repair ongoing damage, a helmsman or two to alternate shifts with the nakhoda, an assistant to the first mate, a cook to prepare meals for the sailors, and sometimes even a skilled navigator (*m'allim*) to assist nakhodas who were unfamiliar with how to navigate beyond sight of the coast. A good singer (*nahhām*), too, formed an essential part of the crew; his music animated the labor that mariners performed on board the dhow and lifted spirits during the long intervals of boredom that invariably accompanied life at sea.[46]

Finding a body of reliable crew members necessitated a considerable amount of social knowledge on the part of the *mjaddimi*. Though some of the sailors would be drawn from the different neighborhoods around the port town, the number of locals on board the dhow shrunk with the growth of Kuwait's trade. The good times attracted waves of seasonal laborers—migrant laborers who came to the town during the summer months to work on the pearling and trading dhows, and who huddled together at the waterfront cafés drinking tea and smoking and who slept out on the beaches at night.[47] These men came from port cities around the Persian Gulf—from Basra, but also Oman, Yemen, Persia, and Balochistan. A good *mjaddimi* had experience with many of the sailors he recruited and would have at least had enough background knowledge of others to make an informed decision about their reliability. His value as a first mate was anchored in his social network as much as it was in his personal qualities.

There was good reason why a *mjaddami* expended so much effort in recruiting good sailors: his success, and the success of the enterprise as a whole, depended on it. Dhow sailors were not the wage-laboring masses of proletarian Jack Tars of the Atlantic.[48] Rather, they were more like the whaling crews of New England in that they worked for a share of the season's net profits, after deducting the cost of provisions and customs duties, and then after handing over half the share to the dhow's owner; a mariner's share was based on his position, though the nakhoda could hold out other inducements. For individual crew members, the incentives to contribute as much to the season's profits as possible were apparent in the profit-sharing system; and for those tasked with managing the enterprise, the necessity of a reliable, hard-working crew was clear.

If the work of finding crew members fell principally to the *mjaddimi*, when it came to binding individual mariners to the dhow enterprise, he had to rely on the help of the nakhoda and merchant. Because the crew worked for a share of the net proceeds, they needed loans in the interim to get by. The nakhoda thus had to open his ledger book to the mariners, giving them advances of money—or more often, rice, sugar, tea, and other provisions on credit, which would then be deducted from the mariners' share of the season's profits. By taking on the advances, the mariner bound himself into an obligation to sail for that nakhoda. The credit, however, rarely came from the nakhoda himself: more often, it came from his merchant employer, to whom he was also bound through some sort of debt-based obligation. Through advances of money and store credit—credit that many of the itinerant sailors would not have otherwise enjoyed—nakhodas bound mariners to himself and his merchant and established the human capital that constituted the dhow.[49]

Not all the sailors were free to choose whether they joined the dhow. At least a few had names that suggested that they were enslaved—or had at least descended from enslaved people: names like Maskin bin Fairuz, Bukhit bin Khamis, or the *nahhām* ʿAntar bin Yaqut. One merchant recalled that at the turn of the century, many of the mariners on a dhow would have been enslaved.[50] It's hard to tell what the proportion of enslaved mariners on a trading dhow may have been, but we do know that enslaved mariners abounded on dhows around the Indian Ocean. In the labor-hungry pearl dive, enslaved individuals from East Africa and Makran toiled as divers and haulers, making up substantial proportions of the labor force.[51] Nakhodas in East Africa also regularly included enslaved mariners among their crewmembers; those from Kuwait did too. Not all these mariners were completely unfree: many were manumitted, and at least some of those who were enslaved managed to hold onto their advances and earnings.[52] Still, the stain of enslavement was a difficult one to scrub off the dhow, and plagued nakhodas for nearly a century by the time Al-Failakawi came to captain the *Crooked*.

The dhow thus formed a space where multiple itineraries and histories intersected—where itinerant and enslaved individuals were combined and rearranged to be sent out circulating into the world once again. But it was also a site in which these itineraries were grounded in local ties of credit and debt—in that complex of ledger books, promissory notes, profit-sharing arrangements, contracts, long-standing relationships, and circulating gunny sacks of rice that bridged together ship, shore, and transregional trade. Under the carapace of the dhow, these mariners were brought into the firm and anchored in the marketplace. In a sense, their routes generated certain kinds of roots.

• • •

The mariners loaded the dhow's cargo hold with sacks of rice, flour, lentils, and dates, and hoisted the masts upright. When the dhow was finally ready, sailors, half of whom brought with them hand drums of different sizes, gathered at the shipyard by the *Crooked*. The *nahhām* called them around, and they would begin to drum slowly, over a long, drawn-out 128-beat cycle. Others clapped their open palms rhythmically and sang in a droning, guttural growl, while one sailor blared a tune into his double-reed oboe. Together, they sang songs of lament—of having to leave loved ones behind for the uncertainties of the sea.[53] And to the crowd of onlookers who would have gathered around to enjoy the music, this performance—the *sangani*—signaled that the time for the crew to set sail had nearly arrived. Over the next day or two, the sailors scrambled to wrap up their business around the town and spend their final moments in the company of family and friends, ensuring they had enough food to last them the year and bidding them tearful goodbyes.

As August turned to September, they got word from Al-Ghanim: it was time to set sail. Al-Failakawi and his crew got onto their longboats and rowed out to the *Crooked*, clambering on board one at a time, lest the dhow list too much. On the deck of the dhow, they hoisted the main sail to the barking orders of the *mjaddimi*, singing shanties to give rhythm to their work. As the winds filled the raised sails, they pushed the *Crooked* out into the water. The shore receded further and further away, until the ship was surrounded by the foamy green sea.

INSCRIPTION

Debts

IMMEDIATELY BELOW THE COLOPHON to his first *ruznamah*, Al-Failakawi copied out two contracts. The first was an acknowledgment of debt from Muhammad bin Hassan to Ahmad bin ʿAli for Rs 400, due in four months; the contract notes the date as 1341 AH (which ran from August 1922 to July 1923) but makes no mention of the day or month. The second is very similar: an acknowledgment of debt, this time for Rs 600, made out from a Hassan bin ʿAli to a Muhammad bin Ahmad, due in six months; the year is the same, and there are no other dates given. Both end with the same statement: "I have given this document as proof, and Allah is the greatest of witnesses."

That Al-Failakawi would have felt compelled to copy contracts into his logbook is unusual; it doesn't belong in the genre as we understand it. That he would copy two contracts that didn't involve him is even more unusual, even odd. But there are at least a few things that are even more curious about the two acknowledgments that Al-Failakawi copied into his logbook. There are no specific dates—at least none that would give a contract its legal specificity (imagine basing a claim on a contract that makes no mention of a month and day!). The names, too, are generic: Hassan, ʿAli, Muhammad, and Ahmad, all in different combinations. One can imagine dozens of Hassan bin ʿAlis or Muhammad bin Ahmads—hundreds, even. Who the contracts refer to—even who they *might* refer to—is entirely unclear.

There is a third contract, copied into another *ruznamah*, that might shed some light on these questions. This is in Al-Khariji's distinctively clear handwriting. It reads: "Yes I, Fulan bin Fulan, agree that I have for the holder of this legal document [*sanad sharʿi*], the honorable gentleman Ahmad bin Muhammad, money to the amount of Rs 1,200, and this is a benevolent loan [*qarḍ al-ḥasanah*], and is the value of rice, and it is due in a year from today, and will be handed over without excuse or delay." The contract relays a specific date (16 Safar 1341) but declares that it was written by Fulan bin Fulan—which translates to "so-and-so, son of so-and-so." The contract, it is clear, is a template for a written acknowledgment of debt; both *fulan* and Ahmad bin Muhammad are placeholders in a formulaic expression of a particular sort of contract. Reading this contract also makes it clear that the first two, with all of the Hassans, Muhammads, Ahmads, and ʿAlis, were most likely generic templates as well. But templates for what, and why?

The clue is in the rice. The one common denominator that ran through all the accounts between merchants and mariners, on and off the dhow, was rice—counted by the 50 kg (110 lb) gunny sack, referred to in the colloquial as *gūniya* or *jūniya*. By the late nineteenth century, rice was being exported from Karachi in mass quantities, becoming a staple of the Gulf diet. It was hardy and filling, and though it lacked much nutritional value, it provided the inhabitants of the region with the energy they needed to work; they called it *ʿaysh*, that which sustains life. Provisioning a dhow for its voyage principally meant securing enough rice for mariners to consume over the course of the season. As a commodity, rice circulated in and out of the bazaar, between merchants, nakhodas, sailors, and their families; it could be stored, cooked and consumed, or simply resold, resuming its life in circulation. And it was in its latter form—as a commodity meant as much for circulation as it was for consumption—that rice acted as credit.

Rice, and other dry goods like it, constituted the lifeblood of the marketplace; they moved back and forth between merchant, mariner, and nakhoda, and often stood in for all other forms of credit. When a mariner took on a loan, it was in rice or flour, not money; those who wanted to convert it into money had to hawk the rice around the marketplace or sell it back to the merchant at a discounted rate. This circulation of sacks of rice into and out of the bazaar generated a visible paper trail: with every movement of a sack of rice, flour, or sugar, a new line emerged in a ledger book and a new receipt was drawn up. The social ties that animated life in the maritime economy were thus interlaced with and punctuated by account sheets, declarations of money owed, receipts, notes, and all manner of paper instruments, much like those that Al-Failakawi copied into his *ruznamah*.

. . .

There was a reason why I was immediately drawn to the acknowledgments of debt in Al-Failakawi's *ruznamah*. Just a couple of years before seeing Al-Failakawi's writings, I had published a book on law and economic life in the Indian Ocean—one that drew primarily on acknowledgments of debt that were very much like the ones inscribed into the *ruznamah*. One could say that I was familiar with the genre.

The acknowledgment of debt was perhaps the most ubiquitous contractual form in the Western Indian Ocean. One finds examples of it everywhere, from Basra to Zanzibar and in different arenas of economic life. It formed the basic bond that linked actors together in a relationship of mutual credit, debt, and obligation. "It was hard for a sailor not to be in debt, and so far as I could see none of them tried very much not to be," observed Alan Villiers, during his time in Kuwait in the late 1930s. "Debt as an accepted thing, and to spend a lifetime owing money was apparently usual. The sailors owed money to the nakhodas, the nakhodas to the merchants, the merchants to other merchants or the shaikh.... It was obvious enough that the whole industry rested on a structure of debt."[1]

To understand the ways in which debt worked in economic life, we have to discard our stereotypes of creditors and debtors and see these individuals in their different capacities. Although professional moneylenders were not rare, they formed the minority of creditors around the Western Indian Ocean. The bulk were everyday economic actors extending credit to others under the regular circumstances of work: shopkeepers extending credit to their customers; nakhodas giving advances to mariners; traders giving loans to owners of farmland; and merchants extending credit to one another as part of the normal course of business. Credit was a means to an end that lay somewhere in the realm of production or commerce—or, from the perspective of the debtor, consumption.

In most cases, the debtor was expected to repay the debt not with money but through some form of service. A mariner who borrowed from a nakhoda was expected to work for him. A planter who borrowed from a trader was often expected to supply him with the sorts of goods the trader did business in—dates, grains, cloves, or coconuts, for example. Very rarely would this be a one-time transaction; in most cases, the loan was extended and reextended over many years, and in some cases over generations of creditors and debtors. Repayment in full would have signaled the end of the relationship—a severance of the bond, and its replacement with a new one, with someone else.

Of course, this wasn't the case all the time; history matters here. Credit and debt might not have always occupied as central a place in economic life as they did in the century leading up to the inscription in Al-Failakawi's *ruznamah*. Port cities around the Western Indian Ocean experienced a commercial boom that unfolded steadily, if not always evenly, over the course of the nineteenth and twentieth centuries. As markets in Europe and the United States opened up to goods from around the Indian Ocean, actors around the Indian Ocean witnessed a surge in demand for luxury items—ivory and pearls—but also mundane goods like dates, coconuts, coffee, cloves, and hides. With the expansion in economic activity that accompanied the boom came an expansion in credit, supplied principally by Indian merchants from Gujarat and Sindh. Acknowledgments of debt were negotiated, drawn up, and argued over around the coasts of Arabia, India, and East Africa, across ethnic, religious, and legal lines.

Nor were times always good. At different moments, commercial troubles, political disturbances, or environmental calamities—small, local occurrences, or more widespread downturns—could strain the credit relationship, turning it into something that looked more like the kind of debt that demanded repayment. One such moment occurred in the 1870s, when a combination of factors—political upheaval in Muscat, a bursting bubble in real estate and cotton speculation in Bombay, and a hurricane in Zanzibar—caused some panic in the regional credit market. Another, more expansive moment of economic upheaval came with the worldwide economic depression in the 1930s, when markets around the world felt more constrained, and creditors began calling in the money they had loaned out.[2] But that takes us slightly ahead of Al-Failakawi and his contractual templates, which bore the date 1341 AH (1922–23).

• • •

Nakhodas like Al-Failakawi issued declarations of debt as a matter of course. When they loaned money or goods to mariners, they issued documents called *barwas*—small, unassuming scraps of paper no larger than 15 cm long and wide, and often even smaller, generated as a matter of course in the maritime economies of the Persian Gulf. A form of debt accounting, a *barwa* was effectively a reverse IOU—something like a "you owe me," or more accurately a "he owes me," coupled with a request for repayment. The terms it laid out were brief, but clear. As one 1889 *barwa* read: "To whomever sees this from the nakhodas, [let it be known that] we have on [*lanā ʿalā*] Faraj, the follower of . . . sixty-five rupees and a half rupee. Whomever wants to take him on [*yuḍimmah*] [is to] hand over to us the aforementioned."[3] Another *barwa* drawn up five years later is virtually identical in its wording. On a sheet of paper barely the size of a matchbox, it reads: "To whomever sees this from the nakhodas, [let it be known] that we have on Hussain Al-Balushi two-hundred rupees and sixty-six rupees. . . . Whomever wants to take him on is to hand over the aforementioned."[4] Some *barwas* directly addressed nakhodas associated with the pearl dive, whereas others specified long-distance trading nakhodas, whom they addressed as *nawākhithat al-safar*—"the voyaging nakhodas." And not all *barwas* demanded full repayment. Many set out alternative terms, allowing for the mariner's new nakhoda to hand over half, a third, or even a quarter of the mariner's earnings until the debt was paid off. Still others, following the same formula, declared that the mariner owed nothing and any nakhoda was free to take him on "in full."[5]

Reading these scraps of paper, it's tempting to imagine the chains of debt that bound sailors to their nakhodas as a form of debt bondage. This was, after all, a monetary figure attached to people and their labor, and one that placed clear limits on their freedom to work. And yet, the labor market for trading dhows was characterized by a considerable amount of mobility, especially among mariners with skills. Mariners did not stay with a nakhoda for more than two or three sailing seasons before finding work with another. The bonds of debt and credit that bound nakhodas to mariners were thus temporary; when the season's profits were good, a mariner made more than enough to cover his advances and move on to a nakhoda who might have been willing to advance him even more.

It might be more useful, perhaps, to think of debt and credit as forming the vertical bonds that structured the dhow enterprise.[6] We might see this as reflecting a form of building wealth in people that animated the maritime economy of the Persian Gulf—a mode of engaging labor in which wealth was imagined as resting in communities of people that actors would form, and particularly in the specialized knowledge they would bring to the collective and the ability of the community to reproduce its social capital over the course of generations.[7] For their part, mariners who took the time to develop the skills they brought to the collective, be they in navigation, carpentry, or music, were given a higher share of the profits—and larger advances throughout the season—than those who did not.

But wealth was not only in people. It was in things as well: in the sacks of rice, flour, coffee, and sugar, and the yards of fabric that went in and out of the marketplace; in the scores of timber and mangrove poles that lay in piles in the shipyard; in the properties—shops, homes, farms—that merchants owned; and in the coins and banknotes that occasionally changed hands. All these sat alongside the people who made up the enterprise and at different times intersected with them. Merchants drew on property to access the capital necessary to outfit ships; they passed goods and money down to the nakhoda, who then used it to bring the mariners into the enterprise.

And this is where the boundary between the free and enslaved could blur. If these inanimate forms of wealth could be used to bring people into the enterprise, people could also be converted into other forms of wealth as well. There are not many times when a merchant or nakhoda would look to liquidate their wealth in people, but the possibility was always baked into the relationship. Those who could no longer meet their financial obligations would look to the forms of capital they held that were the most expendable; first, money and goods, and then humans and the debts they owed. When times were hard, people could, with little difficulty, come to be seen as property; it would be the mariner's body—by way of his *barwa*—that would ultimately absorb the shocks of the market, when they came.

. . .

It is no wonder that Al-Failakawi included the two acknowledgments of debt where he did, on the first page of the logbook, immediately beneath his colophon. The debt acknowledgment was, in so many ways, an obvious candidate. Though it was short and basic, the acknowledgment of debt stood in for a whole world of transactions, and of transacting. It signaled the ties of credit and debt that shaped the maritime economy of the Western Indian Ocean. It formed the links in the chains of obligation that stretched from the shipyards of the Gulf to the marketplaces of India, the mangrove deltas of East Africa, and beyond. It constituted the basic building block of the dhow enterprise that Al-Failakawi managed: through it, he brought mariners into his crew, provisioned his dhow, and wove the enterprise into the marketplace. If there was one contract to know how to generate, this was it. And so he wrote down not one, but two—or, counting Al-Khariji's, three.

TWO

The Shatt Al-ʿArab

THE FIRST PORT AL-FAILAKAWI and the other Kuwaiti dhows that sailed alongside him called at was not far across the water. It was about as close to Kuwait as one could get: the Shatt Al-ʿArab waterway, outside of Basra, downriver from where the Tigris and Euphrates rivers met. It took less than a day for dhows to reach the Shatt, and nakhodas knew the route there so intimately that they wasted no energy describing it. Few nakhodas would have gone all the way to Basra, which, despite its prominence as a port city, actually lay more than sixty miles upriver from the Persian Gulf. Instead, they sailed to the many small towns and villages that dotted the distance between the city and the sea—a world in the orbit of Basra, to be sure, but distinct from the city itself.

Before they made their way up the Shatt, all dhows gathered at Al-Faw, a small stretch near the mouth of the waterway that ran around eight miles upriver from the sea. There, the *Crooked* bobbed in the water, alongside dhows from around the Western Indian Ocean. "Dhows sailed and dhows came in," observed Alan Villiers. "All kinds of dhows lay off Fao, or moved slowly on the water—sweet-lined great *booms* from nearby Kuwait, shapely *baggalas* and sturdy *sambuks* from Sur, *kotiyas* from Bombay" began arriving in late August, and continued to sail into the Shatt until late October; only a few would come in November, as the weather began to cool.

Like Al-Failakawi, all who came to the Shatt came for one reason and one reason only. "Dates brought them all," Villiers wrote; "they were the forerunners of the season's date fleet, to distribute the produce of Iraq over the Eastern seas." For Kuwait's merchants, Shatt dates constituted the currency through which they bought into the markets of India and South Arabia. The date plantations of the Shatt left a deep imprint on the dhow itself, too: a dhow was measured not in abstract tonnage, but in its capacity to hold packages of dates. A dhow's masts, too, were called its *digal*—a term that also referred to the seedlings that sprouted from the tree's base; dates and dhows converged in the minds of shipwrights and mariners.[1]

The Shatt basin produced an astonishingly expansive date crop. One observer of date agriculture around the Middle East called the region "the largest single area" of date palm cultivation in the world, covering roughly 138,000 acres and comprising some fifteen to sixteen million palms. Another noted that as a date-growing region, the Shatt was "probably the most prolific and extensive in the world." Whereas the dates much further up the Tigris and Euphrates served nearby markets in Northern Arabia, Syria, and Northern Persia, those of the Shatt supplied "the United Kingdom, India, the United States of America, Southern and Central Arabia, and the rest of the world."[2]

Sailing up the Shatt, mariners on Al-Failakawi's dhow would have been treated to a view unlike any other: rows and rows of date palms, as far as their eyes could see, abutting the sweet river water—a veritable feast for the eyes. "The sight of the river's banks," wrote one traveler in 1827, "the freshness of the air, and the smoothness of the water combined to produce sensations of the most exquisite description." More than a century later, Villiers echoed the sentiment: "This green and this fresh water made up the Kuwaiti's Paradise," he declared, "this land where the water flowed and all things grew, and all a good Arab need do was to sleep and take his ease while the dates ripened and the fruits of the earth came to his table."[3]

The crew passed hamlet after village—settlements so small it would be difficult to locate them on a map. From Al-Faw, they sailed past one after the other: Al-Qaṣba, Al-Manyūḥi, Al-Maʿāmir, Al-Dora, Al-Duwaib, Al-Dawasir, Al-Ziyadiah, and many more, running all the way up to Basra and then another forty-five miles past it. From the deck of the dhow, there would have been little to distinguish one from the other; together they formed an unbroken chain of date groves running roughly forty miles on the Shatt's eastern bank, and more than one hundred miles on the opposite side.

Al-Failakawi knew, though, that this was anything but an undifferentiated landscape of date palms; the nakhoda knew every one of the places his dhow passed. To him, the chain of date groves broke down into towns and villages, and even further, into discrete date farms owned by known individuals. He knew, too, that despite the deep roots that the palms thrust into the soil and the bucolic warmth of the scene surrounding him, there was hardly anything local about the place. Many of the plantations he sailed past were owned by someone from elsewhere: there were many Iraqi landowners, yes, but there were just as many plantations owned by merchants from Kuwait, Bahrain, Bushehr, Muscat, and beyond.

The dates that hung from the branches of the palms in hued bunches of yellow, brown, and red—those, too, were almost all destined for markets elsewhere. Around the Western Indian Ocean, consumers had their own preferences—and there were many choices to be had, given the varieties of dates that plantations around the Shatt produced. By the 1920s, there were 132 different types of dates produced in Iraq, and Basra merchants exported growing quantities of them. Between 1899 and 1906, Basra exported an average of around forty thousand tons of dates every year, while reports from 1918 and 1919 put the number at around eighty thousand tons.[4] Four varieties dominated the Indian Ocean market. The most popular was *istaʿmarān*, which merchants called *sāyer*—a hardy variety that accounted for nearly half of the date palm population in the Shatt. By the early twentieth century, the *ḥalāwi* variety, whose light brown color at ripening appealed to buyers in the United States, grew to constitute roughly a third of all dates produced in the region. The richly flavored *khaḍrāwi* constituted roughly a tenth of the Shatt's date palms, and along with *zāhidi* dates (which accounted for less than 5 percent of all date palms) found ready markets around India.[5] Other varieties, like *dīri*, were also shipped out of the Shatt, though in much smaller quantities.

The specific date variety was only part of the trade calculus. Just as important was its stage of ripeness. Generally, dates ripened in three distinct stages: at the earliest stage, the *khalal*, the fruit's skin was still firm and largely inedible due to its bitter tannins. With time, the tannins broke down and the hard skin softened into moist, sweet flesh. This was called the *riṭab* stage, and merchants knew well that as *riṭab* the dates could not withstand the journey overseas, as they would quickly rot; virtually all of the harvested *riṭab* was sold locally. Very soon after, though, the dates fully ripened and began to turn into the sweet and juicy fruit that most consumers knew them as being: *tamar*.

The picking season lasted roughly four months from the earliest stage of the *khalal* harvest in August to the last stage of the *tamar* harvest in late December. Though there was no market for *khalal* due to its bitterness, demand for *raṭab* dates was steady. To bring *raṭab* to overseas markets, farm workers had to pick them at the *khalal* stage and then boil them to help them keep for longer periods; they would then pack them into baskets for loading onto the dhows. The heat and moisture that the packed *khalal* was subjected to over the course of the journey hastened its transition into *raṭab*, which could be sold in nearby port cities. The other varieties destined for overseas markets—that is, the *sayer, ḥalawi*, and *zahidi* dates—were all shipped at the *tamar* stage.[6] Thus, an enterprising nakhoda had to factor in different preferences among consumers around the Indian Ocean, the time of the year in which the dates began to ripen, and the length of the journey to its final destination.

Dhows coming to the Shatt from Kuwait and other ports, then, timed their arrival: those who wanted to supply the market with the cooked *khalal* dates arrived in late August, just as the harvest was beginning. These were known as the *harfi* nakhodas: the first batch of the season, who came to load their cargoes and leave in the hopes of being able to make at least two, if not three, voyages to India over the year.[7] More dhows began to arrive in September, and by October the process of harvesting and loading was in full swing, stretching into November and even early December.

• • •

Letters, receipts, telegraphs, manifests—Hajji Muhammad bin ʿAbdullah Al-Matrook could hardly keep up with it all. Day in and day out, paper streamed into and out of his office in the Al-ʿAshar neighborhood of Basra. Things were always busy, but the four months between August and December were exceptionally frantic for him and other date merchants. When the dhows came in, they had to receive the nakhodas in their homes and offices and arrange to load their ship with its cargo; notes had to be signed, and customs forms had to be filled out. At the same time, mail poured in from their business partners from around the Gulf and Indian Ocean—letters detailing orders of specific goods to be loaded onto the dhows, notices of goods that had been sent or were received, sheets advertising the prices that different varieties of dates commanded in markets like Karachi and Bombay, wire transfers of money, and more. Those who wanted to succeed had to dedicate themselves to coordinating this circulation of goods, money, information, and people, all while keeping an eye on a ticking meteorological clock. A merchant like Al-Matrook thus had to straddle multiple geographies and temporalities simultaneously, contending with pressures from suppliers and buyers, all while ensuring the timely circulation of information, goods, and money.

By 1924, Al-Matrook had been in the date business for a long time. He was born in Kuwait in 1885, and at a young age began working with his father

FIGURE 6. The merchant Hajji Muhammad bin ʿAbdullah Al-Matrook. Photo from Aslan Al-Matrook.

ʿAbdullah, who ran business offices in Kuwait, Basra, and Mohammerah, alongside dozens of other Kuwaiti merchant-landowners. After his father passed away at the turn of the century, Al-Matrook took over the family office and oversaw the expansion of the business's operations to include port cities around the Western Indian Ocean. The earliest document from his collection is an account sheet dated January 7, 1903, when Al-Matrook would have been just eighteen years old.[8]

The geographical scope of Al-Matrook's business dealings was broad, to say the least. He received weekly letters from traders in Kuwait, Bandar ʿAbbas, Lingeh, Bahrain, Karachi, Bombay, Mangalore, and Calicut. His correspondents wrote mostly in Arabic, but some also sent him letters in Persian and occasionally in Gujarati and English as well. Most of these were regular trading associates with whom he kept open accounts, but he did have some particularly durable engagements. He appointed a regular agent in Mohammerah, Muhammad Jaʿfar Al-Kaʿbi, and found a long-term business partner in Nasser bin ʿAbdulmuhsin Al-Kharafi, who operated the Karachi branch of the firm's business; Al-Kharafi's son ʿAbdulmuhsin joined in the partnership as well and mostly worked out of Bahrain. All three were peripatetic, though to different degrees.

As part of a transregional community of merchants, Al-Matrook was in the business of circulating goods. Much of his work was in dates: in receiving orders from abroad and ensuring there were enough cargoes of the dried fruit on hand to load onto the dhows and steamers that sat waiting. However, dates formed only part of his business portfolio. Al-Matrook received shipments of goods from India and Persia, which he would then sell locally or reexport to markets around the Gulf. Most of the incoming cargoes he dealt with involved grains—barley (*ḥinṭa*) and rice—and different varieties of sugar. In shipping goods, Al-Matrook took advantage of Basra's status as a port of call for steamships from India: he regularly sent and received goods on steamers run by the Peninsular & Oriental Company (P&O) and the British India Steam Navigation Company, which P&O bought in 1914 and which had established a regular shipping line between the Gulf and Western India since the 1860s.[9] As with dhows, Al-Matrook used steamships to send a variety of goods to port cities in India—principally dates, which he shipped under the mark "MK" (which presumably stood for Matrook-Kharafi).[10]

To coordinate the movement of goods into and out of Basra, Al-Matrook utilized a few different technologies that were within arm's reach in the port city. In addition to the written correspondence that he made sure to keep up, he regularly sent and received cables via telegraph (which they referred to as simply *tayl*) from his business partners, often for similar purposes: to coordinate orders, to circulate information on prices, and to conduct all manner of mundane business. Gulf merchants like Al-Matrook seamlessly joined together their cables and epistolary exchanges: virtually all of his correspondence mentioned cables sent and received alongside letters that had come and gone, all usually within the first few lines.[11]

Running alongside the goods, letters, and telegraphs that circulated into and out of Al-Matrook's office in Basra were flows of money. Al-Matrook was a regular visitor to the Basra branch of the Eastern Bank Ltd., which had set up in the port city in 1915, just after the British invasion of Iraq during the First World War. By 1923, it had branches in several port cities where he had business associates, including Karachi, Bombay, and Bahrain.[12] Bank transfers formed a critical part of Al-Matrook's business operations: wiring money allowed him to settle accounts with his associates on an ongoing basis, but also allowed him to send ready cash for large orders so that his business partners would be able to meet their obligations.

All of this was part of Al-Matrook's daily operations, regardless of the time of the year. However, between August and December, when the dhows came in, the lion's share of his work involved receiving nakhodas sent by trading partners from around the Western Indian Ocean to bring dates to markets around the region. Many had sailed to Basra from Kuwait, but just as many came from other ports: one of Al-Matrook's business partners in Karachi regularly sent over dhows, *kotiyas*, and other sailing ships carrying rice, flour, sugar, coal, and other goods, and requested that he arrange for them to be loaded with cargoes of dates for the return journey.[13] A partner in Bombay once dispatched three dhows, all owned by different Gujarati merchants, to load dates for markets in Bharuch, Bombay, Mukalla, Aden, and Berbera.[14]

For all of the Shatt's natural bounty, there was a lot that merchants like Al-Matrook had to track in the months leading up to the summer so that they'd be able to supply the steady stream of dhows coming to load dates. Landowners, many of whom did not live along the Shatt, had to hire seasonal laborers to till the land around the date palms, removing roots and adding manure that was supplied from the city's markets and stables. They also had to ensure the palms were constantly being irrigated, either via irrigation channels that directed water from the river or via nearby wells. By April, planters had to also ensure all of their palms had been fertilized, which required peasants (*fellāḥs*) to manually place male date palms' pollen in the female palms' stigmae. And when the palms finally began to bear fruit, merchants had to secure additional *fellāḥs* who would harvest and pack the dates—for dhows, into baskets woven from palm fronds, and for European and American merchants, into wooden boxes—in time for them to be loaded onto the ships.[15]

The delicate timing of these processes was made all the more precarious by the annual fluctuations in the date crop itself, which could affect supply, and by extension, price. Sometimes, these were beyond the immediate control of anyone: in 1896, for example, prolonged flooding of the Shatt Al-ʿArab followed by excessive heat damaged nearby date groves, leading to a drop in both the quality and the supply of the fruit. In a few years, growers had brought supplies back up, but then weather fluctuations—alternately, cool, dry winds in the summer of 1900 and hot, dry winds during the ripening stage in 1901—once again reduced the quality of the crop. At other times, growers faced the opposite problem: an oversupply of dates as a result of speculation following a particularly lucrative season. The result was a precipitous drop in date prices.[16]

Even when all the climatological factors went their way, merchants like Al-Matrook still had to contend with people. After all, the supply chain did not begin with him: there were landowners, *fellāḥs*, and other laborers, all of whom could claim different rights in the series of stages that dates had to pass through before they could be loaded onto a dhow.[17] One Bahraini landowner in the Shatt heard complaints from his *fellāḥs* that his agent, whom he had hired to manage his farms while he was away, had taken their share of the produce; they threatened to leave work altogether. One of his correspondents noted that the agent had also left the family's gardens in such a state of ruin that they were no longer profitable to run.[18] Problems like this weren't uncommon: absentee landlords along the Shatt who were forced to rely on agents to manage their properties for them often found themselves embroiled in similar disputes.

To manage the human dimensions of the date business, Al-Matrook turned to paper. With every batch of dates that arrived, he drew up a brief receipt, noting that dates arrived on a small boat sent by one of his suppliers, and attesting to the weight and variety of the dates, as well as the price paid. Al-Matrook did not write the notes himself; a few dozen named other people as the authors—most likely middlemen that Al-Matrook contracted to help secure supplies of dates. He collected the receipts and compiled them into longer lists, tallying up the cargoes of dates that came in from different suppliers around Basra. Atop each chit, he inscribed a number that corresponded to an entry in an accounting sheet. A list of numbers corresponding to the chits ran down a column on the left, and on the right, the name of the sender, the variety sent, and the weight of the cargo. An example, read from right to left: "[Collected] 65 || Arrived from Shaikh Saleh Al-Khaled Al-Saʿdun, *zahdi* || the weight of one is 30 || [Receipt] No. 3." Another read "83 || Arrived from Ahmad Al-ʿAbdulnabi, *zahdi* || the weight of one is 30 || No. 45." In a column on the right, he kept a running total of *manns* received.[19]

Through these sheets of paper, Al-Matrook monitored his investments and the flow of goods into and out of his office: he transformed qualities into quantities, lending visibility to processes that unfolded over broad distances and across multiple hands. Small sheets of paper poured into larger ones that compiled into ledgers designed to track the flow of goods and money back and forth between the date garden, the business office, and the bank. Like the goods and money that it attested to, the paper trail spread itself out along a dense web of people who moved along the land and its waterways, carving channels for the movement of goods into and out of the Shatt.

. . .

As dhows arrived at the Shatt, they sailed past the rows of trees, hamlets, and creeks that began at Al-Faw. And there, shortly after they entered the waterway, should they have looked to the western side of the riverbank, they would have been just barely able to make out the remnants of a light-colored stone structure, surrounded by thickets of date palms. Those who had been sailing for long enough might have remembered that this was what remained of an Ottoman fort that had once housed forty-five soldiers and a medical officer.[20] Though the fort was hardly very old—the Ottoman government built it in 1886—it stood as a reminder of a much longer history, of the tensions that ran between the Ottomans, the successive Persian empires, and the tribesmen who inhabited the lands surrounding the Shatt.

The Shatt had arguably always been a borderlands area—a frontier—of one sort or another throughout early modern history, from the fall of the ʿAbbasid empires onward. Situated as it was at the confluence of the historic Tigris and Euphrates rivers, at the mouth of the Karun river and at the head of the Persian Gulf, the region was frequently in the crosshairs of different expanding empires. Over the course of the sixteenth century, as the Ottomans and Safavids successively thrust themselves across the Tigris-Euphrates valley with equal vigor, the inhabitants of Basra and the towns near it found themselves at the receiving end of overtures and threats from a changing group of overlords. By the middle of the century, the two empires had largely settled on a tense frontier zone—far from an established boundary, but less of a recurring battleground—that ran down the eastern edge of what is today Iraq, all the way down to the Shatt itself.[21]

But Basra and the Shatt were not within the sightlines of the Ottomans and Persians alone. From the mid-1500s onward, the towns formed part of a changing web of European expansion in the Indian Ocean, starting with the Portuguese—who, from their fort in Hormuz, saw in Basra an important trading partner—and then, over the 1600s and 1700s, the Dutch and English East India Companies, who traded extensively with Basra from their factories in India and the Gulf. Both the Dutch and the English gradually shifted their Gulf headquarters from Bandar ʿAbbas to Basra in the 1720s, following the fall of the Safavids and the political uncertainties that trade in Persia presented.[22] And as the European presence gained traction, the Ottomans attempted to further incorporate Basra into the sprawling imperial administration; during the eighteenth century, the Ottoman government began appointing members of the local elite as tax farmers, weaving them into the imperial fabric.[23]

For the shaikhdoms and the tribal confederations that established themselves at the head of the Gulf, the ebb and flow of imperial power thus came from multiple directions: from Istanbul in the northwest, Tehran in the northeast, and then from Hormuz and Bombay within quick succession of one another. Theirs was an inter-imperial borderland world, one in which neither the Ottomans nor the Persians in their various guises had the means or energy to invest in an entrenched presence on the Gulf's shores, and in which none of the Europeans saw much to be gained from embroiling themselves in local politics. These groups also frequently clashed with one another and aligned themselves with whichever imperial power would give them an advantage over their rivals. And as they fought among themselves, trade suffered: merchants, fearing loss of property and life, sent their goods along different, more secure routes, much to the chagrin of the central governments that hoped to tax them. The scattershot nature of the political geography of the Shatt and the northern Gulf, then, emerged out of attempts by different local groups to capture the gains of shifting trade routes.[24]

Gradually, the central governments in Tehran and Istanbul came to realize that the most effective way to subdue these groups was to incorporate them. Over the eighteenth and nineteenth centuries, the Ottomans designated many of the political elite around Basra as members of a sprawling Ottoman imperial bureaucracy—as tax farmers, district governors, and more—creating a class of merchants and tribal heads with strong ties to the central government. This extended beyond Basra itself and included Ottoman overtures further down the Gulf, to the rulers of Kuwait and Hasa. The Kaʿb rulers of Mohammerah similarly received a series of investitures from the Qajar government of Persia but acted independently of them. By the middle of the nineteenth century, these actors had come to a sometimes tense but often manageable arrangement whereby the local political elite paid lip service—and occasional taxes—to their nominal overlords in Istanbul or Tehran and acknowledged their suzerainty but exercised substantive autonomy over their affairs. For their part, officials in Tehran and Istanbul gradually formalized their boundaries along the Shatt in two treaties, signed in 1823 and 1847.[25]

The advent of a more energetic British Indian imperial policy in the Gulf during the nineteenth century further reconfigured the calculus among political actors along the entire stretch of the littoral, from the Shatt to Muscat. Prompted by the threat that the Napoleonic invasion of Egypt in 1798 posed to India, officials from the Government of India set about establishing a firmer British presence in the Gulf, one of the principal routes to India. British officials found themselves further drawn into regional politics out of a desire to stamp out slave trading and piracy in the region but also an obligation to protect the mercantile interests of their Indian subjects in the region. During the second half of the nineteenth century, investments in a submarine telegraph system that ran through the Gulf, coupled with the threat that a renewed Ottoman interest in the region posed, pushed British officials in the Gulf to take a more proactive role in regional politics.[26]

For rulers trying to balance obligations to their imperial patrons with their desire for political autonomy, the British presence offered a useful counterweight. The diplomatically savvy could exploit fears by imperial officials on either side of them to negotiate terms that would leave them with the greatest degree of autonomy. In Kuwait, it was Mubarak Al-Sabah who actively courted the British, playing them off against Ottoman officials who saw in Kuwait an extension of the province of Basra. In Mohammerah, the ruler Jaber bin Mirdaw Al-Ka'bi had tried to court the British government in the mid-nineteenth century, but to no avail; they were not yet interested in establishing a strong presence in the area, and Mohammerah's sensitive position on the border between the Persians and Ottomans made it too much of a liability to interfere with.[27] By the time Jaber's son Khaz'al came to power in 1897, the situation had changed: Britain was more deeply entrenched in the affairs of Southern Iraq and the Gulf. Though Khaz'al entertained a close relationship with the Qajar dynasty, he saw in the British an opportunity to establish autonomous rule over the tribes in his area and sought assurances from them that he could pursue his independence if the Qajars should ever fall.[28]

But Gulf rulers did not only draw strength from larger imperial protectors; they actively forged relationships and alliances with one another as well. By the beginning of the twentieth century, a stable alliance emerged between Mubarak and Khaz'al. The two developed their friendship in the mid-1890s, as they met in different towns along the Shatt in their capacity as agents for their brothers, the rulers of Mohammerah and Kuwait. They quickly formed a bond with one another—one that only grew denser as they both came to assume positions of leadership within their respective emirates. "The two regions," wrote one historian, "had never witnessed denser ties or relations of trust so bare" than they did under the two rulers.[29] They even went so far as to build palaces in each other's emirates; the ruins of Khaz'al's palace still stand in Kuwait.

By the turn of the century, members of the Al-Sabah family came to possess large tracts of land along the Shatt.[30] The ruler, Mubarak Al-Sabah, owned much of the land at Al-Faw, which by then included some ten thousand date palms and fifty hamlets of more than three hundred households.

The land came into his possession only after a protracted dispute among members of his family surrounding how estates on the banks of the Shatt were to be divided. Mubarak also owned extensive date gardens on the banks that were opposite Mohammerah, on the Turkish side of the Shatt, including two sizable properties in the village of Fadaghiya, which he had bought from the families of a handful of different Ottoman political notables.[31] Like other landowners along the Shatt, Mubarak's date farms supplied the dhows that plied the waters between the Gulf and the Western Indian Ocean. Every year, before the dates even ripened, Mubarak would summon Kuwaiti merchants and contract out cargoes of dates to them, extending lines of credit that could last as long as three years.[32] Khazʿal, too, marketed his produce through Kuwaiti merchants in India, to whom he would sell dates on credit.

But for a ruler like Mubarak, date plantations were as much a matter of politics as they were of profit. By cultivating an independent source of wealth, the Kuwaiti ruler was able to shift a considerable amount of political power away from the merchants, whom his predecessors had depended on for voluntary contributions.[33] At times, too, Khazʿal assigned Mubarak additional produce from his own date palms so that the Kuwaiti shaikh might buy the armaments necessary to fight battles in the northern Gulf.[34] But more than that, he was able to use the credit he extended to Kuwaiti merchants as a way of securing a base of clients: officials noted that the ruler continually reextended the period of repayment on loans, particularly in lean years, "for his policy is literally to be the 'father of his people.'"[35] Wealth in the Shatt thus constituted a resource that rulers could circulate, convert, and more generally draw upon to shore up their political positions.

Iraq's political landscape changed dramatically in the decade before the *Crooked* arrived there with Al-Failakawi and his crew. As Al-Khariji wrote: "The English entered into and bombed Al-Faw on 17 Dhu Al-Hijja 1332 [November 6, 1914] on a Friday, at 5 minutes until 5 [10:55] in the day. That day, heavy winds blew from the south, and the next day the ships sailed into the creek and there erupted an intense battle between them, lasting an hour and a half. We saw this with our own eyes, as we loaded dates from Al-Qaṣba."[36] From his dhow, Al-Khariji watched as the British landed Indian soldiers and began bombarding the Ottoman fort—the very one Al-Failakawi sailed by on his way into the Shatt. The Mesopotamia Campaign, one of the main theaters of the First World War, lasted another four years. After the guns fell silent, a new round of violence erupted as Iraqis sounded calls for revolt against the British Mandate; thousands died on both sides of the conflict. In 1921, officials agreed to install Faisal bin Hussain Al-Hashemi, the king of Syria, as king in Iraq. For the next decade, they would rule Iraq through him.[37]

Like his teacher, Al-Failakawi worked on dhows going into and out of the Shatt throughout the armed violence. The waterway that he sailed into as captain might have moved the same as it did when he was a young sailor, but it bore the scars of a decade's violence.

• • •

Gulf merchants positioned themselves at the interstices of the global, regional, and local political formations and alliances along the Shatt. The fragmented nature of political authority in the imperial borderlands was balanced by the substantive autonomy of the local shaikhs, and merchants with enough means could convert wealth from the Indian Ocean world into landholdings along the Shatt. And for Gulf merchants, date-growing land on the Shatt constituted the bedrock of the family's wealth. Many would have had investments in at least a few different sectors: the pearl trade, real estate, and other local enterprises. Those involved in overseas trade, however, seldom failed to establish a foothold in the date gardens of Southern Iraq. Profits generated from the trade in pearls, buoyed by global demand, invariably made their way back to landholdings, and the Shatt was undoubtedly the best place to own land.[38]

Those who profited from the pearl trade took the opportunity to invest their earnings along the Shatt, diversifying their business portfolios to include dates as well as pearls. And as they converted at least some of their pearl wealth into agricultural land, they met with yet another swell in profits. Starting in the 1870s, a growing number of American and European mercantile firms established shipping operations in the area surrounding Basra and began to export large quantities of dates from the Shatt. Over the next few decades, they shipped hundreds of thousands of tons of the dried fruit to markets in India, England, continental Europe, and the United States. By the 1890s, the American market for dates had become so extensive that merchants launched a small fleet of steamers to sail the nearly ten thousand miles from Basra to New York directly, stopping only in Muscat to pick up even more dates; one firm, the Hills Bros, even bought its own date farm and riverfront building, which it called Beit Hills, or the Hills' House.[39] The growing trade buoyed the profits of date growers and local merchants, adding to the wealth that the Shatt already generated.

It was no wonder, then, that rulers, notables, and the regional well-to-do sought to amass property in the region. Large landholders abounded in the villages along the Shatt, and many marked their wealth with prominent houses. The settlement of Bait Naʿmah, for example, was a small hamlet of 250 people, but was distinguished by "a palatial mansion with a frontage of about 400 feet by the late Hajj Ahmad an-Naʿamah," whose family owned a plantation of 15,000 palms. In Sabiliyat, a village of about 5,500 inhabitants and 55,000 palms, the *naqīb* of Basra—the shaikh of nobles—resided in a "well built and fairly large house," and owned date groves in another village, Kut Al-Shaikh, where he partnered with the Lynch Bros firm in shipping and the date trade. Other villages were wholly owned by certain merchants—for example, the village of Majmaʿ, which consisted of 160 inhabitants and 6,000 palms and "belong[ed] to M. Asfar, who has bought the whole of it."[40] Merchants from around the Gulf—from Kuwait, Bahrain, Persia, and Iraq itself—all sought to own land in the Shatt and establish themselves among the region's landed aristocracy during Basra's "Gilded Age."[41]

The story of the Al-Ibrahim family vividly illustrates the changing possibilities that the Shatt presented the Gulf's entrepreneurs. Its scion, Muhammad bin Ibrahim, settled in Kuwait in the mid-1700s, engaging in the transit trade between India and the northern Gulf by the end of the century. When Muhammad passed away, his four sons took on an aspect of the family's business. ʿAbdullah began investing the family's wealth in farms near the Shatt in the early 1800s, leasing a sizable plot of land from the governor of Basra in 1814. His brother ʿIsa migrated to Surat, on the northwestern coast of India, where he continued to do business on behalf of the family for the rest of his life, while ʿAbdulwahhab oversaw transportation, directing the family-owned dhows that plied the waters between the Gulf and the Western Indian Ocean. Of the four brothers, only ʿAli chose to remain in Kuwait, representing the family interests there; his two sons Muhammad and ʿAbdulaziz joined their uncle ʿIsa in Surat.[42] The Al-Ibrahims thus illustrate well what historians call the "family firm"—a business enterprise principally, though not exclusively, comprising members who share a blood relation—often for reasons of trust, but also to manage wealth.[43]

Women played an important role in developing the wealth and connections that nurtured these family firms. In 1815, just as the family business was gaining momentum, the eldest brother ʿAbdullah died. When his brother ʿIsa in Surat fathered a son later that year, he named him ʿAbdullah. ʿAbdullah bin ʿIsa inherited the mercantile mantle of the family, moving back to the Gulf after his father died in Surat in 1839.[44] There, he had a head start: on his father's side, his uncle had already established the family presence in the date gardens along the Shatt. Meanwhile, his mother Fatima bint ʿIsa Al-ʿAbdulkarim had left him some choice property as well: her own mother had established a *waqf* (or endowment) of property that she had received from her brother in Al-Saraji, a small village on the eastern bank of the Shatt, close to Basra; she designated her two daughters, Fatima and Lulwah, along with their children, as beneficiaries.[45] Lulwah had also purchased property in Al-Saraji, and in 1826, she established her own *waqf*, designating her son Ahmad and her sister Fatima, along with their descendants, as beneficiaries. By 1839, ʿAbdullah bin ʿIsa Al-Ibrahim was the only designated heir left, and received the proceeds from both *waqfs*—the one established by his aunt and the other by his grandmother, all through his mother.[46] When it came time to marry, he chose his paternal cousin Mariam, who herself had inherited family wealth along the Shatt.

Though he was already flush with property that had accumulated on both sides of his family and through his wife, ʿAbdullah quickly set about expanding his family's holdings, establishing himself as one of the principal landholders in the Shatt. During the 1850s and 1860s, he bought up plots of land both large and small, all around the village of Al-Dora—mostly from small landholders, but occasionally from Ottoman officials who owned date farms in the area, sometimes in partnership with members of the Al-Sabah family.[47] Through purchases like these, ʿAbdullah slowly accumulated enormous landholdings around the Shatt, anchoring himself principally in the towns of Al-Dora and, in the 1860s, Abu Al-Khasib. By one estimate, he owned roughly seven thousand acres of land containing roughly seven hundred thousand date palms.[48] A presence in Abu Al-Khasib was a strategic choice: the annual meeting between the date growers and exporters, held early in September to fix the prices for the season, sometimes took place there.[49] By investing in land there, ʿAbdullah forcefully established the family's presence in the transregional date trade.

When ʿAbdullah died in 1875, the bulk of the family wealth passed down to his son Yousef, who quickly emerged as a prominent member of the Gulf merchant community.[50] Meanwhile, Yousef's first cousin, Jassim—the son of one of the brothers who moved to Surat—had established himself among Bombay's mercantile elite. Together, the cousins benefited from a regional commercial boom that unfolded just as they took on larger roles within the family business. Jassim benefited directly from the boom in the Gulf pearl trade, which reached its heights shortly after he was born and extended well into his young adulthood. And it was just as Yousef took over the family's Shatt properties that there emerged a growing interest among European and American traders in marketing Basra dates. The Al-Ibrahims adapted quickly to the new market and even began packing their dates in branded wooden boxes so as to appeal to the new buyers.[51]

The swell in profits and deeper integration into global markets allowed the new generation of Al-Ibrahim merchants to reinforce the family's position in regional politics. The family had so deeply entrenched themselves in Shatt society that by the beginning of the twentieth century, British intelligence officers would refer to the town of Al-Dora as "Dorat Bin-Ibrahim"; Yousef himself came to be known as "the Shaikh of Al-Dora."[52] And through the female members of the Al-Ibrahim family, he further inserted himself in regional politics. His maternal cousins had married Mubarak Al-Sabah's half brothers Muhammad and Jarrah, both of whom ruled Kuwait in the early 1890s, and Yousef was known to have very close ties to the siblings.[53]

The Al-Ibrahims were unique in their visibility, but there were many other merchant families who amassed agricultural land around the Shatt before and during the date boom. The Al-Sager family bought up equally large tracts of land, focusing their energies on Al-Faw and Mohammerah. The head of the family, Yousef bin Sager, invested in date farms along the Shatt Al-ʿArab, like many other merchants. An 1829 deed in the family's possession details his purchase of shares in two different Shatt farms. Nearly a decade later, he bought a one-third interest in two different properties. Al-Sager's luck was short-lived: he died in a shipwreck off the Malabar coast, and his son ended up permanently migrating to the East African coast. But by then, the family was set up for success. The Basra properties would form the core of the family's wealth for generations.

By the late nineteenth century, Yousef's grandsons Sager and Hamad and their sister Bazza emerged among the large land-owning merchants of the Shatt. The brothers often divided up the work of the family business between themselves. Hamad usually held down the Basra branch of the family firm, while Sager, who was principally based in Kuwait, was often (though not always) the one who traveled between the Gulf, India, and Yemen, maintaining the family network of business associates who would buy their dates and send them other goods. All the while, the two continued to add to the family's holdings in Basra, purchasing date farms until late into the 1920s.[54] Sager's son Yousef, too, ended up establishing a business office in Calicut—

not far from where his great-grandfather's ship once foundered. And their sister Bazza married another date-trading merchant, Thunayyan Al-Ghanim; their son Muhammad Thunayyan Al-Ghanim would become one of the biggest date traders in Karachi and Gujarat and would employ countless nakhodas in his lifetime.

The history of the Al-Sager family—their slow accumulation of wealth in date farms, their gradual development of a network of associates, and their marriage into families with similar interests—was typical of the Gulf-based Indian Ocean family firm; the Al-Matrooks, the Al-Ibrahims, and many others did as much. The Al-Sagers, though, distinguished themselves in the amount they invested in ships. The brothers owned one of the largest fleets of dhows in Kuwait: in 1920, they owned seven ocean-going *booms*, seven *balams* for fishing and short-distance trading, two pearling *sambuks*, two *shūʿis*, and two other vessels designed to navigate the inner creeks of the Shatt. They also owned anywhere between a one-third to a three-quarters interest in nine other vessels.[55] Theirs was a family that was committed to building an integrated enterprise—from farm to ship to distant shore.

As Al-Failakawi navigated the *Crooked* along the Shatt Al-ʿArab, he traced the outlines of this long history. On September 20, he and his crewmembers loaded dates from Al-Dora, where the Al-Ibrahims had spent decades amassing farmland and weaving themselves into the fabric of regional politics. From there, the crew sailed on to the date gardens of Al-Fadaghiya, which Mubarak Al-Sabah had purchased just fifteen years before, and which were the subject of extensive litigation between the Al-Sabah family and the Kingdom of Iraq over whether the Kuwaiti rulers could continue to enjoy the tax exemptions they had received from the Ottoman government when they bought the gardens. From his dhow, the nakhoda could see the sediments of the past—of rulers, merchant-princes, and sultans, of fortunes made and contested—gather along the banks of the Shatt.

• • •

But nakhodas along the Shatt had to grapple with more immediate temporal horizons. Roughly eighty miles downriver from Al-Matrook's office, the nakhoda Rashid bin ʿAli Al-ʿAsʿousi dropped anchor at the village of Qasba, just up from the mouth of the Shatt. It was a small village with few inhabitants, but boasted date plantations stretching twenty miles along the river banks and about two to three miles inland; it produced about one hundred

FIGURE 7. Dhows loading dates at the Shatt Al-ʿArab waterway. Photo from Pitt Rivers Museum, Oxford, UK.

thousand baskets of dates every season, enough to fully load at least thirty average-sized dhows.[56] That year, the most pressing issue Rashid faced was timing: how long he spent waiting to load his cargo, when he could leave, and what it meant for his ability to reach overseas markets on time.

Loading the dhow took up precious time. For Kuwait's nakhodas, a wait of two to three weeks was normal; at times, it took as long as a month and a half or even two months. Rashid was losing patience. In a letter he wrote to Kuwait in mid-October, he complained that he had only been able to load two hundred *qoṣras* of dates so far; most of the dates had been reserved for the shaikhs. More than two weeks later, he still hadn't loaded a full cargo. There were many dates in Qasba, he grumbled, but they were in the hands of the *fellāḥs*: "May Allah ease our affairs."[57]

The more pressing obstacle, though, was weather. He complained about the *kaws*—localized, gale-force winds that blew from the southeast, which brought rain and created weather conditions that made it difficult to safely sail out of the Shatt. Every day that the dhows had to spend waiting at the Shatt was another day of consumed provisions, forcing nakhodas to replenish their stores. Merchants and shipowners in the Shatt provided the sailors with provisions during the loading period and even put on a feast at the end, but that came to an end after the last basket had been loaded.[58] In a letter he wrote from Qasba in late November, Rashid grumbled that the *kaws* winds had delayed their departure from the Shatt, and asked that 'Abdulrahman Al-'As'ousi send the ship "coffee and sugar, and some tea" from Kuwait; he also let him know that he was forced to replenish the ship's water supplies.[59] Rashid also took food on credit from Al-Matrook at least once—just over Rs 550 worth of provisions for the outward voyage.[60]

Mariners might also grow restless while waiting in the Shatt, or have a change of heart altogether. One year, Rashid wrote to 'Abdulrahman to let him know that one of his crew members, Muhammad bin Harban, had absconded into Iraq after disembarking from the dhow two days earlier. His dhow, he wrote, was short on sailors; would 'Abdulrahman be able to send some—at least a helmsman (*rā'i sikkān*) from Kuwait? 'Abdulrahman responded quickly: less than a week later, he let Rashid know that he had hired "Fairuz, the slave (*'abd*) of Bin Fahad," who was both a helmsman and a "good man"; he advanced him Rs 120 and sent him along with another nakhoda and his men. The nakhoda was also going to bring supplies of rice and flour for Rashid's dhow. 'Abdulrahman apologized for not being able to send more men: "We asked around," he said, "but people are out fishing or have gone for the winter pearl dive (*al-radda*)."[61]

A delayed departure could mean the difference between arriving with the first fleet of dhows in northwestern India and selling at a good price or finding the market oversaturated and having to sail elsewhere to dispose of their goods. Here, having access to someone with a strong network of correspondents helped enormously. As Rashid waited for the *kaws* winds to ease, he received a last-minute set of instructions that would change his course slightly: 'Abdulrahman wrote to let him know that although he initially instructed Rashid to sail to Karachi, he had recently heard that freightage there was low—perhaps due to an oversupply of dhows—and that the market was oversaturated. He suggested a different, more profitable route: to first sell at Kathiawar, and then sell the remainder of the cargo at Porbandar (which he and others referred to as *Khormiyan*) and Veravel. "And if Allah wills it and you sell your dates," he wrote, "there should remain in the *boom* 300–400 *qoṣras* for Al-Naibar [the nakhodas' name for the Malabar coast], and if you hear that there is money to be made in Al-Naibar [in shipping goods along the coast of India], or if you sell what you have, load the *boom* with ballast, and make sure to stop for repairs in Nawā Bombay [Kannur] or Mangalore."[62]

. . .

When the dhow was finally loaded, the crew accounted for, and the winds favorable, nakhodas could start preparing their dhows to sail out of the Shatt. It was early October when Al-Failakawi was finally ready to leave. But before crossing the mouth of the Shatt into the Persian Gulf, he had to make one final stop, back at Al-Faw, where British officials had recently set up a post office and telegraph station near the site of the Ottoman fort they had once bombarded. There, he and other nakhodas checked the incoming wires for any last-minute changes to their plans, and then called at the customs house, where they presented their receipts to be verified and stamped.[63] With his papers finally in hand, Al-Failakawi was cleared to leave. As he inched toward the Gulf and prepared to sail out, he took care to note the dhows that left the Shatt with him: that year he left with the *boom Marzouq* belonging to the nakhoda Mansur—perhaps his teacher, Al-Khariji.[64] He'd see his fellow nakhodas all again, if not on the Persian coast then in India or elsewhere in the Western Indian Ocean.

As he looked out onto the water from the edge of the Shatt, Al-Failakawi ordered the *Crooked*'s mainsails to be hoisted. He watched as the winds filled them out, propelling the dhow forward with them. Heavy with dates, the *Crooked* crossed the mouth of the waterway and sailed out into the Gulf's open waters.

INSCRIPTION

Freightage

THE TIME OF LOADING was also a time of paperwork. As dates moved from the gardens that lined the Shatt Al-ʿArab waterway to the dhow's hold, a stream of documents moved alongside them. As the cargo changed hands from one person to another, pen was put to paper: a receipt, a contract, a line in an account ledger.

The nakhodas knew these well. Of the six different contracts that Al-Khariji inscribed into two pages of the notebook that he passed on to Al-Failakawi, three involved the transport of dates. The longest one read:

> Carried, with the blessings of Allah and his good fortune, in the *boom* called *Fateḥ Al-Khair*, owned by ʿAbdulnabi bin Al-Hajj Qasim, under the guidance of the nakhoda Mansur bin Al-Hajj Ibrahim [Al-Khariji] are 3,000 *qoṣras* of *sayer* dates, equivalent to 75 large *kāras*, and the freightage for one *kāra* is Rs 45 from Basra Creek to Karachi and the ports of Kathiawar, the final port being Bombay. This is sent from Hamad bin ʿAbdullah, transferred, in the aforementioned ports, to Muhammad Thunayyan, and the total freightage is Rs 3,375, and it will be handed over after the safe arrival of the goods to the port and after their unloading, and without negligence. He [the nakhoda] has two manifests, one original and the other a copy, and if one is present the other is null, and in what we say, we place our trust in Allah.

The other two were much shorter. They narrated the contract from the standpoint of different parties: one was dictated by a merchant who was contracting with the nakhoda to deliver one thousand *qoṣras* of *sayer* dates, also in *kāras* that each carried forty *qoṣras*, at a rate of Rs 50 apiece, while the other narrated the same contract in the nakhoda's voice. All three contracts bore the same names, and all included the same date: 22 Muharram 1341, or September 14, 1922.

Like the acknowledgments of debt that the nakhodas had inscribed before, all three of these were model contracts—templates for the prospective nakhoda rather than actual contracts the nakhoda had entered into. These were three versions of the same contract, and there was good reason for it. This may have been the most important contract a nakhoda drew up all season. It was there, along the banks of the Shatt Al-ʿArab, that he staked out the legal and financial contours of the voyage: the amount that he and his mariners would earn for transporting the cargoes to their destinations, the negotiations surrounding how far they could travel within the framework of the arrangement ("the final port being Bombay"), and the bureaucratic expectations that accompanied them.

That Al-Failakawi included three different versions of the same type of contract suggests the nakhoda's attentiveness to writing and to the routine kinds of paperwork that underpinned the movement of goods around the Western Indian Ocean. The contract was one of many pieces of paper that moved dates onto and along with the dhow; and for Al-Failakawi and Al-Khariji, this was one of the many moments of accounting that punctuated the seasonal voyage. Here, we see how concerns surrounding risk and uncertainty were baked into the agreement itself and how they accounted for the instability of contractual terms as they moved from one language to another. In short, we see a multidimensional world of law, grappling with a far-flung arena of commerce, in a single piece of paper.

• • •

Al-Matrook's archive teems with different kinds of paperwork related to the loading and transport of dates. Among them are many contracts with nakhodas—seventy-six of them, to be exact, all from between 1935 and 1936. The majority of Al-Matrook's contracts arrange for deliveries of dates within the Gulf itself, from Basra and Mohammerah to his associates in Lingeh or Bandar ʿAbbas, on the Persian coast; only a few go as far as Karachi, Kathiawar, and Bombay.

Almost all of Al-Matrook's contracts are in the third person, noting that an agreement had taken place between him and the nakhoda for the transport of dates; only three are dictated by nakhodas. Their wording is consistent across the collection and is very similar to the template that Al-Khariji inscribed into the *ruznamah*; the differences mostly relate to units of measure. On the whole, the materials suggest closely shared conventions when it came to the lexicon and syntax of contracts. And in the absence of a manual that supplied merchants with the contractual form, we can only surmise that they learned to write them by observing others do the same—a horizontal form of legal knowledge transmission. That Al-Khariji's look so similar suggests that he distilled the template from extant contracts that he saw or that he himself had entered into.

On the nakhoda's end, what mattered was where they were going and on what terms. Let us take a contract for ferrying dates to India from Basra, written on December 8, 1924, just two months after Al-Failakawi left that port city. That month, an Indian sailing vessel—a *kotiya*, captained by a nakhoda from Kachchh—arrived in Basra to load a cargo of dates from the merchant Hamad bin ʿAbdullah Al-Sager—the same Hamad bin ʿAbdullah mentioned in Al-Khariji's contract. The freightage contract they drew up was virtually identical to those that Al-Failakawi and Al-Matrook had seen over the course of their business, but with telling differences. It is worth quoting in full:

> Carried, with the blessings of Allah and his good fortune, in the *kotiya* called *Trikam Pasa*, owned by Lakhmidas Premji and in the company of the nakhoda ʿAbdullah bin Saḥāq Al-Hindi, of the people of Kachchh, Anjar, from the port of Basra to the ports of India—Kathiawar and Bombay, according to what has been described above—in the amount of 1,004 *mann* of zahdi dates of Najaf; 703 *mann* and 396 *mann* of sayer dates; and seven *mann* of zahdi dates weighing 60; all property and cargo of Hajji Hamad Al-ʿAbdullah Al-Sager, and he will transfer it to whomever he pleases. The freightage will be dispensed in Basra, and it will be 10 Annas per *mann*, and it was received in cash in Basra, [a total of] Rs 200 in cash, and this will be a loan for which I am liable, and [will constitute] my freightage after I arrive safely in port and hand over the property in full. This offer and acceptance took place between us in the Arabic form, and not the Hindi, and I am obligated to pass Porbandar and its likes by instruction of the agent of the aforementioned Hajji Hamad, to wherever he may direct me from the ports of Kathiawar and Bombay, and if I do not follow his instructions I will be responsible for any delay and damage, and I have given two manifests, an original and a copy, and we will depend on one of them.[1]

What is striking is how Al-Sager and the Kachchhi nakhoda inscribed the uncertainties of the voyage and the vagaries of the date market into the contract, and with such detail. Merchants like Al-Sager grounded their calculus in the time and money they spent securing and collecting dates from their suppliers and middlemen—the piles of receipts that Al-Matrook kept, all of which led up to the freightage contract. They based what they were willing to pay on how much they had already paid and what they might expect to receive for their dates abroad. For nakhodas, though, the question was, in part, whether they could expect to net enough at the end of the voyage to pay themselves and their mariners. The freightage earned on a single voyage to India was never enough; they would either have to do it twice in a season or find other cargoes to ferry around after unloading the dates. At its simplest, the freightage rate stated in the contract—10 annas per *mann*—reflected a point of equilibrium: the meeting point between what merchants were willing to pay for the mariners' labor and what nakhodas were willing to take for it.

But the rate was not just a question of supply and demand; it was a matter of liability as well. The terms of the contract speak to the different ways in which merchants sought to protect their capital from the risks of the voyage. Al-Matrook's freightage contracts stipulated that nakhodas would receive their freightage upon delivering the cargo, making no mention of who bore the liability if the dates were lost at sea. The version Al-Khariji inscribed into Al-Failakawi's notebook does the same but stipulates that it all was to take place "without negligence"—a loaded term if there ever was one. Al-Sager insulated himself even further, stipulating the freightage as a loan—a debt on the part of the nakhoda that he'd only consider as fulfilled when the dates were delivered. All of them shifted the risk of losing the goods at sea and the freightage itself onto the nakhoda, circumventing the protections against the loss or damage of cargoes that he otherwise enjoyed.[2] Freightage was thus not simply a matter of haggling over price; the rate, or price, was always shaped by tradeoffs between risk and responsibility.

The parties also wove the risk into the route itself. The itineraries that the contracts laid out bore the weight of the contingencies of the date market. Most freightage contracts specified where the nakhoda was supposed to sail to—Lingeh, Karachi, or some other port. Others described the general itinerary that the nakhoda was expected to follow but left the rest up to him. Like Al-Khariji's inscription, they asked that he stop at Karachi, the ports of the Kathiawar Peninsula—presumably leaving the nakhoda to decide which ports—and end the journey at Bombay. In a few of Al-Matrook's contracts, the merchant enumerated different freightage rates for different ports: one contract, for example, states that the parties agreed to a rate of 13 annas per *mann* of dates to Karachi, and 13.5 annas per *mann* to the ports of Kathiawar. Different markets entailed different risks and costs and thus called forth different equilibria.

By contrast, the contract between Al-Sager and nakhoda ʿAbdullah tethered the nakhoda's itinerary to Al-Sager's agents, of which there were principally two: one who did business out of Karachi and Kathiawar, and another in Bombay.[3] The work of responding to the changing market fell not on the nakhoda but on the agent, who would have had access to more precise information as to which markets were saturated and which were not. By shifting the work of finding markets onto the agent, Al-Sager alleviated the nakhoda of the burden of searching for information on markets but also kept him from being able to react on the spot. In the tradeoff between direction and discretion, the date trader opted for the former.

There was one other point that Al-Sager wanted to ensure there would be no confusion over. The agreement, he stated explicitly, took place in Arabic, not in Hindi. There was good reason for this caveat: contracts that involved parties from different language communities often included different languages on the space of the contract itself. Al-Sager's contract included a few lines in Kachchhi (not Hindi, as the contract suggests) below the Arabic text; one of Al-Matrook's contracts, with the Kachchhi nakhoda Muhammad Osman, did the same.[4] In both cases, the Kachchhi text was considerably shorter, listing only details on the cargo and the rate. In Al-Sager's contract, the nakhoda noted in Kachchhi that the freightage, which amounted to Rs 200 in cash, was "his own."

It is tempting to think of these as moments of translation from one contractual lexicon to another, but the brevity of the Kachchhi text militates against it. Instead, we might think of these as annotations of the sort we see in Al-Matrook's date receipts, which linked up to a much broader system of accounting and filing. The details weren't missing from the Kachchhi end, they just ended up in different genres of the finance economy.[5] These are moments in which we see the palimpsests of a multilingual world of contracting, accounting, and filing that structured the movement of goods around the Indian Ocean. And while Al-Sager was part of this layered linguistic world, he wanted to leave no doubt as to what transpired when he and the nakhoda ʿAbdullah struck up an agreement: "The offer and acceptance," he wanted to make clear, "took place between us in the Arabic form, and not the Hindi." If there were many ways of narrating the same moment of loading, only one telling would carry weight.

• • •

As he set sail for India, Al-Failakawi left in his wake a trail of documentation that carved the path from the date palm to the dhow—a trail that moved through different systems of accounting and contracting, different sheets of paper, and sometimes different languages. He also held in his chest, on the deck of the dhow, a document that looked ahead to the marketplaces of coastal India that he would have to shuttle between to find takers for his cargo—as the other contracts laid out, Karachi, the Kathiawar Peninsula, and ultimately Bombay. The negotiations that he engaged in over the details of the contract would recur again, on this voyage and others to come, as he loaded and unloaded goods in different port cities around the Western Indian Ocean.

The contract that Al-Failakawi entered into when he finished loading dates in the autumn of 1924 has not survived. There is no way of telling exactly who he loaded dates from and on what terms he agreed to ferry them to India that season. But we do know what these contracts looked like more generally, and we know that they varied only slightly from one nakhoda to another. Moreover, we know that Al-Failakawi would have known that, too: in his logbook, he carried a reminder with him of what a freightage contract was supposed to look like—a model passed down to him from Al-Khariji. Learning to voyage, then, was not just about learning to sail; navigation was a necessary component of the voyage, but not a sufficient one. To set out for markets abroad, a nakhoda needed a grasp of the different negotiations and contracts—the paperwork—that he had to manage along the way.

THREE

The Gulf

LEAVING THE MOUTH of the Shatt Al-ʿArab waterway, Al-Failakawi and his crew set out into the Persian Gulf. The *Crooked* was heavy with dates, its hull submerged in the briny green waters of the northern Gulf. It would take just over three, largely uneventful weeks of sailing along the Persian coast for Al-Failakawi to reach the first major stop along his voyage, the port city of Karachi.

But life on board the dhow was anything but quiet. Crew members constantly listened for orders from the nakhoda and the *mjaddimi* to raise or lower the different sails and manipulate the halyards. Each sailor knew which mast and which yard he was responsible for and reached get to it quickly, leaping over chests, coils of rope, and other stray items on the deck. When they weren't actively maneuvering their dhow, they kept their hands busy mending sails and ropes. Even relieving oneself entailed work: to defecate, a sailor had to step over the rail and onto a hanging platform with a hole cut into the floor, over which he squatted precariously as the dhow lurched in all different directions. The only relief from work was when sailors ate or assembled for the five daily prayers, led by the nakhoda himself. Even sleep only came in shifts, with half the crew staying awake while the other half found a part of the deck where they could curl up and get a few hours' rest.

There was an order to life on the ship –not a strict hierarchy, but a clear division of labor grounded in a structure of occupational difference.[1] At the bottom of this social structure was the *tabbāb*—a boy as young as eight years old, whose job was to assist in the preparation of the daily meals and to take on the more menial tasks on board the dhow like cleaning the deck and toilet, prepping the tea and hookah pipes, and cleaning up after meals. He earned nothing aside from whatever tips he received from the nakhoda and

FIGURE 8. Sailors maneuvering the mainsail on a Kuwaiti dhow, 1939. Photo from National Maritime Museum, Greenwich, UK.

sailors, but he gained work experience on the dhow and got to eat for free. Over time, he might hope to be assigned more tasks and be promoted to half-share sailor (*baḥḥār nuṣṣ glāṭa*), which entitled him to half what a normal sailor would earn from the net profits over the course of the season.

The dhow's rank and file sailors all had specific tasks assigned to them as well. At the beginning of the voyage, the *mjaddimi* assigned each sailor a particular halyard, such that when he barked out his orders, the sailor knew where to be and what to do. True to his title, the *mjaddimi*—the driver—pushed them forward with every shout. At times, he was assisted by another *mjaddimi*, chosen from among the sailors, whom they called the *mjaddimi al-ṣgheer*, or assistant first mate; he would be promoted to *mjaddimi* if his senior took a position on another dhow.

Aside from the *mjaddimi*, there were only a handful of crew members who were exempted from the regular tasks: the *sukuni*, or helmsman, and the *m'allim*, or navigator, if there was one on board the vessel. These two specialized in guiding the dhow along the currents, shoals, and reefs, beneath the sun and stars. Regular sailors and the *mjaddimi* all worked on the lower deck, whereas the *sukūni*, the *m'allim*, and the nakhoda spent most of the voyage up on the poop deck. They, too, slept in shifts, so that there was always at least one or two people awake and alert to what was happening in the dhow's immediate vicinity.

Al-Failakawi, the nakhoda, was less an imperious, authoritarian figure than he was primus inter pares. He occasionally shouted out commands from the poop deck but just as often joined the sailors in their work and always ate with them. Still, he distinguished himself from them in his authority and comportment: his word was law, and it was in the sailors' interest to follow it, as their lives and livelihoods depended on his skill and good judgment. In the wooden world of the dhow, he sat atop the hierarchy.

• • •

From his perch on the upper deck, Al-Failakawi watched the sun and shorelines as the *Crooked* bobbed along the Persian coast, moving with the surface currents that would take it past one coastal town to the next. Day after day, Al-Failakawi entered into his logbook the names of places only a mariner might attach value to: Al-Maṭāf, Al-Ṭāhiriyya, Shaheen Kuh, and Hendorabi Island.

Every morning, the nakhoda or his *mʿallim* would engage in the same ritual: he took his telescope and sextant out of his chest and walked out onto the main deck. With his eye pressed firmly against the sextant's eyepiece, he looked out onto the shore, setting the horizon squarely within his sight. He then moved the instrument's index arm until the sun appeared close to the horizon, gradually making adjustments until it just barely touched the horizon. He then noted his reading of the sun's angle from the horizon, correcting for its declination—the angular distance of its rays from the equator on that particular day. Al-Failakawi had this information handy several years before and after the voyage, in neatly copied tables in the notebook he received from Al-Khariji. With these data in hand, he could determine the ship's true latitude, which he then checked against the landmarks he saw through his telescope—coastal markers that confirmed to nakhodas their dhow's location.

Within five days of leaving the Shatt, Al-Failakawi noted in his logbook that the dhow had reached the island of Qais—a small island of roughly thirty-five square miles, ruled by the shaikh of Charak. At the turn of the century, it was home to just around 2,250 people, though that would change during the summer pearl dive. There was little trade on the island—a small market and a handful of dhows.[2] As Al-Failakawi sailed past the island's

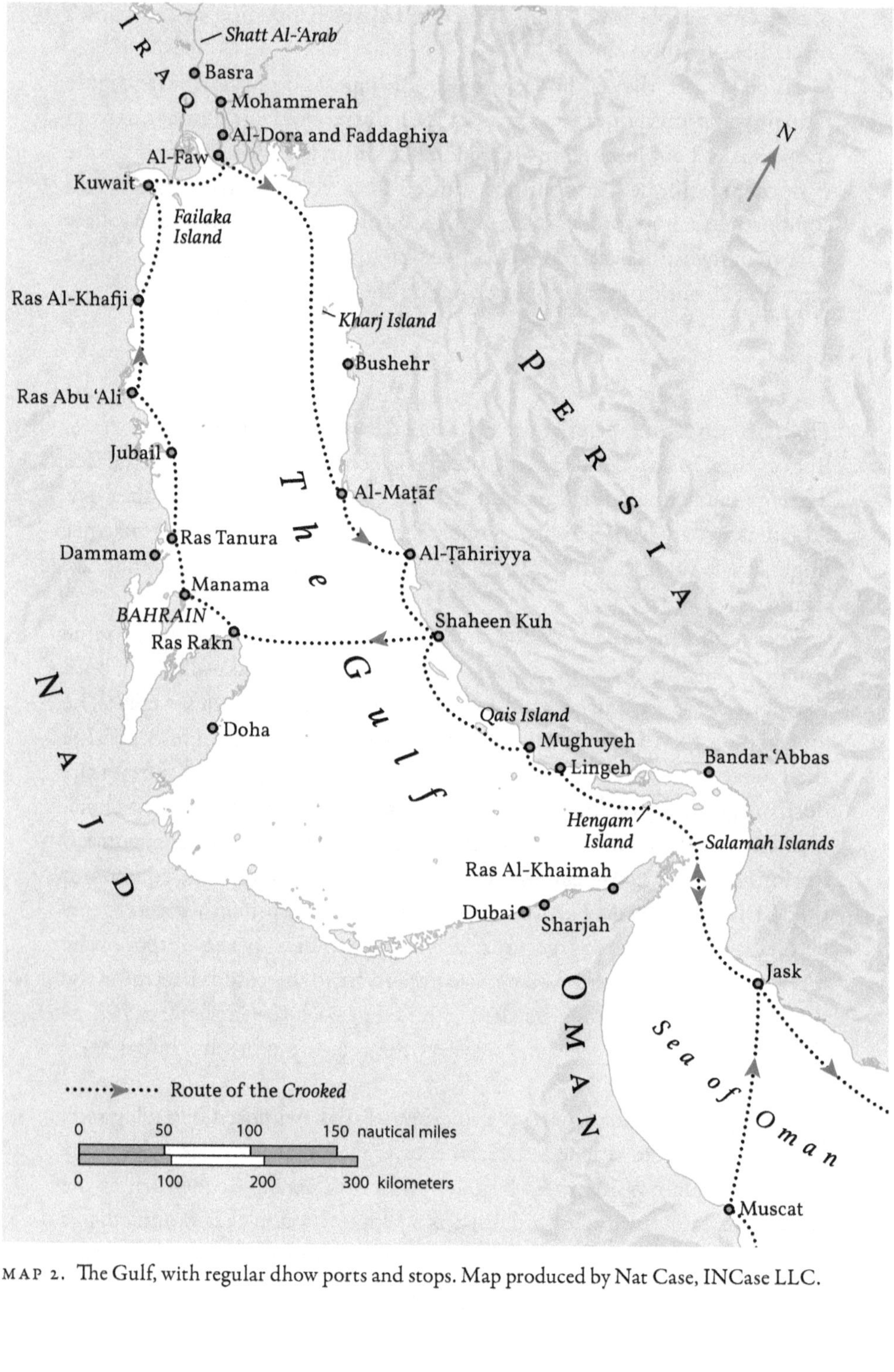

MAP 2. The Gulf, with regular dhow ports and stops. Map produced by Nat Case, INCase LLC.

north coast, he would have been able to make out the outlines of what appeared to be historical ruins—mounds of stone, fallen pillars, and a crumbled minaret. Avid reader that he was, Al-Failakawi knew that long ago the island once held the Gulf's keys to the riches of the Indian Ocean world. The ruins of Qais Island, like those that peppered many of the other Gulf ports the *Crooked* sailed past, stood as testaments to a Gulf medieval maritime golden age, when the riches of India, China, and Africa flowed into the Middle East through port cities like Siraf, Hormuz, and Bandar ʿAbbas, each of which succeeded the other as the Gulf's chief entrepôt.

But Qais also had a more sordid past. In the twelfth century, the island had become notorious as a place of maritime predation. The ruler of Qais frequently sent out ships to attack rivals, and plundered goods from ships that sailed nearby. He was no anomaly: port cities in the medieval Gulf and Indian Ocean often sent out ships to capture the fleets of their rivals and regularly attempted to direct traffic to their marketplaces by way of force. This was hardly the peaceful trading zone historians had long thought it to be. Rather, in the medieval Indian Ocean, economic violence constituted a form of statecraft. Commercial wealth and armed raiding were not at all at odds with one another; they coexisted quite comfortably.[3]

Things had changed since then, but not completely. A cargo-laden dhow like the *Crooked* could often attract the wrong kind of attention—especially as it sailed out of the Shatt carrying heavy cargoes of dates. Less than fifteen years before Al-Failakawi's voyage, in June 1910, a Kuwaiti merchant leaving the Shatt found his dhow being fired upon by a small boat. His crew returned fire but soon ran out of ammunition. The attackers then boarded the dhow, killed everyone on board, and robbed the vessel of all its goods; those who jumped into the sea were pursued and shot in the water.[4] Just a few years earlier, a group of Arabs on the Persian coast launched a series of attacks on dhows leaving Basra. From their base in Dayyer, just south of Bushehr, they would wait for dhows to leave the Shatt and, when the dhows would find anchorage on their voyage down the Gulf, they would raid the vessel and kill the crewmembers and passengers on board. In one of their more gruesome episodes, in 1907, they attacked a Kuwaiti vessel leaving Basra. The dhow was later found stranded on an island near Bahrain; its cargo had been carried off, and all thirty-nine crewmembers and passengers had been killed.[5]

These outbursts of violence were just as likely to emerge from the deck of the dhow itself as they were from attackers beyond its rails. In one incident, a group of passengers on a dhow sailing out of Basra overpowered the nakhoda and his crew, throwing them overboard and wresting control of the vessel. More common, though, were mutinous plots by the crew members themselves. On one Kuwaiti dhow sailing the Gulf in 1912 with a cargo of dates, a group of sailors—hired hands from among the subjects of Shaikh Khazʿal—conspired to take over the dhow and make off with its cargo. When the nakhoda and his son were asleep, the mutineers slit their throats and cast their bodies into the sea; they also threw an uncooperative crew member from Najd overboard. The following year, a group of Somali sailors murdered their Kuwaiti nakhoda and made off with the cash he carried on his ship.[6] If Al-Failakawi felt the slightest bit nervous as he watched out from the poop deck over a dhow heavy with baskets of dates, it was with good reason.

What all these incidents shared, whether on or off the dhow, was that they could all plausibly be labeled acts of piracy. For by the early twentieth century, "piracy" had become a broad tent. It could encompass all sorts of different acts of violence, whether among coworkers or complete strangers. In virtually all these cases, rulers disavowed any connection to the so-called pirates and even went so far as to arrest and punish the perpetrators. There existed a membrane, if not a clear barrier, between piracy and politics. Politics was the arena of diplomacy and legitimate—albeit sometimes questionable—warfare. Piracy, at least the forms Al-Failakawi witnessed and heard of, was a largely private endeavor. That, however, was not always the case.

• • •

By all accounts, political life in the eighteenth-century Gulf was mired in violence. Arab chroniclers from the period described one battle after another, as though little else happened in the region. And in bound volumes of letters, European company officials sent reports detailing robberies, attacks, and reprisals—sometimes involving Europeans, but often not.

But these were not wanton acts of violence between different coastal groups. Rather, through violence, these communities sought to forge the contours of political authority on the crumbling frontiers of regional empires—the Ottomans and the Safavids. The Ottoman Empire had by the early seventeenth century largely given up direct control of Basra, handing the reigns over to Afrasiyab, one of the city's notables; his descendants ruled the city for nearly seventy years, building alliances with local tribes. Even when the Ottomans retook control of the city in the eighteenth century, they delegated control to its governor, who in turn farmed out defense and revenue collection responsibilities to the city's notables. All the while, the tribes that inhabited the vicinity of Basra practically ceased to answer to their Turkish overlords and in some cases stopped paying taxes altogether.[7]

Of equal, if not greater consequence for the Arab tribes of the Gulf was the collapse of two successive Persian regimes in the eighteenth century. The first were the Safavids, who seldom exhibited any interest in projecting a strong naval presence. They were followed by Nader Shah, who ruled from 1736 to 1747 and strongly asserted himself in the Persian Gulf: he developed shipbuilding facilities in Bushehr and other ports, conquered Bahrain, and even briefly invaded Oman.[8] His legacy was not to last, though, as his would-be dynasty fell apart in the hands of his sons, who were locked in internecine conflict with one another. Nader Shah's successor, Karim Khan Zand (1751–1779), courted European companies in Persian ports, and even at one point invaded Basra, but was ultimately unable to sustain a government presence on the coast.

The sputtering attempts by these Persian dynasties to assert themselves in the Gulf, coupled with the absence of any meaningful Ottoman presence in the region, left something of a power vacuum on the coast. The gates of the political arena on the Persian and Arab littorals were thrust wide open, and local actors from around the coasts jostled for control of coastlines and resources. Up and down the Persian and Arabian coasts, political entrepreneurs clashed with one another over access to, if not outright control of, the rich date gardens of the Shatt and Eastern Arabia and devoted much of their energy toward situating themselves close to the Gulf's pearl banks. But more than that, they fought for the control of particular routes along the Gulf—for the right to send their ships out without harassment and to exact tolls on the ships that passed through the areas they controlled.

On the Persian coast, the transformations in the rickety central government gave the Arab governors of the coastal towns the opportunity to assert themselves as independent rulers. Successive rulers of Persia would occasionally have rogue governors arrested but would release them after a short period and could only sometimes force them to pay taxes. The overwhelming majority of these governors maintained their own armies and navies, and those who did not invariably fell prey to those who did.[9]

The inhabitants of Kuwait—the ʿUtub—were deeply imbricated in this changing political world. Soon after their port's establishment, the ʿUtub were locked into armed conflict with their most immediate neighbors: the Bani Kaʿb of Mohammerah, as well as the inhabitants of Bushehr and Bandar

Rig.[10] Elsewhere, the ʿUtub went on the offensive. A handful of families left Kuwait in the mid-1700s, returning to their earlier settlement in Zubara, on the western coast of the Qatari peninsula.[11] From there, they launched a series of raids against the island of Bahrain, then under the authority of the shaikh of Bushehr, culminating in a full-fledged invasion, to which the ʿUtub of Kuwait lent naval support. By 1783, the ʿUtub succeeded in wresting control of the island from Bushehr and wasted no time in enjoying the fruits of their military victory. By the century's end, they were sending out a thousand boats every summer to the pearl banks near Bahrain—at least twice as many as any other community. Each of these paid a tax to the island's ruler; they also paid him to furnish armed vessels to protect the pearling fleet.[12]

There was good reason for the armed escort. The capture of Bahrain engendered as much tension as it resolved. The Al-Jalahmi family, which had participated in the Bahrain offensive, found themselves left out of the spoils of victory; the Al-Khalifa seated themselves at the island's helm. Thus scorned, the Al-Jalahmi leader, a man by the name of Rahmah bin Jaber, took his boats and followers and established an independent base in the small inlet of Khor Hassan, later spreading to the nearby port of Dammam. Rahmah would devote the rest of his life to attacking ships sailing out of Bahrain and nearby ports, all while forging fleeting allegiances with inland powers who offered vague promises to help him capture Bahrain from his enemies. The lore of his swashbuckling life is perhaps only outdone by the story of his death: In an 1826 naval battle with the Al-Khalifa of Bahrain, Rahmah found himself in a situation of near-certain capture. Rather than allow himself to be taken alive, the Jalahmi chief disappeared into his dhow's hold. With his eight-year-old son clutched in his arms, he lit the dhow's powder keg—"and in a few seconds," a later report recounted, "the sea was covered in the scattered timbers of the exploded vessel, and the miserable remains of Rahmah bin Jaubir and his devoted followers."[13]

Though he ended his life scattered at sea, Rahmah left his mark on the shore. In Dammam, as in Khor Hassan, he had built a fort, which he used to guard his stretch of the coastline, inlets and all, and to shelter his dependents. Forts like Rahmah's dotted the coast, on both shores. Forts were often the starting point of a coastal settlement, attracting client tribes and other protégés, and serving as the foundation for expansion later on.[14] Kuwait itself began that way: the name itself, we might recall, is a diminutive form of *kūt*, or fort—a reference to the summer fort the Bani Khaled maintained in the area.

The stakes in the maritime economy were changing, and with it, too, the political geography of the region. Settlements oriented toward the pearl banks in the Gulf had sprung up in rapid succession over the second half of the 1700s; these fed into major markets in the area—principally Basra, and later, Muscat, as the Indian merchant community there grew in the last quarter of the century. However, by the 1800s a number of these settlements began to grow, attracting greater numbers of merchants and their dependents—mariners, shopkeepers, artisans, and their families. They had begun to develop into full-fledged towns with the capacity to trade their pearls directly with mercantile centers on the western coast of India—chiefly Surat at first, and then Bombay. And their growth invariably came at the expense of others, as they siphoned off manpower, capital, and ultimately revenue.[15]

Nowhere was this a smooth process. As a settlement grew, its presence became a matter of politics: the claims it made to shorelines, pearl banks, and routes invariably butted up against similar claims made by its neighbors. For officials from the European companies trying to move their goods through the Gulf, the establishment of one of these sorts of settlements could cause alarm. "You will be pleased to endeavor by all means to obtain early intelligence of any attempts that may be made to build forts on the seacoast, to construct vessels for warlike purposes, and generally all proceedings indicative of a tendency to the renewal of Piracy," wrote one East India Company official in Bombay to their representative on the island of Qeshm. Company officers knew that a fort often signaled the establishment of an independent potentate or the expansion of an existing one. New forts and ships could herald a fresh wave of maritime violence, which would ultimately hurt Company commerce in the area.[16]

But commerce, raiding, and politics were not necessarily opposed to one another; in fact, they often fit together quite neatly. The marriage of trade and violence was less an anomaly than a regular aspect of economic and political life in the Gulf—and less a permanent feature than a seasonal activity. Those who took part in raiding also engaged in other aspects of the regional economy: in trade, pearl diving, in shipbuilding, and in provisioning markets and laborers. The goods they carried off the ships they plundered—and sometimes, even the ships themselves—often made their way into the marketplaces of the Gulf, where they moved through the regular channels of commerce. At other times, they might be distributed locally, as a way of building up a base of dependents. And as opportunities for trade expanded, so, too, did opportunities for raiding ships at sea. Trade and raid, then, were not opposites; they were cousins.

What, then, of raiding's more troublesome sibling, piracy?

• • •

Francis Erskine Loch, the commander of the East India Company brig *Eden*, was sailing the stretch of water between Gujarat and Muscat in the company of the *Psyche* when, on Christmas morning 1818, he saw three sailing vessels, one of which was being towed by the other two. "The superior manner in which the sails were cut and set, as well as the rig of the masts and form of the hull of the two latter," he wrote, "bespoke them at once to be Pirate vessels." The *Eden* and *Psyche* immediately chased the dhows, whose "shoulder of mutton" sails, bulging with wind, would have skipped them across the water like flying fish—but for the ship they towed. The dhows cut their prize loose and raced off. Loch spent the next three days chasing the pirates around the Sea of Oman, stopping briefly in Gwadar only to give his sailors a chance to partake in a belated Christmas celebration. On December 28, they spotted a convoy of seven dhows—"immediately observed to be" pirates, for they had a vessel in tow—and pursued them. The ships exchanged gunfire and cannon shot, and most of the dhows managed to escape; Loch was only able to capture the towed dhow, which had been cut adrift, and one of the other vessels. He sent an interpreter on board the captured dhow, "who by dint of threats and persuasion" managed to produce thirteen pirates, "most uncouth, athletic, and almost naked wretches."[17]

Loch had come to the Gulf from England, by way of India, to take command of a planned naval expedition against the Qawasim—a tribal confederation based in Ras Al-Khaimah and Sharjah (both in what is today the United Arab Emirates) that had been associated with "piratical" activities around the Gulf. He stopped in Muscat on his way up the Gulf to call on Saʿid bin Sultan Al-Busaʿidi, the Omani sultan, and hand over the captured men to him so that Saʿid might take them to India on his own ship. The sultan was eager to transport the Qawasim: their followers had killed his father, Sultan bin Ahmad, in 1804, when he and a small crew of sailors lay anchored off the coast of Persia.

The assassination was part of a regional offensive by the Qawasim that mimicked, but also departed from, the regional pattern. In the early 1760s, during the period of political upheaval that came after the fall of the Safavids, the Qawasim slowly established themselves on the Persian coast, creeping into Qeshm Island, and from there into the areas surrounding the port of Bandar ʿAbbas. Though they faced a stiff resistance from local powerholders and their allies, the Qawasim managed to mount a robust offensive, asserting themselves as more than just a fly-by-night band of traders-cum-raiders.[18]

When the Qawasim expanded in the Gulf, they did so in the shadow of two much larger polities, both of which would shape the destiny of the small but growing confederation. The first was a growing empire based out of Muscat, headed by the energetic Al-Busaʿidi dynasty. The Al-Busaʿidis were relative newcomers to regional politics, having precariously established themselves as Oman's leaders in the 1740s, after the previous dynasty's collapse gave way to a years-long civil war. By the 1780s, the Busaʿidis had embarked upon a vigorous campaign of empire-building: they consolidated their holdings on the coast of Oman, had spread onto the Makran coast on the opposite side of the Gulf of Oman, and had established themselves in port cities along the coast of East Africa. From the island of Zanzibar, they exercised suzerainty over large swaths of East Africa, from the Benadir coast of Somalia to Cape Delgado in Mozambique, appointing governors and tax farmers along the coast.[19]

The Busaʿidis also sought to extend their power further into the Gulf. By the time he met Loch, Saʿid bin Sultan had tried to take Bahrain several times, and on a few occasions actually succeeded, albeit briefly.[20] His father had done the same, but also subjugated the islands of Qeshm and Hormuz and the Makran coast.[21] The Yaʿrubi dynasty that preceded them had also spread themselves out along the coasts of the Gulf, at different times taking Qeshm, Larak, and Bahrain.[22] Busaʿidi attempts at expansion into the Gulf was part of much broader and longer policy of empire-building on the part of different rulers of Muscat.

But it wasn't the Busaʿidis who were on the minds of Loch and other British officials in the Gulf when they thought of the history of the Qawasim. Their more pressing concern emerged from a rather unexpected place. Deep in the interior of the Arabian Peninsula, an alliance had formed in the 1740s between a cleric, Muhammad ibn ʿAbdulwahhab, and the emir of a small oasis town. Ibn ʿAbdulwahhab had been preaching a fundamentalist vision of a monotheism shorn of any idolatrous practices: saintly intercession, Sufism, Shiʿism, and even lax practices by Sunni Muslims were all deemed repugnant to the true faith. Because of the hostile nature of his message and the sort of action he pushed local rulers to engage in, he struggled to find a patron—that is, until he found Muhammad ibn Saʿud, the ruler of Dirʿiyya, who perhaps saw in the preacher an opening to pursue an expansionist agenda and enlarge his revenue base.

And expand they did: with Ibn ʿAbdulwahhab by his side, Muhammad ibn Saʿud raided nearby towns and villages, forcing the inhabitants to pay taxes to him. As they moved through one town after the other, they would offer their enemies a chance to submit and join the movement; those who did not were dealt with accordingly.[23] After his death in 1765, his son ʿAbdulaziz, who married Ibn ʿAbdulwahhab's daughter, vigorously pursued his father's policies. By the beginning of the 1800s, the Saudi-Wahhabi alliance had control of virtually all of Central and Eastern Arabia and had gone so far as to raid Shiʿite shrines in southern Iraq. And as they brought large swaths of territory under their control, they redirected the overland caravan traffic through towns they occupied, taxing pilgrims and merchants alike.[24] Much like the emerging polities on the coast, then, the Saudis sought to control routes—the arteries through which people and goods moved across the Arabian Peninsula.

Why the Qawasim ended up in the Wahhabis' service is unclear. Officials like Loch attributed it to the Wahhabis' missionary zeal and their military prowess. As the Wahhabis expanded onto the coast, the Qawasim became "strict adherents" of their creed and found in it a justification for their activities, "claiming the right and holy work of extermination as following the dictates of the Founder of their sect."[25] Loch's contemporary, the English traveler James Silk Buckingham, agreed: the Qawasim were once peaceful traders, but "at the same moment that they received the conquerors within their gates, they bowed submission to the new doctrines which they taught, and swore fidelity to such laws and injunctions as the most learned and holy of the leaders might pronounce these doctrines to impose." The Qawasim "directed their views to war and conquest . . . war and plunder was the universal cry, and destruction to infidels was vowed in the same breath that uttered the name of their merciful creator"—and they directed their activities to the sea, "the great high-way of nations on which men of every faith and denomination had hitherto passed unmolested."[26]

If the Qawasim allied themselves with the Saudis for reasons of self-preservation, they would have been no different from the countless other towns in the eighteenth-century Arabian Peninsula that faced an unenviable choice between submission or death. But at the same time, the Qawasim may have seen in the Saudis reliable partners to help them fight the Busaʿidi dynasty, who stood in the way of their quest for regional supremacy. The

Saudis offered military support: they marched soldiers all the way to the capital at Muscat, occupying much of northern Oman along the way. Moreover, the Wahhabi creed furnished an ideological justification for going to war with the Omanis: the Busaʿidis were members of the Ibadi sect of Islam and thus were anathema to the Wahhabi creed. On at least one occasion, the Saudi emir ʿAbdulaziz Al-Saʿud gave the Busaʿidis a chance to submit to him, even going so far as to send them a copy of a tract written by Muhammad ibn ʿAbdulwahhab on the articles of the faith.[27]

For the Saudis, though, there may have been more to their alliance with the Qawasim than a desire to win over adherents and raid those they considered unbelievers; there was material benefit too. That the Saudis were interested in a coastal presence is certain enough: in the late 1780s, they laid siege to the former ʿUtubi stronghold of Zubara, sending its inhabitants fleeing to Bahrain and precipitating the town's ultimate demise.[28] The Qawasim may have seemed like good candidates to help expand the Saudi presence onto the shores, banks, and routes of the Gulf. They also might have facilitated their allies' access to supplies of goods—consumer goods, but also arms, for which the Saudis exhibited an unquenchable thirst and for which the port cities of the Gulf constituted an important source from early on. Whether by way of trade—after all, they never stopped trading, even in their "pirating" years—or by raid, the Qawasim helped their inland allies acquire the victuals and weapons they needed to sustain their territorial expansion.[29] There was no tension here between trade and raid; they both fed the same agenda.

As far as the British were concerned, what tipped the balance from raiding to piracy was the decision to attack East India Company vessels—good sources of provisions and weapons, but ones that came with an enormous tactical cost. Early raids on Company vessels involved only minor looting and were treated with indifference by Company officials in India.[30] Over time—and perhaps because of the early signs of indifference—the attacks grew bolder and more violent, some bordering on the macabre. Officials on Company ships wrote of their ships being attacked and crewmembers killed, maimed, and even decapitated, the attackers making off with goods, guns, provisions, and equipment—even the clothes the crew wore. In one attack in May 1809, the Qawasim reportedly carried off the Armenian wife of a British official, along with her infant son and servants, and forcibly circumcised the Christian men on the ship.[31]

After reports of Qasimi depredations against their shipping began to circulate in England and the public outcry reached a crescendo, Company officials had heard enough. In the early hours of November 11, 1809, a joint Anglo-Omani flotilla descended on the Qasimi port of Ras Al-Khaimah. For the next three days, they fired cannonballs into the town as waves of bayonet-wielding sepoys set fire to one building after another. After capturing the town, soldiers set fire to the fifty vessels they found in the harbor, burned the town's warehouses, and destroyed whatever stores of ammunition they could find. They then burned Lingeh and took the village of Luft, which they handed back to the Sultan of Muscat, who had farmed it from the Persian government.[32]

For all its dramatic flair, the expedition proved only momentarily effective; the bulk of the Qasimi fleet lay safely anchored elsewhere. By 1814, there were regular reports of Qasimi vessels—no longer harassing Company ships in the Gulf, but instead capturing Indian-owned ships off the coasts of northwestern India and attacking Omani shipping in the Arabian Sea.[33] Faced with a resurgence of Qasimi attacks, the Company authorized a second expedition against Ras Al-Khaimah in 1819—one to be headed by Loch.

The timing of the 1819 attack was perhaps more felicitous than the first: Ibrahim Pasha, the son of the Ottoman viceroy of Egyptian Mehmet Ali, had just led a successful campaign against the Saudis, who had overplayed their hand in expanding into the Hijaz and threatening Ottoman control of Mecca and Medina. Over the course of several months, the Turkish and Egyptian troops decimated the Wahhabi army and razed the Saudi capital at Dirʿiyya. Seizing on the opportunity to take action, Company officials approached Ibrahim to join them in fighting the Qawasim. When he demurred, Saʿid stepped in, offering both land and sea support. The joint force arrived at Ras Al-Khaimah in early December and relentlessly bombarded the town, simultaneously landing artillery and troops on shore to pursue the Qawasim and their allies. Within five days, the battle was over, and two weeks later the Qasimi rulers, who had retreated to a nearby village, surrendered themselves to Company soldiers. Loch left Ras Al-Khaimah in the hands of its ruler Hassan bin Rahmah, but the settlement had been reduced to a smoldering heap of ruins; only the fort and its surrounding stone buildings were left intact.[34]

Loch's voyages around the Gulf in preparation for the 1819 expedition mapped out British understandings of the boundary between piracy and licit acts of plunder. The distinction had little to do with the act itself: it was fundamentally political. Piracy, insofar as Loch and other Company officials understood it, was a form of violence and plunder enacted by private actors—

that is, actors who were not engaged in plunder on behalf of states. His understanding drew from a long history of British encounters with "pirates"—in the Atlantic and the Mediterranean, but also around the Indian Ocean, where British encounters with different coastal political actors gave rise to a discourse of piracy. In India and Southeast Asia, British officials used the term "piracy" to separate what they understood to be legitimate forms of littoral statecraft from those that fell short.[35]

Company expeditions in the Gulf reflected an elastic definition of piracy, one that could conjure up the very political realities they wanted to see. There was, after all, nothing private about the sorts of violence the Qawasim engaged in; they were very much actions by a state, however small that state may have been. And yet, Loch and others categorically denied them the possibility of legitimation by way of statehood. Indeed, there seemed to be only a few states—the Ottomans, the Persians, and perhaps the Omanis—that they were willing to recognize as such, even as they made the Qasimi chiefs answer for the actions of their subjects. The field of licit violence on the high seas was thus subject to serious constraint; the terms British officialdom operated on denied the possibility of raiding as a means of staking out political authority.[36]

There were, however, other forms of "private" political violence that British officials were more than willing to countenance. As he prepared for his expedition against the Qawasim, Loch looked the other way as Rahmah bin Jabir, "the terror of the Gulph," attacked ships belonging to the port of Lingeh, claiming that people there had been selling dates and other provisions to the Qawasim. What distinguished Rahmah was that he knew where to draw his limits: "He protected the British trade and was at peace with Bushir and Bussora," wrote Loch, "but was at war with every other part of the Gulph."[37] In the immediate wake of the expedition, though, Loch held a firm line with Rahmah when he asked whether he had carte blanche to attack Bahrain.[38] Certain actors, then, could weave between the lines British officials drew between licit and illicit forms of maritime violence, and saw clearly that those distinctions were less about the substance of that violence than they were about who might hold the right to legitimately plunder others.

Others may have helped draw those same lines. Saʿid bin Sultan Al-Busaʿidi, who joined in both expeditions against the Qawasim, may have been the only clear winner in that moment of upheaval, emerging both with more powerful allies on his side and fewer powerful enemies to check his expansion. The Qawasim were no longer a threat, and the Wahhabis, who had been harassing him in Oman since the first expedition, had fallen to

Ibrahim Pasha's troops. The state-sponsored violence they engaged in against their neighbors was not only successful, it was legitimated. It is telling, then, that in the aftermath of the 1819 expedition, when Loch visited Bahrain to destroy whatever Qasimi vessels remained there, the rulers of Bahrain expressed a concern that he was there to give up the island to the Sultan of Muscat.[39] Loch, who read much of the Gulf seascape through the lens of the recently published translations of the voyages Alexander the Great's Admiral Nearchus, may have been familiar with the dialogue related by Augustine of Hippo between Alexander and a pirate that had been captured. "For when that king had asked the man what he meant by keeping hostile possession of the sea," related Augustine, "he answered with bold pride, 'What thou meanest by seizing the whole earth; but because I do it with a petty ship, I am called a robber, whilst thou who dost it with a great fleet art styled emperor.'"[40]

• • •

If the 1819 expedition forced a military definition of piracy onto the inhabitants of the Gulf, it left open the legal definition; this brought in a wider range of actors. Very soon after the expedition, British officials called the rulers of the Ras Al-Khaimah and the nearby port towns to a meeting, where they presented them with a draft peace treaty designed to prevent future acts of maritime aggression.[41] The treaty's first article called for "a cessation of plunder and piracy by land and sea on the part of the Arabs," while the second clarified that whoever attacked someone who passed through by land and sea outside of the context of "acknowledged war" was to be considered "an enemy of all mankind," forfeiting "both life and goods." It went on to clarify: "Acknowledged war is that which is proclaimed, avowed, and ordered by government against government; and the killing of men and taking of goods without proclamation, avowal, and the order of a government, is plunder and piracy." The treaty's language drew on a deep well of ideas from the Law of Nations, particularly in the distinction it drew between acts of legitimate warfare and those deemed piratical. For officials tasked with enforcing the treaty's terms, a sense of where different rulers in the Gulf stood vis-à-vis one another was critical.

But the language of the 1820 treaty suggested there was more to piracy than unauthorized violence; commerce formed part of it as well. It made distinctions between licit and illicit trade, as even engaging in slave trafficking could be considered piracy. To give shape to its vision for a post-piratical

Gulf, the treaty laid down practical foundations for commerce in the time of peace. The third and fourth articles required that parties to the treaty outfit their dhows with a red and white flag, which was to signal their treaty relationship to the British: "This shall be the flag of the friendly Arabs," it proclaimed, "and they shall use it and no other." The two articles that followed required that dhows sailing from treaty ports carry papers—registers and port clearances—describing the ship, its crew, its contents, and its destination, each signed by the port's ruler and an envoy he was to send to the British Resident. Legitimate commerce, then, was not simply a matter of staying on the right side of the line between war and piracy; it was also about conforming to British-imposed bureaucratic expectations. Article 10 made this clear, suggesting that only Arab dhows that met these requirements "shall enter into all the British ports and into the ports of the allies of the British so far as they shall be able to affect it."[42]

The 1820 treaty thus captured another impulse: the hope that imposing a maritime truce among the Gulf Arabs would encourage them the pursuit of *doux commerce*, the sort that rehabilitated pirates and remade them into proper merchants. Company officials in Bombay strongly believed that the mercantile spirit of the Qawasim and their neighbors ought to be revived, and that they should be exposed to civilizing influences by being allowed to trade in India.[43] The treaty emerged out of a worldview that naturalized the human desire to want to engage in commerce without hindrance. It was in part for this reason that the rulers of other nearby ports were encouraged to join the treaty; by March 1820, it included the rulers of six other towns, in addition to Ras Al-Khaimah.[44]

But the treaty also bore the personal imprint of its authors. The expedition's Arabic interpreter and drafter, Perronet Thompson, had served as a governor of Sierra Leone and was a staunch abolitionist. He found in the treaty a chance to create the first public document in which the slave trade was called piracy; "the opportunity was an excellent one, and much too good to miss," he wrote to a colleague.[45] He was also an enthusiastic reader of work in political economy and espoused many of the free-market ideas that had been recently popularized by Adam Smith. When he stayed on as temporary political agent in Ras Al-Khaimah after the treaty, he sought to reorder the military provisions bazaar on Smithian terms. The "military bazar ought to be regulated by the principles common to markets in general . . . that of leaving the buyer and seller to settle their own terms, trusting entirely their mutual necessities to produce the most favorable result, and confining the interference of the public authority to the prevention of monopoly," he wrote to the incoming bazaar master. Rather than regulate the prices for different goods, his successor ought "to repress all monopolies or exclusive advantages of sale . . . and to encourage competition to the greatest possible extent."[46] In his work both as a political agent and as a drafter of the treaty, then, Thompson found an outlet for his belief in the salutary effects of free trade.

One British survey of the Gulf, penned in the wake of the 1820 treaty, reflects some of the treaty-writers' aspirations. Describing Ras Al-Khaimah, the author noted it had been reduced to roughly twenty-five hundred inhabitants, most of whom lived in a new settlement and distinguished themselves from their predecessors by occupation. "They now participate in the trade of the Gulf, and have a number of boats thus employed," he declared, adding that "they trade to Bombay and the Malabar Coast during the north-east monsoon, and to the Red Sea; they also take a large share in the pearl fishing." Kuwait, too, was "a place of much importance, owing to the maritime spirit of its inhabitants"—the main reason why Kuwaitis "enjoyed peace while all other parts of the Gulf have been embroiled."[47]

And yet, for all of its ambitions, the specific terms of the 1820 treaty wouldn't last very long. The governor of Bombay worried that the maritime regulations that the treaty imposed offered no guarantee of good behavior and included no provision for the punishment of those who tried to game its provisions—by, say, falsifying papers—and did not secure for British vessels the right to seize and search suspicious dhows. Moreover, the extent to which they stretched the definition of piracy so as to include virtually all forms of maritime violence and human trafficking might have established too broad of an arena for them to police. There were, of course, clear acts of piracy and plunder that the government prosecuted and punished, but they were largely drowned out by more ambiguous forms of maritime violence.[48]

Officials were concerned, too, that routine acts of maritime violence, piratical or not, threatened the annual pearl dive and would prove damaging to the trade of the Gulf more generally. It was thus that in 1835, the Gulf Resident negotiated a temporary truce for the duration of the pearling season, and stipulated penalties for anyone who violated it. The experiment proved a success, and the truce was reextended—this time, among the rulers of the so-called Trucial Coast, excluding Kuwait and Bahrain, who were not nearly as closely associated with piracy.[49] In 1843, British officials negotiated another extension to the truce, one lasting a decade. The language of the truce was shorn of the broad ambitions that characterized the 1820 treaty.

The Gulf chiefs, the preamble declared, assented to the truce out of a desire to prevent violence from interrupting the pearl dive, along with a recognition of a "general advantage to be derived" from maritime peace—a distant nod to the salutary effects of commerce. "Acts of aggression at sea," the treaty declared, were to be reported to the British Resident.[50] A decade later, a peace treaty "in perpetuity" repeated many of the same points, but ended with a declaration that the maintenance of the peace "shall be watched over by the British Government, who will take steps to ensure at all times the due observance of the above Articles."[51]

For the tribes of the Gulf and their rulers, this more permanent investment by the British reconfigured the calculus of maritime competition but did nothing to eliminate it. What the treaties did offer, though, was a new language for talking about those sorts of contests—the possibility of using the language of piracy to reclothe an old phenomenon in a new garb, and one with major political ramifications. Groups around the Gulf played on the changing boundaries between legitimate warfare and commerce on the one hand and piracy on the other, and to good effect. Those who drew on maritime violence and raiding to establish alternative bases of power could easily be branded pirates by their rivals, who could then legitimately mobilize violence against them.

More broadly, the treaties local rulers signed with British officials allowed them the possibility of recognition as sovereigns by an outside power. For rulers whose authority was perpetually under challenge from within their families and from the outside, this was a highly attractive prospect. Company officials signed the treaty with specific rulers and depended on their cooperation for its success; parties to the treaty thus had the backing of the East India Company and all its firepower, and they could rely on British support when they needed it. Legitimate violence, successive treaties made clear, was limited to actors who were parties to the truce; everyone else lay beyond the pale. The cycles of plunder and claim-making that had long established the rhythm of politics in the region now moved along a different axis.[52]

We shouldn't overstate the transformations wrought by the British and their treaty-making. After all, Gulf rulers had long sought to expand their web of alliances to include regional and global powers; at different times, they sought alliances with the Omanis, the Persians, the Ottomans, the Portuguese, and the Dutch. For small city-states whose political fortunes often depended on the protection afforded to them by larger, more powerful polities, this was a long-established fact of political life. Much of this, too, involved attempts at delegitimating rivals in the eyes of their protectors. The precise mechanics and language they drew on might have changed, but the British were otherwise just the latest in a long chain of polities that the rulers of port cities in the Gulf would weave into the fabric of regional politics.

• • •

The treaty of 1820 and the truces that followed ushered in a new era of Gulf history: the time of the Pax Brittanica, in which the Government of India became increasingly involved in matters of law and order, both on the water and on the coasts. The merchants whose goods were being carried on Al-Failakawi's dhow overwhelmingly approved of this transformation. For them, the British Empire brought security and lent them the freedom to pursue their commercial interests with relatively few hindrances. Their goods, their ships, and their money were all secure, and the political landscape suffered few, if any, of the sorts of convulsions that characterized much of the pre-British era.

But counterintuitively, maritime peace came at a cost to dhows like the *Crooked*. In the minds of British officials, dhows had traversed the boundary between peacetime commerce and wartime plunder enough that one might justifiably cast doubt on them. For nakhodas, it was precisely the converse. Their activities hadn't changed, but the lines between public and private violence, and between commerce and piracy, had been redrawn; the dhow became associated with the kinds of illicit activity the treaties sought to stamp out. This was not a chapter that had come to a close, either: the tensions remained unresolved. As Al-Failakawi sailed past the remnants of old trading centers and forts strewn across the Persian littoral, he may have been reminded of how these lines continued to shift, even during his time, and just how difficult it could be to remain on the right side of history.

INSCRIPTION

Passage

PASSING THROUGH THE GULF and into the Arabian Sea involved more than just navigating through the Straits of Hormuz and passing from one body of water—one coastline—to another. The sea teemed with polities, small and large, and a seasoned nakhoda knew that a successful voyage was one that could navigate different claims to shorelines and waterways.

Although Al-Failakawi's logbook tells us little about the political dimensions of the voyage, he did carry with him a document that signaled something of the world he sailed through. The small sheet of paper, bearing the letterhead of the Kuwaiti Emir Ahmad Al-Jabir Al-Sabah, was something of a safe-conduct pass. It read: "Let it be known among those who inspect this decree of ours, pertaining to the nakhoda ʿAbdulmajeed bin Ahmad, [that] he is of our people, the inhabitants of Kuwait. We ask that officials of nearby governments ease his path, treat him in accordance with the ties of friendship between the two, so that there is no confusion. The Ruler of Kuwait, Ahmad Al-Jaber, 23 Rabiʿ Al-Thani, 1358; 11 June 1939."

There is nothing in Al-Failakawi's pass that makes mention of the sea, aside from referring to him as the nakhoda; it simply asked that those who inspected the pass allow him to move through their jurisdiction. But buried deep inside Al-Khariji's notebook are copies of two similar passes, both of which reveal a strong sense of the claims that different polities had made to the seascape. "In the name of Allah, the most gracious, the most merciful," he wrote, "this is a copy of a ship's *qawl* for this year, 1358 AH [1939]":

> Let it be known among those who inspect this decree of ours, from those who traverse the seas and inhabit the ports, from all the great and eminent states: the dhow named *Fateḥ Al-Khair* is the property of Fulan bin Fulan, and he is of our people, the inhabitants of Kuwait, and is our subject. We pray that if

> the leaders of the known great and friendly states stop and inspect him, they treat him in accordance with the principles, laws, treaties, and ties of friendly states. This is what we have decreed, and we have given this to its carrier.

The second document, inscribed onto the same page, was much like the first. The nakhoda wrote "this is a ship's *qawl* from the Persian Government, in sample form," using the Persian term for template, *namūna*. The document he copied into the notebook was in Arabic—telling, for a document that would have originated from the Persian government. The gist, however, was the same: the document declared that its carrier was a subject of Reza Shah (who is named after several lines of honorifics), and asked that "those with whom we enjoy the long friendship of those who go back and forth in the seas and to the ports, do not treat him but with the usual friendship, hospitality, dignity, and honor."[1]

Al-Failakawi and his mentor's documents came more than a century after the expedition against the Qawasim and the maritime treaties that followed, but they followed the conventions of longer-standing genre. The term Al-Khariji used to describe the documents, *qawl*, might link back to the *qawlnāma*—a treaty written up for the safe passage of different groups through the territory of a particular sovereign. However, the declarations the nakhodas carried were more outward looking, addressing themselves to friendly governments whose shorelines and port cities the nakhoda might pass through. In a seascape that teemed with different kinds of political authority, from the local to the imperial, the *qawl* was something of a passport—an attempt to thrust a political presence onto the water.

And it was there, on the *qawls* in Al-Failakawi's chest and Al-Khariji's notebook, that a changing world of imperial politics on land and at sea most visibly spilled over into the dhow's archives.

• • •

This requirement that a nakhoda carry with him a pass signed by both his ruler and the British Agent likely has its roots in the fifth article of the 1820 treaty, which stipulated that "the vessels of the Friendly Arabs" were to carry "a paper ('Register')" signed by the ruler, describing the dhow, and another "Port-Clearance" document from the same, detailing its owner, nakhoda, crew, origin, destination, and arms carried. "And if a British or other vessel meets, them," the clause continued, "they shall produce the Register and the

Clearance."[2] As the cannon smoke cleared from the coast of the Gulf, the treaty sought to impose order on a new basis. The Pax Brittanica was to be preserved less by guns than by paper—a maritime rule by writing desk.

But was it so new? The Arabic text of the 1820 treaty, which stipulated that dhows carry a pass, referred to it as a *girṭāṣ*—the same term that produced the *cartaz*, another safe-conduct pass, one that Portuguese imperial officials forced onto local carriers in the Indian Ocean as a means of directing maritime traffic to their ports. There were clear differences between the regime the British proposed in their treaty and the *cartazes* issued by the Portuguese, which effectively sought to protect the holder from the Portuguese themselves. The stated goals of the British Empire in the nineteenth century, of suppressing piracy and the slave trade, and those of the Portuguese Empire of the sixteenth century—pepper and proselytization—were simply not the same. But the idea that an empire might project its power over routes and sea lanes by way of paper was not in itself anything new.

For the nineteenth century, we might read the pass regime against the backdrop of British regulatory endeavors across a much broader Indian Ocean arena. Documents confiscated from one dhow captured off the coast of Zanzibar in 1867 included three *qawls* issued by Sultan Majid bin Saʿid of Zanzibar, the son of Saʿid bin Sultan Al-Busaʿidi and the brother to the ruler of Muscat at the time—and so in the orbit of Gulf maritime politics. "To all who see this from our dear respected friends who inhabit the warships that sail East and West, may Allah protect you," read one; "this is the dhow of Sayyid Abubaker Ahmad [and it is] leaving Zanzibar and headed towards Mwali [in the Comoros] and Madagascar [*Bukīn*], and it is captained by Ahmad Adam, and he has with him seven sailors and other passengers. We ask those who encounter them at sea to provide them the protection they deserve, and to treat them in a manner appropriate to the agreements of amity and bonds of unity."[3] The *qawls* addressed a broad audience of imperial powers that had entered into treaty relations with Zanzibar, but which had also begun to engage in surveillance at sea. It reflects a recognition that one could no longer move freely through the waters—that even the short voyage from Zanzibar to the Comoros and Madagascar involved encounters with European warships, and that a dhow that could not signal attachments to a recognized sovereign risked harassment.

Nakhodas could carry *qawls* from places other than their home ports. An 1859 *qawl* issued to the nakhoda Hussain Muhammad Al-ʿAsʿousi and another issued to the nakhoda ʿAbdullah Yousef Al-Sager five years later both drew on Ottoman protection.[4] The *qawls* declare that their holders were inhabitants of Kuwait, traveling by ship to India and the Swahili coast, and that they were subjects of Ottoman state; they ask that "illustrious foreign nations" to "maintain him [i.e., the holder], protect him, and treat him in accordance with the principles, laws, bonds, and treaties of friendly states"—a phrase that seems to have withstood the vagaries of space and time.[5]

On the face of it, the Ottomans seem to be an odd choice: the Ottoman naval presence in the Gulf, much less the Indian Ocean, was minimal—though this was not for their lack of trying. In any case, at issue was not which states could plausibly send gunships out into the Indian Ocean; rather, it was whether the state in question was imagined to be strong enough to lodge impactful diplomatic claims against another. And here, the Ottoman state was not at any disadvantage: the *qawls*, countersigned by the British Agent in Basra, constituted a recognition on the part of the preeminent maritime power in the Indian Ocean that the nakhodas were both Kuwaiti *and* Ottoman subjects, and would enjoy protections as such. Subjecthood in the nineteenth-century Indian Ocean could be nested, and *qawls* could invoke multiple layers of political belonging at once.

At some point, perhaps as the British presence in the Gulf became more locally entrenched, the two different documents collapsed into one, issued by the ruler and countersigned by the British Agent. And by Al-Failakawi's time, they came to be issued as a matter of course. One register of contracts from the British Political Agency in Kuwait that runs between 1927 and 1932 includes dozens of *qawls*, all of which follow exactly the same wording and syntax that Al-Khariji includes in his template; same, too, for a *qawl* that was in Al-ʿAsʿousi's possession in 1916.[6] All asked the same thing: that those who inspected the document treat its holder, an inhabitant of Kuwait, with the dignity he deserved as a subject of a friendly ruler.

And yet, the fact that Al-Khariji carried templates to *qawls* from two different polities, Kuwait and Persia, speaks to the long histories of connection that continued to animate the Arab and Persian littorals of the Gulf well into the twentieth century. But more importantly, it signals that nakhodas were able to engage in multiscalar legal thinking and interlace their travels through multiple political formations with enormous facility. They were able to recognize when different polities appeared on the horizon and were cognizant of the shifting balance of power in the region and how that could impact their ability to move from one port to another and to carry different sorts of cargoes. And they knew how to situate themselves advantageously when they needed to. This is in part why so many Kuwaiti nakhodas registered their *qawls* at the British Political Agency: there was a gap between the pleas of the Kuwaiti ruler that "the great friendly nations" treat his subjects with respect and the realities of the encounter with British naval officers at sea, but it might plausibly be bridged on the pages of the consular registry. With the British Agent's countersignature on a Kuwaiti *qawl*, nakhodas like Al-Khariji, Al-Failakawi, and the scores of others like them drew the language and the spirit of treaties past into their repertoires.

• • •

For nakhodas who spent their days at sea, empire manifested itself in very particular ways. The most obvious manifestation of imperial power was in the red tape that ran through the ports and harbors of the Western Indian Ocean—in the customs houses, quarantine stations, and other points of inspection that punctuated the voyage. But empire also manifested itself in small things, too: in signed pieces of paper that nakhodas could fold up and place in a chest on their dhows. They understood empire in the abstract, as an entity they intersected with in different ways over the course of centuries. But they also knew it in its particular formations, as one polity among many, that they could then draw into their own world. Through the *qawl* and the signatures on it, nakhodas like Al-Khariji and Al-Failakawi could thread the needle of the dhow through the world of empire and enrobe themselves in its fabric. This act, as we shall see, called forth a much longer history.

FOUR

The Sea of Oman

AS THE *CROOKED* CRUISED DOWN the Persian coast, it would eventually pass through the narrowest point in the Gulf, the Straits of Hormuz. The moment came just four days after Al-Failakawi and his crew left Qais Island: from there, they sailed due south to the Greater Tunb island and across to Hengam Island, where they anchored for the night. The wind had died down on their way to Hengam, and they had to wait for it to pick back up; "we took water," Al-Failakawi noted, adding his usual ending, "May Allah ease matters upon us and upon all Muslims, Amen o Master of the World."[1] The next morning, on October 16, he and his crew set out again, the winds bearing down onto the coast from the sea. By the late afternoon, they sighted the Salamah Islands—really, a cluster of volcanic rock jutting out from the Gulf, so small they barely appear on even the most detailed maps.

Like other dhows making the voyage to India, the *Crooked* sailed eastward past the ports of Kuh Mubarak and Jask, along the coast of Balochistan. The coastline didn't change much at all: long stretches of yellow-brown sand that gradually crept up to a long ridge of rocky, dirt-brown mountains. One might be forgiven for thinking there was nothing different about this stretch of the journey. It spent nearly two weeks sailing past one coastal landmark after another—Kuh Kalat, the mountain that marked Bandar Tang, and the port of Chabahar, where Al-Failakawi noted that they "stopped to take water and sailed out." As far as the logbook went, it was the least remarkable stretch of the entire voyage.

And yet, much had changed. Passing the Salamah Islands, the *Crooked* was now in the more expansive waters of the Sea of Oman. The water was nearly ten times deeper here, and the fish were plentiful.[2] No longer contained by the Gulf littoral, Al-Failakawi and his crew could now easily reach

Aden and the Red Sea, East Africa, and India. But nakhodas sailing to these destinations would rarely, if ever, sail across the open water to get there. They invariably preferred to stay within sight of the land, within reach of provisions and safe anchorage. At this point in the voyage, there was too little to be gained from crossing the open waters, and too much to lose.

In the past, this stretch of the voyage would have been the riskiest. From at least the beginning of the 1500s until the mid-nineteenth century, entry into and out of the Gulf was a highly regulated affair, as empire after empire sought to control the mouth of the Gulf and the shipping that ran through it. If the small polities of the inner Gulf sought possession of shorelines and pearl banks, these empires—some regional, others global—were in the business of controlling routes. They zealously guarded their right to ship goods into and out of the Gulf from around the Indian Ocean, and they forced others to either buy from them, pay a toll for safe passage, or both. Whoever controlled of the mouth of the Gulf—the Straits of Hormuz and what is today called the Sea of Oman—had their hands on the throat of Basra, one of the Ottoman Empire's only outlets to the Indian Ocean world, and held the keys to the riches of India, Yemen, and Africa.

As Al-Failakawi and his crew sailed past the Salamah Islands and across the open sea toward the Makran coast, they would have, wittingly or not, been tracing lines in a seascape that had been hotly contested for at least four hundred years before their voyage.

• • •

The fifteenth-century navigator Shihab Al-Din Ahmad ibn Majid knew at least a little about what competition over routes was like. Ibn Majid was born into a family of Najdi migrants-cum-mariners sometime in the 1420s, in the town of Julfar, which would later become known as Ras Al-Khaimah—the same port town that fielded the Qawasim roughly three hundred years later. Over the course of his life, Ibn Majid accumulated a staggering amount of knowledge on celestial navigation, but also on the particularities of sailing around the Indian Ocean—all expressed in combinations of verse and prose, compiled and recompiled over the course of his life and in the centuries beyond it. His most famous work is his treatise, *Kitāb Al-Fawāʾid fī Uṣūl ʿIlm Al-Biḥār wa Al-Qawāʿid*, or *The Book of Useful Information on the Principles and Rules of the Science of the Seas*, which contained descriptions of routes around the Indian Ocean, extensive analyses of the movements of stars over the course of different lunar mansions, and helpful maxims for the aspiring navigator.

Ibn Majid taught his readers to recognize the patterns by which stars rise and fall in the sky and how to plot courses of movement by them. The moon and stars had their own histories, animated by myths that he and others spun around them by way of poetry. He joined a mythopoetic world of stars, routes, and courses onto a material world of seas, ports, and coastlines. "Route is my mother and latitude measurement, my father," said the course to its parents, "and I am the child which the skillful, clear-thinking navigator has brought forth from both of you."[3] He instructed his reader on how to identify different coasts by their features, where to locate safe harbors and fresh water, and places to avoid—reefs, coals, and sandbanks that would jeopardize the ship. He dedicated an entire chapter to the monsoon winds, detailing all windows within which a nakhoda might be able to safely leave one part of the Indian Ocean for another. Reading his work, one might walk away with a sense of a frictionless Indian Ocean world, one through which a ship—accounting for the winds and direction—might easily glide. Such is the genre of the navigational treatise, which is, at its core, a text on route-making.

There were people around this Indian Ocean world, too—groups that inhabited the islands and coastlines. Most simply went about their business, trading, fishing, or pearl diving. But some, Ibn Majid hinted, were interested in the business of others. Peppered throughout the text are references to places where different groups made claims to points, often islands, along the route. In the archipelagos past the Straits of Malacca, different sultans, Muslim and non-Muslim alike, waged war on one another. In the Gulf, the ruler of Hormuz took Bahrain, Qatif, and swaths of coastal Oman as well. The island of Socotra, off the tip of the Horn of Africa, fell to the rulers of Mahra, in South Arabia, after a long period of jostling with other contenders. Ibn Majid warned his readers, too, that off the coast of Malabar, they ought to beware of the "Al-Kābkūrī," who are "a people ruled by their own rulers," numbering about one thousand men with canoes.[4] One can guess as to why the prospective navigator would want to be careful. Around Ibn Majid's Indian Ocean world, routes of trade could easily shade into vectors of power.

In the Arab navigational tradition, Ibn Majid is a towering figure. As a historical actor, though, he is far more elusive and has been asked to bear more than his fair share of the load. Perhaps the most pernicious myth surrounding him has him linked to the Portuguese admiral Vasco da Gama: Ibn Majid is erroneously thought to be the Moorish pilot who helped Da Gama sail to India from East Africa, thus opening the floodgates to Portuguese imperialism in the Indian Ocean.[5] And if Villiers is to be believed, even the nakhodas of the twentieth century had bought into the idea.[6] Historians have since pointed out that Ibn Majid was an unlikely candidate for the dishonor: not only had Ibn Majid most likely already passed away at the time of Da Gama's voyage, but the Portuguese commander himself referred to the pilot in question as a Gujarati.[7]

As is often the case, these myths tell us more about ourselves than they do the past. Until very recently, historians had little grasp of how different communities extended their authority over the water in the pre-Portuguese Indian Ocean. The prevailing understanding has been one of a competitive but peaceful trading arena, in which "the idea of 'sovereignty over the sea' except in narrow straits was unknown to Asian conception." It was the Portuguese, historians asserted, who first mounted cannons onto their ships and militarized trade in the region.[8] As historians grasped for a way to mark out one epoch from another, the contrast between Ibn Majid's world and that of the Portuguese seemed clear—and the possibility of their paths having intersected maybe seemed too deliciously ironic to pass up.

• • •

Inaccurate as it is, the association of Ibn Majid with the Portuguese does illustrate the changing tides that the navigator saw. He knew that in the Mediterranean, Muslims and Christians jostled for authority, and he wrote of Portuguese fleets violently thrusting themselves into the Indian Ocean world at the outset of the sixteenth century.[9] Within a decade, the Portuguese established a forceful presence around the Indian Ocean, as admirals and commanders took port cities and built factories and forts along the coasts of Africa, India, and the Gulf; they tried to take Aden, at the mouth of the Red Sea, too, but failed.[10] Their machinations over the course of the 1500s marked a moment of transition in the history of the Indian Ocean world—one in which trade and violence were woven into a broader fabric of empire-making on a scale that had not been seen before. In this history, the Sea of Oman would constitute one of the principal battlegrounds.

In the Gulf, it was a squadron under the command of Afonso de Albuquerque that would plant the Portuguese flag. In 1507, De Albuquerque's fleet sailed along the southern coasts of the Arabian Peninsula, sacking one town after another. The commander extracted a promise from Hormuz's young ruler, then only twelve years old, to pay an annual tribute of gold, and returned in 1515 to press for the establishment of a Portuguese fortress on the island. There, he found the young ruler under the influence of a Persian minister, whom he promptly had stabbed. Hormuz's king quickly assented to De Albuquerque's demand, and construction on the fortress began; it was one of many that the Portuguese would build around the Gulf and Indian Ocean.[11]

The capture of Hormuz was a key piece of the Portuguese's grand strategy in the Indian Ocean world. Although their main material interest was in gaining a monopoly on the trade in pepper from southwestern India, a secondary—and necessary corollary—aim was control over the movement of local ships around the region. By directing traffic toward ports under their control, they would benefit from the customs receipts that trade would generate. They gradually learned that for all the expansiveness of the Indian Ocean, shipping moved along predictable routes and passed through a very limited number of chokepoints. By situating themselves in port cities and islands close to these chokepoints, they could lodge themselves into the throat of Indian Ocean commerce.

The plan rested on the issuance of *cartazes*, safe-conduct passes that they issued to local vessels to guarantee they could sail without being hassled by ships from pirate states and smaller polities around the Indian Ocean, but also from the Portuguese men-of-war themselves.[12] Other polities had made use of similar passes before but seldom invested in the naval might necessary to police movement on such a large scale.[13] In the hands of the Portuguese, the *cartaz* effectively became a tool of organized violence on the high seas. Those who did not purchase them risked harassment from the very entity that was claiming to protect them from being harassed, and those who did were obligated to call at Portuguese factories in Hormuz, Goa, or Malacca, thus ensuring Portuguese profits on both ends of the racket.

What underpinned the *cartaz* system, though, was not just simply the blunt force of violence, but a theory of legal personhood, or the right to have rights. The Portuguese chronicler João de Barros expressed it most fully in his mid-sixteenth-century history of Portuguese exploits in Asia: the right to freely navigate the seas "does not extend beyond Europe," he wrote, "and therefore the Portuguese by the strengths of their fleets are justified in compelling Moors to take out safe-conducts under pain of confiscation and death." This, he argued, was fundamentally a question of whether one's religion disbarred them from participation in the community of men: "The Moors and Gentiles are outside the laws of Jesus Christ, which is the true law which everyone has to keep under pain of damnation to eternal fire. If the soul is thus condemned, what right has the body to the privileges of our laws?"[14] The Portuguese also drew on papal authority, resting their claims on a series of proclamations issued over the course of the fifteenth century in which the Vatican sanctioned the seaborne expansion of the two Iberian kingdoms of Spain and Portugal, ultimately partitioning the world between them in the 1494 Treaty of Tordesillas. With the papal winds at their back, the Portuguese sailed into the Indian Ocean as representatives of Christianity, enjoying sovereignty over the Eastern seas as its "true lords."[15]

Their self-assuredness would not last long. Almost immediately, the Portuguese found themselves challenged by an Ottoman fleet that sought to dislodge them from both the Red Sea and the Gulf, both of which the Ottoman sultan had vested interests in. Through a combination of corsairs, privateers, and appointed admirals, the Ottomans repelled the Portuguese from the Gulf of Aden and attacked their holdings in Muscat and the Straits of Hormuz. The Ottoman fleet even successfully lay siege to the Portuguese in Diu in 1538. They failed to capture the city, though, and ultimately could not mount a successful offensive against the Portuguese in the Indian Ocean, in part because of a lack of interest among policymakers in Istanbul.[16]

Others challenged the Portuguese position, too. In 1603, three Dutch East India Company (VOC) ships captured a Portuguese merchant carrack near the Straits of Malacca and declared the ship and its cargo a fair prize, as the Dutch were formally at war with the Portuguese. When the Portuguese demanded the return of their cargo, the VOC hired the young jurist Hugo Grotius to defend the seizure. The jurist penned a long treatise on the law of prize, one chapter of which challenged the notion that the Portuguese held any claims to sovereignty over the seas; the sea, he wrote, was open to all and the property of none, and the Portuguese enjoyed no exclusive right to it. In 1609, the chapter was published as a pamphlet entitled *Mare Liberum*, or *The Free Sea*.[17]

Grotius's treatise had many dimensions to it but rested on a few main arguments. The first was the idea that the sea did not meet the criteria for any workable definition of property; it was held in common by humankind. The use of the sea by the Portuguese could not impede others' rights to use it. The forts made no difference here: though people build and use them to lay claim to coastlines, shores, rivers, and even inland sea, the open sea, by its nature, "seemeth to resist possession."[18] Even if the sea could be held as property, Grotius argued, the pope had no authority to give it away. The pope was "not a temporal lord of the whole world," and thus had no authority over the sea. The sea and the right to sail "respect gain and mere profit, and not the affairs of piety," and thus the pope had no business interfering in them.[19] Moreover, trade in the Indian Ocean preceded the Portuguese, and even the Ancient Greeks and Romans, who sailed their ships into the region and found dynamic markets for their merchandise. Indians, declared Grotius, enjoyed "authority over their own substance and possessions, which without just cause could not be taken from them."[20] The people of the Indian Ocean enjoyed the right to conduct their trade as they pleased, and the Portuguese could not deny them that right.

Though they enjoyed the veneer of egalitarianism, Grotius's arguments were largely self-serving: he sought to justify the Dutch's capture of a Portuguese vessel and legitimize their entry into the Indian Ocean trade. His arguments for a free sea ultimately emerged from a commitment to furthering Dutch imperial expansion. His contemporaries in Western Europe understood as much. In a dense treatise published sixteen years after Grotius's pamphlet, the Portuguese jurist Seraphim de Freitas offered a point-by-point refutation of the Dutchman's arguments. These mostly revolved around two main points: first, that the Portuguese ruled certain regions of India by conquest, and thus had the right to restrict the trade of foreigners in those lands; and second, that the sea could not be possessed, but certain rights within it could be—such as the right to preserve the security of sea routes that led to specified territories.[21] Others took up these lines of argumentation more vigorously. The Scottish jurist William Welwod argued that while the high seas might be free to all, there existed closer waterways that states had to police access to—those that housed fisheries, for example.[22] Meanwhile, in 1635 the English jurist John Selden published his treatise *Mare Clausum*, in which he argued that waters contiguous to a coastline should fall under the relevant state's dominion—a principle that would later form the core of the idea of territorial waters.[23] The high seas, virtually all agreed, could not be subject to unqualified dominion, but there was otherwise still plenty left for states to claim.

Though these seem like esoteric debates, they animated the history of imperial intervention in the waterways that Al-Failakawi sailed through. As Europeans entered this Indian Ocean world, they thrust themselves into a history of claim-making over routes, shores, and sea lanes. These would later sediment into a discourse on movement at sea that took aim at dhows just like the *Crooked*.

But it was not just Europeans alone who thought about these waterways. Inhabitants of the Gulf and Indian Ocean did, too—and they often converged on issues that were strikingly similar to those that preoccupied their European counterparts.

. . .

Far from the lecture halls of Europe, in the coasts adjoining the Sea of Oman, change was afoot. In 1624, the year before Freitas published his defense of Portuguese practices in the Indian Ocean, a new imam was elected in the interior of Oman: Nasser bin Murshid Al-Yaʿrubi, whom tribal leaders and scholars hoped would reunify the country following its disintegration under the previous dynasty, the Nabhanis. The country that Nasser bin Murshid came to rule had seen better times. Over the preceding century, much of coastal Oman fell under the suzerainty of the Kingdom of Hormuz—and by extension, of the Portuguese. Muscat, Qalhat, and Sohar became the subject of on-and-off naval warfare between the Portuguese and the Ottomans—and at times, even the Persians were able to make claims to parts of the Omani coast.[24]

According to an eighteenth-century chronicler, Nasser bin Murshid advanced through Oman, town by town and province by province, and gradually—often forcefully—reunited the tribesmen of Oman under a single banner. Along with his troops, he arrived in Muscat, where he encountered the Portuguese garrison. "There," the chronicler wrote, "the millstones of death turned between the Muslims and the polytheists, and Allah granted victory to the Muslims, who demolished Muscat's high towers and tall buildings, and took many polytheist lives." The chronicle went on to detail further clashes between the forces of the imam and the Portuguese at Sohar and Sur, each of which ended predictably, with the Omanis vanquishing their foes. "Allah granted the Imam of the Muslims victory over all wrongdoers, and he expelled them from their lands and holdings . . . and Allah supported him against them, and aided him in his victories, and extended his good fortune until Islam triumphed and evil went into hiding."[25]

The truth was perhaps a little more mundane—or at least a little less divine. By the time of Nasser bin Murshid's ascension to the Omani helm, the Portuguese position in the Western Indian Ocean was already under attack. In the Gulf, the most serious blow was the joint campaign that the East India Company undertook with the Safavids against the Portuguese stronghold of Hormuz in 1622, just two years before Nasser bin Murshid's accession. By the time the imam began his campaign against the Portuguese, the path to success had already been made easier—not by God, as the chronicler might have it, but by the ascendant forces of merchant capitalism in the shape of the militarized joint stock company.

Even with Nasser's victories over the Portuguese in Oman, the project was far from complete. The imam died soon after ejecting the Portuguese from Muscat. It was to his cousin Sultan bin Saif (1649–1688) and nephews Bilʿarab (1688–1692) and Saif (1692–1711), both sons of Sultan, that the work of continuing the mission would fall. Over the second half of the 1600s, the Yaʿrubi rulers pursued the fight against the Portuguese overseas, alternatively making use of strategies of plunder, imposition of protection costs, or incorporation into a growing network of port cities.[26] They sacked Zanzibar in 1652, raided Mombasa in 1661, attacked Diu in 1668, sacked Mozambique in 1671, plundered Bassein in 1674, and pillaged and destroyed the Portuguese factory at Kong in 1684. One Yaʿrubi siege of Mombasa in the final years of the seventeenth century lasted nearly two and a half years and decimated the Portuguese presence in the Swahili coast.[27]

As the Yaʿrubis asserted their authority over Portuguese holdings, they enacted a powerful, yet familiar combination of violence and trade, all staged at the hands of the imams and their admirals. This mode of statecraft extended to holdings around the Persian and Arabian coasts and islands of the Gulf. On the whole, they seemed to have left India alone, though they indulged in occasional raids against Portuguese forts, with varying degrees of success. Over time, they turned to other prizes as well: Armenian traders, Persian ports, and the occasional European ship.[28]

The Omani jurist Saʿid bin Khamis Al-Shaqsi understood well the shape that the Yaʿrubi imamate was taking, and he could ground the imamate's campaigns in a compelling legal framework—one that even his contemporaries in Europe might have appreciated. For the jurist, the mounting campaign against the Portuguese raised a simple question: could Muslims justifiably engage in war against them? And if so, by what rationale? This was not wanton plunder: it answered a higher calling and had to be both legitimate and moral.

Al-Shaqsi was well-positioned to reflect on these questions for the Yaʿrubis. He bore witness to much of the imamate's expansions; indeed, he was there from its beginnings. He was born in the interior of Oman, in the town of Al-Rustaq, and later moved to the town of Nizwa, where he married a widow—the mother of Nasser bin Murshid Al-Ya'rubi, whom Al-Shaqsi raised in his own home. Nasser and his cousin Sultan bin Saif, who succeeded him, both studied under Al-Shaqsi, who confirmed his stepson's election as imam, and served as chief qadi to both him and his successor. As qadi and teacher, Al-Shaqsi acted as the moral and intellectual compass for Ya'rubi expansion.

Like many other Muslim jurists, Al-Shaqsi liked to write in hypotheticals, using "what if" situations to reflect on different questions. The fight against the Portuguese, of course, was no hypothetical; it was happening. But what if, he thought—what if a Muslim's ship were to come upon warships from India, which they then raided? And what if that Muslim was confident that these were enemies, and could see among them signs of the polytheists (no doubt a reference to the cross on Portuguese ships) and the raid happened to be in a place where combatants from the polytheists were known to cut off and plunder Muslim shipping? Would that be permissible?[29]

There could be no objection to that sort of behavior, reasoned Al-Shaqsi. In this corner of the Indian Ocean, through which goods regularly passed, brigands were common: some came from the coasts of Yemen, but those that bothered Omani shipping were, these days, from India. These could be raided if they were warships that were known to cut off Omani shipping at sea. They were belligerents, "known for their enmity and injustice, for cutting off shipping and for plundering people," and one could justly attack them. Muslims could even go so far as to burn the trading ships that belonged to their subjects, since these supplied the combatants.[30] All of this fell under the broad heading of *jihad*—in this case, armed struggle against the enemies of Islam.

But *jihad* was no free-for-all. For it to be legitimate, rules had to be followed. In most cases, *jihad* had to be preceded by a call to Islam, giving belligerents a chance to submit—though when it came to known enemies who cut off shipping, the requirement could be suspended. In all cases, there were rules for engaging in combat with the enemy, for raiding them, for capturing, killing, or maiming them, and for dividing the captured booty. On the last issue, there was a litany of rules about how much went to whom, and under what circumstances—lots and lots of what-ifs, all of which Al-Shaqsi unpacked.[31] If there was going to be war, there would be rules that would render that war legal.

Underpinning the rationale for *jihad*—Al-Shaqsi's *casus belli*, as it were—was the notion that a Muslim's right to traverse the sea had been violated. Al-Shaqsi repeated over and over again that all of his rulings on just war hinged on the idea that enemies had "cut off the paths" of Muslims, in what amounted to an act of brigandage; it was what distinguished them from regular combatants. By impeding people's ability to move through spaces held in common, the Portuguese had transgressed the boundary from workaday polytheists to faithless infidels, tipping the balance toward the declaration of all-out war.

The right to move along different kinds of paths on land and at sea held something of a special place in Al-Shaqsi's thinking. Paths fell under the heading of public goods that could not be considered anyone's property—a commons to be enjoyed by all, without exclusion.[32] These included public roads, but also water resources: rivers and irrigation canals, among others. Jurists established protected areas (called *ḥarīm*) around these resources, which prevented any one person from constructing anything that would impede anyone else's ability to enjoy those resources. Different resources called for *ḥarīm* of different extents: irrigation canals were protected to the extent of three hundred *dhirāʿ* (roughly, cubits) on either side; for roads, it was forty *dhirāʿ* in any direction; and for rivers, it was anywhere between two hundred and five hundred *dhirāʿ*, depending on the jurist and context.

Among the resources that called for protection was the sea itself. Omani jurists made it clear that there were to be no property rights in the sea. Depending on whom one consulted, the sea's *ḥarīm* extended anywhere between forty and five hundred *dhirāʿ* inland from where the water reached the shore at high tide; nobody was to build within its vicinity, much less lay any claims to it. Even if one were to lay a fishing weir—a fence-like obstruction placed in tidal waters to trap fish—they could claim no rights to it. If the weir stood on a part of the shore that was only occasionally submerged in water, its owner could prevent others from using it; but if it were wholly submerged, the owner had no right to it, as it prevented others from enjoying the commons of the sea. "The sea is particular," wrote one jurist, "nobody can enclose it, nor can they prevent others from it . . . there is no enclosure nor inheritance in the sea."[33]

If these sound similar to points that Grotius and his colleagues made, it is because they are. In reading the writings of Al-Shaqsi and his Omani interlocutors, one is struck by just how much they converge on the same issues as their European contemporaries. In all cases, the discussions revolved around how to justify violence: for Grotius, how to justify the VOC capture of a Portuguese vessel; for Freitas, how to justify Portuguese violent tactics against Indian Ocean traders, whom they understood as inimical to their imperial project; and for Al-Shaqsi, how to justify armed conflict with the Portuguese. And for all the writers, the core of the matter was whether one had the right to prevent another from movement through the sea for trade or any other purpose. Together, they all reached back into the past for authoritative discourses: legal treatises and injunctions from religious texts, often from far-flung places, animated with notions of the common good—all to place the other beyond the pale of the law. In their subject matter and method—and in some cases, in their conclusions—Oman's jurists were more like their European contemporaries than not.

The discourses that shaped movement through the seas in the early modern period were much more expansive than historians have accounted for. The genealogy that has thus far highlighted the contributions of Grotius, Freitas, Selden, and their European contemporaries to the emergence of the "law of the sea" has come at the expense of thinkers in other legal traditions who articulated similar thoughts. The global shifts that took place in the sixteenth- and seventeenth-century Indian Ocean set in motion a legal conversation with many sources, roots, and actors, and which took different forms and meanings—and yet clearly converged at different points, each in response to the other.

It's unclear how much of this a nakhoda like Al-Failakawi would have been privy to, or even appreciated. He may have understood the gist of the rulings, especially the Omani ones on *ḥarīm* along the seashore, which he may have seen enforced in his lifetime. And he would have likely grasped that the sea was a commons, and that passage through it was a right—though perhaps, as even Ibn Majid knew, it was seldom an unqualified right. But the conversations between Grotius, Freitas, Selden, and Al-Shaqsi took place long before Al-Failakawi's time, and the nuances had long since folded into the froth of the waves that washed up along the shores of the Sea of Oman. In any case, there were other, more proximate conversations on the topic that were on the minds of nakhodas of his generation.

• • •

In the middle of the summer, on July 10, 1905, a group of men gathered in a small room in the city of The Hague, in the Netherlands—just short train ride from where Grotius penned his screed against the Portuguese. Among them were the Dutch law professor Alexander de Savornin Lohman, the Austrian jurist Heinrich Lammasch, and the United States Supreme

Court Justice Melville Fuller. The three had been nominated by the governments of Great Briain and France to settle a dispute between the two states. And the question they grappled with echoed that which had occupied the minds of jurists centuries before: just what right did dhows have to sail freely across the waters of the Western Indian Ocean, without search or seizure?

The Indian Ocean world that Fuller, de Savornin Lohman, and Lammasch discussed in The Hague was not the same as the one that Al-Shaqsi, Grotius, and their contemporaries reflected on, much less the one that Ibn Majid sailed through. It was hardly even the same Indian Ocean that Loch and his East India Company colleagues sought to shape in the earlier years of the nineteenth century. In the decades after the General Treaty of 1820, the political geography of the region underwent several important transformations. In 1856, the Omani Sultan Saʿid bin Sultan Al-Busaʿidi died, leaving an empire of far-flung ports and tributaries with no designated heir. Rather than let Saʿid's sons determine the question of succession among themselves, officials from the Government of India quickly intervened in the matter. A group of arbitrators led by Charles Canning, the Governor-General of India, decided to partition Saʿid's empire into two. Muscat and its dependencies went to one of Saʿid's sons, and Zanzibar and its African dominions went to another.[34] Of course, the division of the Omani Empire into two did nothing to stymie movement between the Arabian Peninsula and East Africa. Inhabitants of Oman continued to travel to East Africa to trade and purchase land, and vice versa; they borrowed money and invested in a future that seemed no less bright than it had been in the past.[35] Meanwhile, the rulership of the sultanates of Muscat and Zanzibar continued to pass between the sons of Saʿid and, later, among their descendants.

What had changed, though, was the European presence in the Western Indian Ocean, which had become more substantial in the years leading up to and following Saʿid's death. From India, British officials began to assert themselves more forcefully in Arabia and East Africa. In 1839, East India Company forces captured Aden and established a settlement there, mostly of Indian merchants, eventually incorporating it into the Bombay Presidency. Their involvement in Zanzibar's affairs ballooned, too, mostly as a result of their desire to suppress the slave trade. Over the 1860s and 1870s, they asserted jurisdiction over greater numbers of Indian merchants on the island, established an active consular court, and played kingmaker as they anointed one Zanzibari ruler after another. By 1890, they established a formal protectorate over the island and over the coasts adjacent to Mombasa, which they designated as the East Africa Protectorate.[36]

It was not only the British that gained ground in the Western Indian Ocean. In the late 1880s, the Germans forced themselves onto the East African coast, declaring a colony in Tanganyika. And France, which had colonies in Pondicherry, Mauritius, and Réunion, gradually extended itself into other Indian Ocean port cities and islands: Nosy Be, an island off northern Madagascar, in 1840; the Comoros (first Mayotte in the 1840s, then a protectorate in Mwali in the 1880s); and the port of Obock in Djibouti in the 1860s. It also maintained a consular presence in Oman and Zanzibar. And though the French never rivaled the British in terms of political influence in the region, they were hardly insignificant; and as their British rivals would find out in the years leading up to the proceedings at The Hague, they could easily throw a wrench in the British oceanic machinery.

To further their ambitions in the Western Indian Ocean, French consular officials took to granting protection to Omani dhows, almost all of which hailed from the port of Sur, roughly one hundred miles south of Muscat. In French ports around the Western Indian Ocean, Suri nakhodas and shipowners applied for flags and passes—*titres des navigation*, which entitled them to the same treatment as French subjects would enjoy. And many of them believed that they were in fact entitled to French subjecthood: they lived and owned property in French possessions like Djibouti and Madagascar, and many had family in those places; some even claimed to have worked for the French in Mayotte. The claims they made to French protection stemmed from a longer history of Omani movement around the coasts and islands of the Western Indian Ocean. And for their part, French officials did nothing to contest those claims.[37]

But it was precisely those claims—to French flags and passes specifically, and to French subjecthood more generally—that were being fought over in The Hague. Did the French have the right to claim extraterritorial jurisdiction over the nakhodas of Sur? The question had been muddied by the Suris' involvement in slave trafficking and their tendency to hide under the cover of the French flag when approached by British vessels. But to the arbitrators at the Permanent Court, it remained, at its core, one of state sovereignty and jurisdiction over mobile subjects. The issue took on heightened importance in the wake of the Brussels Act of 1890, in which Europeans agreed to suppress the slave trade; the legislation necessitated enforcement by states with jurisdiction over a clearly defined body of subjects.

At issue was how to pin down jurisdiction over the Suri nakhodas, given their history of mobility and transregional belonging. For the British-nominated arbitrator, US Supreme Court Justice Melville Fuller, the more ambiguous forms of subjecthood that Suris exercised would have been familiar. Just a year before, Fuller had ruled on a case involving the right of Puerto Ricans to immigrate to the United States; he had to work out what Puerto Rico's status was in relation to the US and what rights its inhabitants might enjoy.[38] The Suris might have been analogous to the Puerto Ricans: they enjoyed a degree of separation from Muscat, but they were still undoubtedly subjects of the sultan, with all the rights and duties that entailed. By contrast, France's arbitrator argued it had the right to grant its flags to whomever it pleased, subject to only its own laws and decisions. In any case, the Suri mariners owed no strict allegiance to the sultan of Muscat; despite his claims over them, they had "no country but the ocean."[39]

The nakhodas viewed the issue differently. Tucked away in the French case files are a dozen petitions by Suri nakhodas, who wrote to the consul in Muscat after the sultan, Faisal bin Turki—Saʿid's grandson—convinced them to hand over their French passes. The nakhodas pleaded the consul to press for their return. The only reason they gave them up, they wrote, was out of fear of the English "men-of-war," Arabized as *manāwīr*. One Suri nakhoda described his exchange with the sultan, telling him that they sought French protection in part because the sultan was unable to shield them from British harassment. Turning his hands, the sultan responded that the French would not be able to come to the Suris' defense should they need them. To this, the defiant nakhoda responded, "Oh Sayyid, do not belittle the French state if you do not know it, for we know that it is not weaker than that state that you came with [i.e., the British] and it is stronger in its cannons and men of war and soldiers. Do you know the state of the Turks? He said yes, and we said it is of that strength and more, and it will be clear to you."[40]

The nakhodas' blending together of the Omanis, British, Ottomans, and French in a single petition, along with their invocation of "men of war," touches on a broader geopolitical imagination, set against a much longer historical backdrop. This was not a question of imperial jurisdiction, as it was for the arbitrators in The Hague; it was simply the latest instance in a long history of encounters with empires in the Western Indian Ocean. But instead of seeing a world of empires that vied with one another for jurisdiction and power over the high seas and the vessels that traversed them, the nakhodas saw a sea of potential protectors whose backing they might selectively draw on at different junctures. In a more distant past, these might have included any number of polities, including the Yaʿrubis, who could challenge European expansion in the region and assert their rights for them. In choosing the French when they did, the Suris hedged against a more recent past in which they saw more of a hierarchy of power, even if not everyone might have agreed with their ranking.

• • •

As the month of October drew to a close, the *Crooked* slowly inched its way along the coasts of Persia, toward India. It sailed past the port of Gwadar on October 26 and made his way along the Makran coast, passing by one coastal formation after another—Ormara (which he called Khormara), a small headland that jutted out into the Arabian Sea, then Ras Malan and its nearby bay. All the while, the wind blew southward, bringing with it the cool air of the mountains of the Makran Coastal Range as the dhow sailed along currents that pushed it further eastward. This stretch of the Makran coast—really, just Gwadar—was virtually all that remained of Oman's empire by the time Al-Failakawi took to the sea, though he may have caught a glimpse of a decaying Portuguese fort just beyond the beach.[41]

It would have been hard for any nakhoda to be ignorant of the sediments of global empires that accumulated along the shores of Sea of Oman. For centuries leading up to the *Crooked*'s voyage, different actors fought over the right to control movement in this corner of the Indian Ocean. Diplomats, jurists, and nakhodas marshaled wide-ranging intellectual resources, mined materials going back to the medieval period, drew comparisons to imperial projects around the world, and mobilized a global repertoire of concepts—property, subjecthood, *jihad*, and jurisdiction—to either regulate maritime traffic or challenge regulations against it. Though these were debates that left little in terms of traces outside of books and archival documents, nakhodas would have been familiar with its broad outlines—if not the concepts themselves, then certainly the men-of-war, the raids, the flags and passes, and passing mentions by their Suri friends, who could remember a world in which a little more was possible.

But that was not Al-Failakawi's world. If the Makran coast remained a testament to Oman's imperial past and the many political configurations that emerged out of its debris, Al-Failakawi's time passing through it was brief. By the afternoon of October 29, the nakhoda spotted Ras Muari, which he called Ras Munayyir—"the illuminating headland," after the lighthouse that stood atop it, one of many that guided ships along the coast. He carefully guided the *Crooked* around the headland and maneuvered it into his first major port of call: Karachi.

INSCRIPTION

Guides

> In the name of Allah, most gracious and merciful: these are the words of the first Shaikh, [who is] appreciative of Allah and his messenger, the navigator of the two seas and two lands, the master Shaikh Ahmad bin Shaikh Mayid bin Muhammad bin Omar bin Yusuf, bin Fadhl bin Hassan bin Hussain ibn Duwaik Al-Saʿdi bin Abi Burkan Al-Najdi, inhabitant of . . . the fort of Ras Al-Khaimah, May Allah have mercy upon him and upon us, and upon all Muslims. . . . Know, o navigator, if you seek accomplishment, you must manage your passions, [know how to] measure by way of sextant, and forgo sleep that is useless, and you are to be aware of the helmsman at all times, even if he is a friend, for he can be greater than your enemies.

So begins a long entry in Al-Failakawi's notebook, written in the hand of Mansur Al-Khariji, his teacher, who inscribed a similar note in his own notebook. The note continues with specific instructions on the ritual of taking solar measurements: first, to recite a combination of two Quranic verses—"Did you imagine that We created you without any purpose, and that you will not be brought back to Us? / When it is Allah who created you and what you do." Then, to perform ablutions, pray two cycles of prostration, and take measurements while reciting the verse: "Ask them (O Muhammad!): 'Who is it that delivers you from dangers in the deep darkness of the land and the sea, and to whom do you call in humility and in the secrecy of your hearts?' To whom do you pray: 'If He will but save us from this distress, we shall most certainly be among the thankful?'"

It is one of the more curious notes in the nakhoda's notebook. Unlike the rest of his notes on navigation and contracting, this one is less practical and more philosophical: it is advice on what a good nakhoda ought to do, not instructions for what a nakhoda normally does. Moreover, it is the only note that explicitly mentions another source—the writings of the fifteenth-century navigator Ahmad ibn Majid, to whom the advice is attributed. The passage does not appear to be a direct quote, but rather a composite of different pieces of advice scattered throughout Ibn Majid's text.

There were many possible sources for Al-Failakawi and his teacher's notes. They had access to a number of different texts on navigation by nakhodas from around the Gulf—books by voyagers on the act of voyaging. Together, these mark out a textual corpus of travel: of writings on the sea, both figuratively (writings about the sea) and literally (writings produced while at sea), opening up potential vistas for the historian seeking to engage with them. Through these genres, we see the deliberate process of charting out a route, and we see how nakhodas suffused this expansive sea with a deep sense of meaning and historical time.

. . .

The nakhoda ʿIsa bin ʿAbdulwahhab Al-Qitami was a contemporary of Al-Khariji's, also from Kuwait. He was a nakhoda of regional eminence: historians and other nakhodas refer to him as "*al-rubbān al-awwal*," the first pilot, and he came from a family that fielded many nakhodas. His reputation, however, was largely a product of his writings: Al-Qitami was a prolific writer and avid consumer of texts. He is best known for authoring a navigation manual, *Dalīl Al-Muḥtār fī ʿIlm Al-Biḥār*, or *The Perplexed's Guide to the Science of the Seas*, first published in 1916. He also produced two other manuals taking on different aspects of the dhow business—a shorter manual on navigating in the Gulf and a manual for weighing and valuing pearls.[1]

Al-Qitami's nautical writings place him within a lineage of writers on navigation and the sea that goes at least as far back as the fifteenth-century navigator Ahmad ibn Majid. Though he didn't specifically name Ibn Majid as one of his sources, he insinuates as much when he declared that in writing the *Dalil*, he "collected [the information in the *Dalil*] from the books of the predecessors from among the scholars of the sea."[2] Like Al-Failakawi and Al-Khariji, Al-Qitami read broadly, and we see echoes of common referents in all their writings—concerns they held in common and phrases that traveled from one text to another.

But even as he praised his predecessors, Al-Qitami knew that his were not the days of Ibn Majid; markets had changed, and routes had adjusted accordingly. There were new nodes: places like Karachi, Bombay, and Aden, all of which had recently emerged as important ports of call, and which called on the nakhoda to rethink the dhow's itineraries altogether. And indeed, for nakhodas like Al-Failakawi and Al-Khariji, Al-Qitami was a more proximate, and maybe even useful, reference than Ibn Majid was. Al-Failakawi's notebook included many notes that were directly copied from the *Dalil*. In Al-Qitami's manual, nakhodas found a very practical guide to navigating the routes they plied.[3]

There were other texts as well, of course. Nakhodas around the Gulf penned navigation manuals, pearling manuals, nautical almanacs on weather patterns, and tables for determining the changing declension of the sun over the course of a year.[4] Some wrote in direct response to Al-Qitami, like the Suri nakhoda Nasser bin ʿAli Al-Khaḍuri, who penned a treatise he called *Maʿdan Al-Asrār fi ʿIlm Al-Biḥār*—a play on Al-Qitami's title. And those were just for the twentieth century. If they are any indication, there was a whole sea of texts that filled the centuries-long gap between Ibn Majid and Al-Qitami, the vast majority of which have either escaped our attention or have been lost to time.[5]

Al-Qitami was aware of who his readers likely were. In the *Dalil*'s opening pages, Al-Qitami—after praising and offering thanks to Allah—declared that he wrote the book after "seeing that some of our brothers from the people of our *waṭan* [the homeland or nation] had many questions about some of the sea routes and currents." He was moved to "serve the *waṭan* specifically and our Muslim brothers more generally." The *Dalil*, he wrote, was to meet "the needs of our Arab brothers, the people of the ships."[6] In a world of contests by empires over coastlines and sea lanes, this was a text by a mariner for other mariners—for those who actually plied the waters of the Indian Ocean.

To address them, he wrote in a distinctly maritime vernacular. In a shorter manual he wrote on navigation within the Gulf, he declared as much. Some people, he wrote, might contend that his writings did not amount to Arabic, "but the grammarians and linguists ought to know that if the prose in the book was not as you see it here, there would be no use for it." The people who would make use of his book "are either illiterate or read very little"; his prose, however faulty it might be, "is enough to convey the message to these readers, many of whom are familiar with this kind of language." Moreover, he wrote, "the Gulf, despite its small size, is home to a diverse array of dialects, and they [i.e., the readers] will see that when it is in print."[7] He deliberately wrote at the margins of an Arab literary tradition, in part to address the needs of his readership, but in part to amplify the world to which he belonged.

• • •

The route was the guiding principle for the structure of Al-Qitami's *Dalil.* Unlike Ibn Majid, who spent the first half of his text describing the movements of the stars and moon before discussing the geography of the Indian Ocean world, Al-Qitami launched into route-making right away. "I begin from the blessed Basra and the Persian coast and the coast of Makran to India and Malabar and what belongs to it . . . along the eastern coast [of the Gulf]," he wrote in his opening lines, "and I follow the western coast, from Basra to the coast of Kuwait and Bahrain and Qatar and Oman and Yemen and the Hejaz, to the coast of Abyssinia and Somalia and the Swahili coast and Zanzibar, and what appertains to it from Madagascar and the Comoros Islands and the Seychelles."[8] He further divided these two broad routes into distinct seas or coastal regions: the coasts of Iran and South Arabia; the Yemeni coast; the Gulf of Aden; the seas of Mocha and Hodeidah, in the southern Red Sea; the Swahili coast (with a separate chapter for Zanzibar); Makran and the western coast of India, with separate chapters for the ports of Goa, Malabar, Mangalore, and Surat; and the sea of ʿAdan, running from Bahrain to Basra.

The stars were still ever-present in Al-Qitami's text, of course. To take one example from an endless sea of them, from a crossing in the southern Persian Gulf: "we cross to the coast of Oman, first from Ras Al-Khaimah to the eastern head of Qeshm Island between the rising *Farqad* [Ursa Minoris] and *Naʿsh* [the Plough of Ursa Majoris], and from Ras Al-Khaimah to Hengam [*Henyam*] Island between the setting *Farqad* and *Al-Jāh* [Polaris], and from Ras Al-Khaimah to Bandar Lingeh set a course [between] the setting *Al-Wāqiʿ* [Vega] to the edge of the *Al-ʿUyūg* [Capella], and watch out for Tunb Island, which will come at you from the bow of the ship."[9] This was not an invitation to stellar navigation: rather, star names stood in for the cardinal and intercardinal points on a compass rose. Nakhodas used these in virtually all their writings, from their description of their currents to their notes on the daily winds bearing down upon their dhows. The wind would be *naʿshī* (i.e., coming from the direction of the *naʿsh* star, which rises north by northeast and sets north by northwest), *ʿuyūgī* (similarly, coming from the *ʿuyūg* star, which rises in the northeast and sets in the northwest), et cetera.

Far more important to the route than the stars was the land. Throughout Al-Qitami's text were descriptions of coastal landmarks (which he called *manātikh*—from the verb *natakha*, "to have spotted") that dotted the Indian Ocean littoral and which guided nakhodas along their routes. Oftentimes, Al-Qitami would describe them in prose. At other times, particularly in his description of the western coast of India, where novice nakhodas were likely to end up sailing to, Al-Qitami supplied his reader with rough hand drawings of notable landmarks—distinct geological formations clearly visible from the sea, such as mountains, craters, and the like. These *manātikh* dotted the nakhoda's mental map of the coastlines stretching from the northern Gulf to South India on the one hand, and down to Zanzibar on the other. They constituted a part of the route, acting as markers rather than endpoints, as we often imagine land to be on a voyage. In a world of changing commercial and political geographies, *manātikh* formed durable anchors to a nakhoda's routes.

But *manātikh* did more than just mark out the route and give nakhodas a sense of where they might be in space. The *manātikh* that Al-Qitami pointed out in his writings, and the routes they directed nakhodas along, marked out his sense of historical time as well. If Al-Qitami's *Dalil* was meant to be a guide for a society formed on the waters of the Indian Ocean, their history was one that was scattered along its routes, washing up against his landmarks like waves crashing on the shore. The routes of the present always bore the sediment of the past.

. . .

One of the more curious footnotes in Al-Qitami's biography is that the nakhoda once penned a manuscript on the history of Oman. By most accounts, the nakhoda likely wrote it during the final years of his life, which he spent in Muscat, where he died and remains buried. And to my knowledge, the manuscript has been lost; nobody has ever seen it. We are left to speculate as to its contents, but at least one thing is certain: Al-Qitami was a nakhoda with an appreciation for the past. Unfortunately, he's left behind very little to go off of: his published writings in the *Dalil* and the *Mukhtaṣar* offer nothing explicit in terms of his historical imagination.

And yet, there are at least a couple of clues. In the third edition of the *Dalil*, published in the early 1960s, Al-Qitami's son ʿAbdulwahhab—himself an accomplished nakhoda—added a few short appendixes on different aspects of the Gulf's maritime economy: on pearling, on fishing, and on different aspects of the dhow trade. He also added a little more: a handful of footnotes peppered throughout a list of ports and headlands, on different *manātikh* a nakhoda might pass during the route.

The first historical marker appears early, just as Al-Qitami sailed his reader past the port of Bushehr on the way to India. He marked the appearance of a landmark: a watchtower that he called *Burj Raḥmah*, adding in a footnote that it was "named for Rahmah bin Jaber Al-Jalahmah"—a reference to the Gulf pirate whose presence loomed so large over the region in the early nineteenth century.[10] The next marker appears on the northwest Indian coast, as nakhodas passed by the Kathiawar Peninsula in Gujarat. As he marked the westernmost headland of the island of Diu—*Ras Al-Diu*—he notes that "there took place a battle between the Ya'rubi Omani and Portuguese fleets, in which the Portuguese navy was destroyed."[11]

Similarly, in describing different landmarks on the coast between Bombay and Goa, Al-Qitami noted a headland he called *Ras Piri* and a "Piri Creek" (*Khor Piri*). A footnote to the entry recalls that the name referred to "Piri Muhyi al-Din Ra'is, who commanded the Ottoman fleet in the Gulf in 1555."[12] The "Piri" here was most likely Ahmad Muhiddin Piri (better known as Piri Reis), the famed Ottoman navigator and admiral, who led a series of campaigns in the Persian Gulf and Indian Ocean during the 1540s and 1550s. The date was a little off: Reis was executed in 1553 for refusing to support the Ottoman *vali* of Basra in another campaign against the Portuguese, after a failed attempt at removing them from Hormuz.[13] Regardless, the historical allusion to a past Ottoman maritime presence in the Indian Ocean was clear, and added to a growing list of contending claimants to a changing seascape.

There were more recent events as well. In a note on the Omani port of Seeb, he recalled that "in it, a treaty was contracted between the Sultan of Muscat and the Imam of Oman, through the mediation of the English"—the Treaty of Seeb, which was concluded in 1920, just four years before Al-Failakawi's voyage.[14] The treaty sought to formalize the political divisions between the sultanate at Muscat and the imamate in the interior of Oman: it established jurisdictions, set duties, and recognized mutual territorial sovereignties. Though it appeared to be a strictly land-based affair, the treaty had much to do with Oman's oceanic past: the imamate, which linked itself back to the Ya'rubis, had always exercised overseas ambitions, and the imam's supporters also regularly moved back and forth between the coasts of Arabia and East Africa and carved out an oceanic world of their own.[15]

Though their provenance is unclear, it is no coincidence that virtually all the notes had something to do with the histories of the Sea of Oman, especially given Al-Qitami's purported project of writing a history of Oman. We might thus read the notes as deliberate hints at a way of reading history from the dhow: an invitation to throw open the gateways that separate past and present, pulling geography and history together into a durable braid that stretched across the Indian Ocean. Al-Qitami's excursions into the past were mapped onto the seascape—onto the ocean's littoral and the landmarks that punctuated dhow routes. Instead of grafting locations in space onto a temporal framework, he gave hints at how one might move in the opposite direction, mapping time onto space.[16] As nakhodas moved from one historical landmark to another, they engaged in a sense of time that did not just pass, but instead accumulated into a textured map of the Indian Ocean's past.[17]

• • •

Al-Qitami might have read the history of the Indian Ocean as one "determined by the triumph of one route, one city, over another route and another city," over the course of several centuries.[18] Routes, like the itineraries of those who traversed them and the places they went, were laden with historical meaning; the places and routes of times past deposited their sediment into the routes of the present. And as Al-Failakawi sailed the *Crooked* along those routes and coasts, he would have perhaps felt the weight of those sedimented pasts—of a sense of time accumulated into different placenames and landmarks.

FIVE

Karachi to Kathiawar

APPROACHING KARACHI, AL-FAILAKAWI COULD SEE from a distance the Manora Lighthouse, which marked the entrance to the harbor. It had been twenty-three days since he and his crew left the Shatt, a full week longer than it normally took them and other dhows to reach Karachi—but they had made it. "The morning began in good health and fortune; we arrived safely in the port of Karachi," he wrote, adding that "we anchored at the lighthouse."[1] In other visits to Karachi, he'd sail further into the harbor. Turning in so that the lighthouse was to the port side of the dhow, he would make his way up the East Channel, sailing past the Kiamari Anchorage where steamships docked and, turning northeast, moving along the New Channel until he reached the Napier Mole wharf—as close to the customs house as one could get by sea. There, along the sixteen-hundred-foot-long wharf, he would dock alongside dhows from all over the Gulf and Western Indian Ocean.[2]

Karachi was the first major date market that Al-Failakawi and other nakhodas visited. Until then, nakhodas kept their cargoes on board the ship, unless expressly directed to unload somewhere on the coasts of Persia and Makran. But for most nakhodas, the most eager buyers were in Karachi and the ports beyond it—and it was in Karachi where they could easily find a good return cargo or something to carry further down the coast.

Karachi was also their first taste of British India. Until then, the voyage had taken them across a string of port cities that were within the ambit of other empires—the Ottomans, the Qajars, and the Omanis, all with the Portuguese, English, and Dutch looming in the background. Even Basra, which had been under British occupation since 1914 and was under British administration since the end of the First World War was still not a British

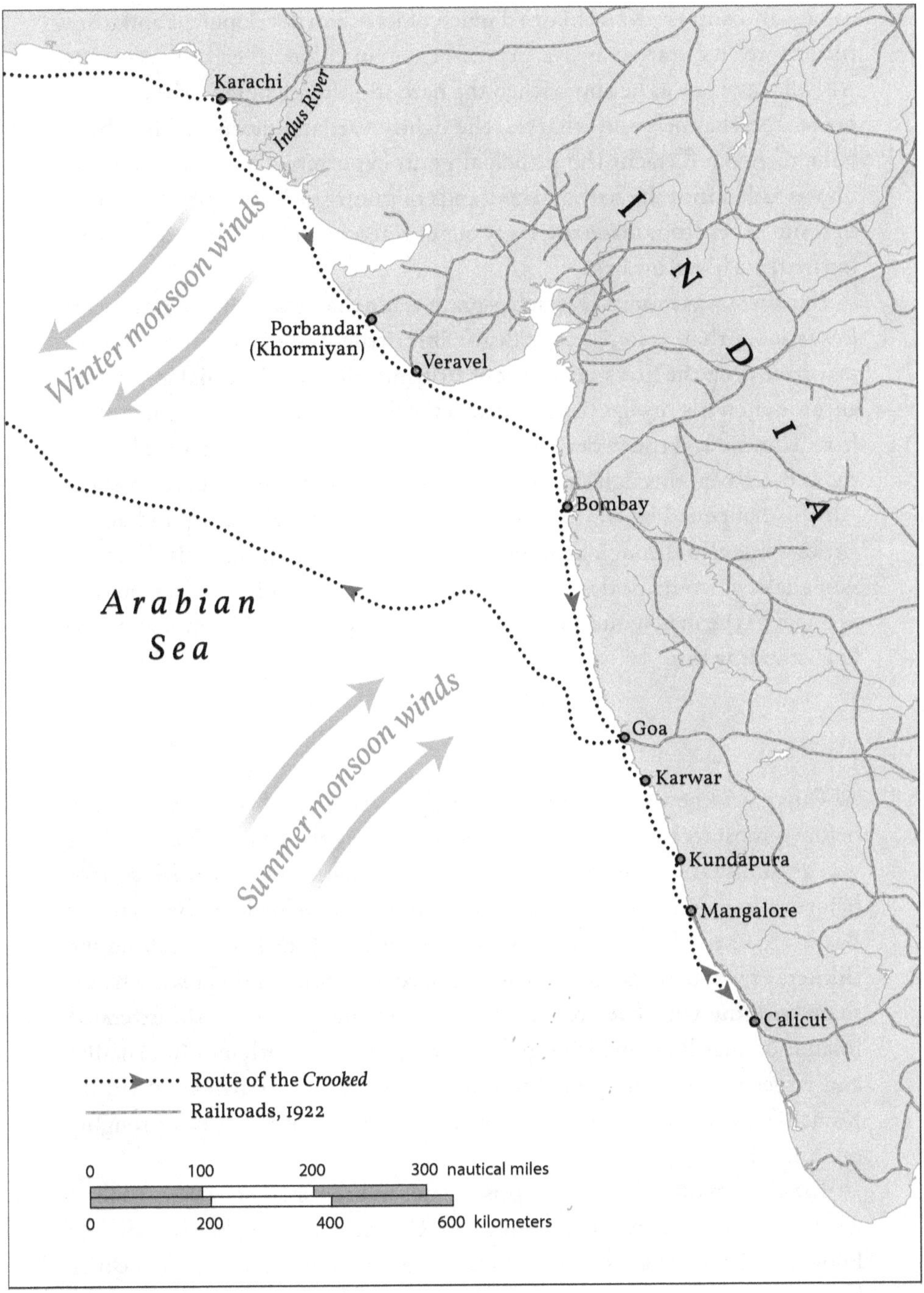

MAP 3. Western India, with the *Crooked*'s route and railways in 1922. Map produced by Nat Case, INCase LLC.

colony. By contrast, Karachi owed much of its urban development and institutions to its status as an export-oriented colonial city. Everything Al-Failakawi saw as he approached the harbor—the anchorage, the customs house, the channels and wharves, the lighthouse, and more—all had been built to make Karachi the principal grain exporter of western India. As dhows sailed into the harbor, thousands of gunny sacks of wheat, rice, cotton, and other crops made their way out of Karachi and across the sea, infusing markets like Kuwait's.

But there was more to Karachi's grain trade than the built environment; it was more than colonial capitalism's infrastructural playground. Karachi may have been the first major port of British India that Al-Failakawi would encounter on his voyage, but it was also the first major outpost of merchants from Kuwait and the wider Gulf. Goods moved from the inland valleys of the Indus River, through the port, and onto the ships waiting in the harbor. All this happened by way of trains and cranes, through canals and across jetties—but also through firms of partners, agents, and brokers. In Karachi, Al-Failakawi would deal with port officials and British bureaucrats, but not as much as the merchants and brokers with whom he and his employers had long-standing ties.

• • •

Al-Failakawi knew how to move through Karachi. He had visited the city before—most recently in 1921, when he spent more than a month there. That was an unusually long stay: other nakhodas would spend a week or two before sailing on to other markets or turning back to load another cargo at Basra. He knew that behind the hustle and bustle of the harbor, behind the thickets of wood, brick, and steel—just beyond the customs house, "a handsome building with five arches *à cheval* upon the road"—a road extended inland: Bandar Road, which ran for nearly two miles, with multiple smaller and larger roads fanning out from it. On either side of Bandar Road lay Karachi's bazaar, which extended outward and took up an area of roughly one and a half square miles.

The city's market district comprised a dozen or so smaller bazaars, each of which suggested a particular orientation. One of these, Bombay Bazaar, was known for its merchants' connections to that other major port city. The link between Bombay and Karachi would have been obvious to those who set up shop in that quarter. In the half century or so leading up to Al-Failakawi's arrival in the city, Karachi had emerged as something of a twin city to Bombay. But where Karachi distinguished itself from its sibling was its riverine access to Sindh and the Indus River Valley, whose agricultural bounty was immediately apparent to its conquerors. Indeed, even to the most casual student of world history, the Indus River Valley and the historical marvels it generated needed no introduction.

To marvel at Sindh's seemingly natural ability to produce was one thing; to marshal that produce and direct it to provision particular markets was, of course, another. As Sindh's main outlet to the sea, Karachi quickly came to be woven into a much broader fabric of colonial capitalism in the region; "Karachi and Alexandria . . . have the family look that becomes brothers," observed the traveler Richard Burton in 1876.[3] It was an apt comparison, at least by the time Burton made it: Alexandria had emerged as the outlet for all the Nile delta's cotton exports, and was the terminus of a network of railways and canals that facilitated the movement of that lucrative crop to the merchants and financiers who eagerly awaited it.[4] Karachi witnessed a similar transformation in the second half of the nineteenth century. Almost immediately after the conquest of Sindh, the colonial government introduced coal-powered steamships onto the Indus River to ferry the valley's produce—principally cotton—up and down the waterway.[5] Downriver, the regularization of steamship traffic into and out of Karachi by the 1840s served to tie the city into developing intraimperial circuits of goods, but also of people—of officials, soldiers, engineers, and regular visitors.

A far more significant development, though, was the emergence of a railroad system that connected the port to the interior—to Sindh, Punjab, and the Indus River Valley more generally. The passage of the Scinde Railway Act in 1855 set in motion two parallel railway projects: the Scinde Railway Company, which connected Karachi to Lahore, and the Punjab Railway, which ran from Multan to Lahore and then to Amritsar. Both were complete by 1861, and together with the Indus Steam Flotilla, a steamship company that ran ships along the Indus and Chernab Rivers, they eased the movement of goods between the interior and Karachi. Over the 1870s and 1880s, the railway network in the interior expanded as smaller railway companies merged, and by 1905 virtually all of it consolidated under the North Western State Railway, which then went on to connect Sindh's railways to Delhi.

The rail connection to Punjab was nothing short of transformative for Karachi and its merchant community. Not long after its annexation in 1849, Punjab had become the principal agrarian province of British India. Its farmers grew cotton, tobacco, sugar cane, and enormous amounts of wheat—a third of all India's wheat crop by the 1920s. With a railroad system in place that connected it to the Punjabi breadbasket, Karachi quickly emerged as a major exporter of grains in the subcontinent, rivaling more established entrepôts like Bombay, Calcutta, and Madras. At the outset of the twentieth century, Karachi became the biggest wheat exporter in the entire east; by 1912, it accounted for 80 percent of all India's wheat exports.[6]

From the port of Karachi, dhows and steamships could tap into the produce of the Indus River Valley and carry it off to the Gulf, the Red Sea, East Africa, and the Bay of Bengal. Karachi's changing status was reflected in the port itself. In 1876, Burton described how "the carcasses of the larger vessels were stranded upon [the] mud banks," at the harbor, alongside the dhows from Muscat, the Gulf, Kachhch, and Bombay that crowded together in the water. But, he noted, things had been changing: engineers were dredging the harbor and building piers, all to accommodate larger vessels.[7] Beginning in the 1880s, they constructed a concrete breakwater, erected stone banks, built a jetty for use by dhows, and erected the Manora Lighthouse that Al-Failakawi stopped at on his way into the harbor.[8] Steamships that traveled through the Indian Ocean could now stop at Karachi and load grains from Punjab and Sindh to take to distant markets.

The roughly concomitant development of a system of telegraph lines linking Karachi to India, Arabia, East Africa, the Mediterranean, and Southeast Asia meant that prices could be communicated quickly and merchants on the spot could act fast. Over the wire, orders came in, price movements on different grains went out, and merchants informed one another of goods purchased and forwarded. As a communications and distribution hub for the regional grain trade, Karachi was less Bombay's twin than it was its sister city. If Bombay, that monied metropolis, was the New York of South Asia, Karachi was its Chicago.

• • •

Many of the Gulf Arabs who set up shop in Karachi established themselves in a small pocket of the bazaar, in the Kharadar neighborhood.[9] By the 1920s, their correspondence from Karachi all bore letterhead that boasted their office addresses, many along Newham Road and Bohri Road in Kharadar. It was in this area, with these merchants, that nakhodas like Al-Failakawi would stay when he visited Karachi. In the early twentieth century, the

Kharadar neighborhood formed the main locus of maritime-oriented activities: it lay just beyond the customs house, and housed communities of Isma'ili merchants that actively participated in commerce with Bombay, Gujarat, Muscat, and East Africa. From Kharadar, astride Jodia Bazaar, Bombay Bazaar, and many of the other markets that populated the old town, they tapped into the flows of sacks of wheat, rice, flour, and other dry goods that moved through the marketplace, directing them to the fleets of sailing vessels that awaited them at the docks, ready to sail on to Bombay, to Aden, to Zanzibar—or back up the Gulf.

From the bazaars of Karachi, these merchants could tap into the enormous agricultural wealth of the Indus River Valley. Chief among these was *balam* rice, a variety known to be "shining white, long, clean, and light," and soft when cooked; Gulf merchants all referred to it as *balam Karachi.*[10] Jute gunny sacks of rice made their way from Punjab into the rail depots, through the marketplace of Karachi and then, via dhow and steamship, to the Gulf. There, they poured into the marketplaces and piled into wholesalers' storage rooms, but also made their way into Gulf rulers' storehouses, which they kept well-stocked to provision their subjects in times of want.[11]

Al-Failakawi and his crew knew those sacks of rice all too well: they formed a staple on board the dhow and in their homes, and they nourished bodies around the Gulf. Indeed, next to dates, there was hardly any food item as common to the Gulf storehouse as rice: rulers and pearl divers alike could access it, and virtually everyone incorporated it into their everyday diet. Indeed, it was so common that it might have seemed timeless, were it not historically inflected. Rice may have been grown in different parts of the Middle East for thousands of years, but it is impossible to overemphasize the impact that the mass production and export of rice from Punjab, by way of Karachi, had on markets in the Gulf. As greater quantities of rice pulsed down the railroads and shipped out of the port on dhows and steamships with greater regularity, it became more widely available, and at cheaper prices. Rice became a commodity a mariner could reasonably consume—albeit on credit from merchants, by way of the nakhoda. As mass quantities of rice moved through the port of Karachi, then, it transformed the foodways of the Gulf, infusing itself into the very bonds between merchants, nakhodas, and mariners.

But rice wasn't the only Karachi export that made its way to Gulf markets. Gulf merchants also oversaw an enormous trade in wheat, both unmilled (which they called *ḥinṭa*) and milled (*ṭaḥīn*, or wheat flour)—but that was just the beginning. Their price lists distinguished between the unmilled wheat of the Punjab and Sindh, and they identified different grades and brands of milled flour. They also exported different pulses: lentils and chickpeas of different varieties, and all manner of beans. From Karachi, too, they coordinated trans-shipments of sugar coming in from Southeast Asia and even Europe, which they directed to Basra and other Gulf ports. All these transformed the Gulf palate, shaping household dishes and redefining what it meant to eat in the region. Whether by rail or steamer, the merchants of Karachi quite literally had their fingers on the pulse of the regional food trade.

To pay for their purchases, Gulf merchants in Karachi looked to Bombay, the regional financial center. After loading their cargoes, they would send bills of lading to Kuwaiti bankers in Bombay—chiefly Muhammad Salem Al-Sudairawi, whom we will come to know more about in the next chapter. Al-Sudairawi would then send back drafts payable on demand: checks that he drew from Bombay banks, which could be cashed or deposited in Karachi. Since not all merchants had direct relations with Al-Sudairawi, the work of coordinating many of these payments fell to Marzouq bin Muhammad Al-Marzouq, who routinely collected orders for payments and bills of lading and forwarded them to his associate in Bombay. Al-Marzouq kept running accounts with his buyers, who would often deposit money into bank accounts he kept in port cities like Basra so that they might pay for goods coming from Karachi.[12]

If payments for Karachi goods came from Bombay, the capital that fueled the commerce came from Basra, in the holds of dhows like the *Crooked*. At Karachi, Al-Failakawi and other nakhodas found their principal market for Gulf dates: no Gulf dhow would sail to India from Basra without first stopping there. In any given year, they unloaded tens of thousands of pounds of the dried fruit to merchants and brokers in the port, who found ready buyers for them both locally and upcountry. The most popular was the slightly cheaper, mass-market *sayer* variety, which in the early 1920s could fetch anywhere between Rs 21 and Rs 28 a *mann*, depending on the market. *Diri* and *zahidi* dates, which were exported in fewer quantities, were more valuable: they could bring anywhere between Rs 31 and Rs 39 a *mann*, and as high as Rs 45.[13]

Gulf merchants in Karachi obsessed over price fluctuations like these and relayed them to one another as quickly as they possibly could. Karachi was a regional communications hub: merchants regularly communicated prices up and down the coast of India, by mail and telegraph. On their letterhead, merchants proudly displayed their telegraph addresses—short handles like "Montasir," "Suhail," "Farid," and "Register," which joined them together with a much broader transregional public of merchants communicating over the wire. Those visiting the port city could tap into information on markets in Gujarat, Bombay, and as far down the coast as Calicut, and could make decisions on the spot as to where to take their cargoes next.

From their offices off Bandar Road in Karachi's bazaar, then, Gulf merchants kept a handle on a much broader transregional food trade. They took the rice, wheat, and pulses that poured from rail cars into the bazaar and funneled them outward, on the steamships and dhows that plied the waters between British India and the Gulf. Meanwhile, they oversaw the thousands of baskets of dates that were being unloaded at the harbor, directing them to merchants and middlemen in the bazaar, and consumers further inland. Karachi was a critical node from which they brokered between different markets, managing the flows that passed through the port city from trading partners upcountry, up and down the coast, and across the sea. Goods, information, and money passed through them, moving through the channels of relationships that they forged with one another.

• • •

In its essence, the Karachi bazaar comprised a society of brokers. Gulf merchants in Karachi, like many others in the port city, worked as commission agents. These were highly flexible arrangements: merchants agreed to assist one another in buying and selling goods, bridging the information gaps between different markets. For their work in securing buyers and sellers, they took a commission. When the Karachi merchant Jassim Boodai wrote to Hamad Al-Sager in Basra in September 1923, he sent Al-Sager a price list with a cover letter that expressed his eagerness to secure whatever it was that the Basra merchant might need. "But our greater purpose," declared Boodai, "is to affirm our love and its exchange [between us], and we hope you will honor us with your communication and your relevant opinions during this blessed year."[14]

Boodai's language seems generic, bearing the same kind of boilerplate honorifics that letters between actors in the Indian Ocean often overflow with. Still, it is telling that he chose to use the language of friendship and mutual love. For although commission agency was a largely individualized form of business organization, it drew from a deep well of sociability. As elsewhere, this language of friendship and love was utilitarian: it implied the reciprocal exchange of favors.[15] To succeed as commission agents, Gulf merchants could not act independently of one another. They operated in a commercial society, scattered as it was, and relied on one another's good offices and services. The bonds of friendship and mutual love that Boodai spoke of reinforced commercial mutualism, and vice versa. Those who could mobilize their social network to pursue commercial gains stood to gain a great deal.

But commission agency was not just a social institution; it was a legal one as well. It was distinct from other principal-agent relationships, especially ones in which merchants specified the duties they expected out of an agent and drew up an agency contract (*wakāla*)—really, a power of attorney—to that end.[16] By contrast, commission agency most often did not involve written contracts with stipulations; it rested simply on offers of assistance and their acceptance.[17] Of course, there were legal ramifications. A commission agent had to guarantee his buyer's solvency: since most did not have the goods on hand, they had to go into the market and purchase them on their buyer's behalf, often on credit.[18] Beyond that, however, the transaction took place on its own, independent of other transactions before or after it. Though the associates were required to keep accounts with one another, accepting someone's offer of goods or services did not legally commit them to anything beyond that exchange. They kept the details of the association incomplete and open-ended. Only the bonds of sociability—of mutual love and friendship—might help fill in the blanks.

More than any other form, commission agency gave Gulf merchants in British India the flexibility they wanted in their dealings with one another—and perhaps the autonomy of action they desired when it came to directing such large volumes of goods—while still embedding them in a commercial society that joined hands across the water in the mutual exchange of goods and services. They did not need to arrange themselves into large-scale organizations with managerial hierarchies; those were already in plentiful supply in Karachi and elsewhere in British India. In any case, the Gulf merchants' enterprise was not one in production; it fell squarely in the realm of circulation. Theirs were firms that could tap into the resources around them—railroads, telegraphs, steamships—and mobilize them without having to manage them.[19] Of course, none of this precluded the possibility of more formal arrangements between them.

• • •

In an office on one of the streets that extended outward from Bandar Road, the merchant Nasser bin ʿAbdulmuhsin Al-Kharafi awaited word from Basra, from his business partner Muhammad Al-Matrook. Al-Kharafi was a relatively recent arrival to Karachi. In 1910, he wrote to a trading partner in Bombay regarding a series of money transfers and asked him to send his reply to the office of an Indian merchant in Bombay Bazaar, suggesting that he had yet to fully situate himself in the marketplace.[20]

How Al-Kharafi and Al-Matrook came to partner with one another is unclear. The two likely began as mutual correspondents and associates, and at some point decided to form a partnership with one another, pooling their contacts and resources into a shared endeavor. In the telling of his grandson Aslan Al-Matrook, Al-Kharafi managed the India side of the business from Karachi, while Al-Matrook was responsible for Basra and the Persian coast. Correspondence between Al-Matrook and Nasser Al-Kharafi first picks up in the fall of 1911; by then, Nasser was already doing business out of Karachi. Al-Kharafi's son ʿAbdulmuhsin also entered the partnership sometime in the late 1910s (his correspondence with Al-Matrook picked up in 1918). While Al-Kharafi was doing business out of Karachi, his son ʿAbdulmuhsin worked as a nakhoda, moving between Gujarat, Bombay, and the Malabar coast and doing business on his father's behalf. In his letters, Al-Kharafi made frequent references to "the boy ʿAbdulmuhsin" engaging in a variety of different trade-related activities. He was the more peripatetic of the three, often moving between Kuwait and Bahrain, but occasionally finding himself in different ports in India, unloading cargoes of dates, picking up other goods, and sending payment orders back and forth between his father in Karachi, Al-Matrook in Basra, and bankers in Bombay.[21] Between the three of them, they managed the movement of ships, goods, and money over virtually all of the Gulf and western India.

At times, partnerships like these emerged out of blood ties, as in the case of the Nasser Al-Kharafi and his son ʿAbdulmushin. Karachi housed at least a few of these merchant families. When the commission agent-cum-banker Marzouq bin Muhammad Al-Marzouq passed away in 1921, his business passed to his four sons, Fahad, Yousef, Jassim, and Muhammad, all of whom joined together to coordinate the movement of ships, grains, and money between India and the Gulf, much as their father did. Fahad immediately split with his brothers and partnered with another Kuwaiti in Karachi, Khalid Al-Ghunaim; and Yousef eventually decided to cash out and return to Kuwait.[22] Indeed, it is unclear that anyone necessarily preferred to partner with their family members. On the whole, family members just as often preferred to set up shop independently, engaging in the same mutual agency with one another that they did with everyone else.

It is most likely this kind of reciprocity that ultimately brought Al-Matrook and Al-Kharafi together, but it is hard to tell. It's also difficult to tell what form the partnership took. Other partnerships were based on contracts that clearly outlined the partners' mutual rights and liabilities, but we have no such document for Al-Matrook and the Al-Kharafis, if they drew one up at all. The accounts that Al-Matrook kept suggest an apportionment of liabilities between the three of them: he designates some transactions as "for us, with him" and others as "for him, with us," and kept separate accounts for Nasser and ʿAbdulmuhsin. The ledger books detail money coming in and going out, all involving goods and services, along with money transfers—all for transactions that took place within the partnership. There is nothing that indicates the division of profits and losses between them, nor any clues as to their mutual contribution of capital and credit. Al-Matrook and the Al-Kharafis designated some transactions as falling under the partnership, but it is not clear how they split the proceeds among themselves. Indeed, the partnership appears mostly as an elevated form of the kind of mutual agency that Gulf merchants already engaged in—a privileged association, but hardly an exclusive one.

Much of the substance of the partnership was in the intense circulation of information between the partners. The paper trail they left behind was enormous. Drafts of outgoing letters kept by Al-Matrook suggest that he wrote more than three thousand letters to them over a period of roughly forty years, a rough average of one to two letters a week if spread evenly—and that is just what has survived! Al-Matrook's outgoing letters, pressed onto carbon paper, are mostly illegible; bleeding ink and water damage make it difficult to decipher much of what Al-Matrook wrote to his partners. Moreover, their letters begin *in medias res*; by the time we see them, it is clear that the coordination between them had already reached higher levels of intensity, which suggests that the partnership was already underway at the time.

The ability to coordinate with a partner proved enormously important when Al-Matrook and Nasser Al-Kharafi tried to find ways to manage government restrictions on wheat exports from Karachi. In December 1917, as the First World War was reaching its crescendo, the Government of India placed an embargo on wheat exports out of fear that food shortages, both from crop failures and the unusually high level of demand brought on by troops, might bring about unrest.[23] Wheat could only be exported with special permission, and Al-Kharafi feared that prices would quickly rise. He was equally concerned about the price of rice, for which there were no controls, but which was becoming more expensive by the day because of high demand and shrinking supplies. Writing to Al-Matrook, he asked whether the market for rice on Al-Matrook's end might justify its cost in Karachi. Shipping had become expensive, he wrote, no doubt because of the enormous demand for dhows created by the recall of steamers for the war effort. The supply of rice itself was uncertain, too. His immediate reaction was to coordinate with Al-Matrook on trying to find alternative supplies of rice, perhaps from Mangalore, in anticipation of a ban on rice exports.[24] It was a prescient move on his part: just two months later, the government prohibited the export of rice, in part to combat domestic shortages and price speculation. "Rice is cheap," wrote Al-Kharafi to Al-Matrook, "but the rail [companies] will not carry it, and nobody can see a way around the ban."[25]

The ban would not last long: by 1920, the Government of India eased the restrictions on grain exports, particularly to the Gulf. Though Karachi's overall exports fell to a fraction of what they were before the war, for the partners it was back to business as usual.[26] In the fall of 1921, Al-Kharafi wrote to Al-Matrook, much as he had done in the past, letting him know the prices that different varieties of dates might fetch in Karachi, and the cost of rice that week. Moroever, a nakhoda whom Al-Matrook had hired to carry dates to Al-Kharafi had just arrived in Karachi and was awaiting further instructions. "We'll wait [another three or four days] for a telegraph from you," Al-Kharafi told his partner, "because the *boom* of [the nakhoda] Ahmad Al-Kharafi will not leave until it has arrived."[27] With the valves of grain and dates loosened once again, the work of coordination was just as critical as ever. A word from his partner could mean the difference between reaching the market at the right time and having to move on to other opportunities—between making a tidy profit and barely covering costs. A partnership, with all its stipulations on sharing profits and losses, was at its core a collaborative enterprise, and one that rested on a solid foundation of timely information.

• • •

The community of Gulf merchants was not self-sustaining. In Karachi and elsewhere, they lived among and worked with a much broader community of merchants. Chief among these were the Gujaratis, who commanded the business of the city: a business directory of Karachi published in 1922 bursts at the seams with Gujarati names. Hardly a sack of rice moved through the port without first passing through the ledger book of one of its Gujarati merchants, be they Hindu or Muslim. To access the wealth of Karachi's hinterland, Gulf merchants relied on Gujarati merchants, who would help them secure goods for export, often on credit, and would also help them dispose of their date cargoes. Gujaratis were thus a critical component of the fabric of Gulf commercial society in British India; there is no understanding Arab trade in the region without them.

Al-Kharafi was no stranger to Gujaratis. The association between Arab and Gujarati merchants was neither new nor limited to Karachi. Gujarati merchants had been doing business out of Gulf ports at least as far back as the 1500s. They were active in Muscat before the arrival of the Portuguese, and over the centuries steadily expanded their commercial activities around coastal Arabia and East Africa. Around the Indian Ocean, they formed transregional firms and actively loaned out money and goods to Arab traders, planters, and consumers; they virtually controlled the regional credit market.[28]

Gulf merchants' business correspondence to and from Karachi suggests that by the outset of the twentieth century, they had developed regular relationships with their Gujarati counterparts in that city and even enjoyed some degree of intimacy with them. Al-Matrook and Al-Kharafi frequently made mention of a "Lalji," whose ships Al-Kharafi often dispatched to Al-Matrook in Basra. This was most likely the Gujarati merchant Lalji Lakhmidas, to whom Al-Kharafi had one of his correspondents address his reply. A directory of Karachi's business community published in 1922 described Lakhmidas as a "Merchant and Commission Agent" with an office in Bombay Bazaar, a quarter with deep connections to that city's money market.[29] Like other Gujaratis in Karachi, Lakhmidas brokered Gulf merchants' access to the grain market, securing for them supplies to send back up the Gulf. He dealt in large volumes: in just one letter to a Kuwaiti banker in Bombay, Lakhmidas sent payment orders amounting to just over Rs 50,000; another letter included two payment orders for Rs 10,000, and yet another for Rs 20,000.[30] The sheer amount of money involved here suggests that Kuwaiti merchants enjoyed close ties to the Gujarati merchant and shipowner.

Many of the Gujarati merchants operating in Karachi had business offices in the Gulf as well. The firm of Ratansi Purshottam kept an office in Karachi but was most renowned in Muscat, where its founder, Ratansi, a Kachchhi from Mandvi, had established himself over the second half of the nineteenth century. From Muscat, Ratansi imported grains from Karachi, but also involved himself in the date trade, packaging the dried fruit for American buyers, but also the arms trade; he even farmed the customs house from the sultan. He was well known to the Arab merchant community of Karachi and frequently did business in various commodities with them.[31]

In their communication with one another, Arab and Gujarati merchants often resorted to Arabic as a lingua franca. Letters from Purshottam and Lakhmidas suggest either a strong command of Arabic or (more likely) that their offices employed seasoned Arabic writers. That wasn't always the case, though. The commission agent Hirji Tulsidas occasionally wrote to Al-Matrook and other Basra merchants, including Al-Sager, to let them know of the prices of various grains on the Karachi market. Unlike the others, Tulsidas struggled to communicate in the language. Among Al-Matrook's letters is one attempt at Arabic from Tulsidas's office—a scrawl by what can only be described as an inexperienced hand—and a typed English letter in which the exasperated merchant writes that "we are always in trouble to get your letters read and so shall thank you if you shall kindly address us letters hereafter in English."[32]

The problem was not usually one of communication, though; there was far more at stake. The Gujaratis that Al-Matrook, Al-Kharafi, and others did business with were themselves embedded in a much broader community of wholesalers and financiers; the concentric circles of rights and liabilities could generate considerable confusion. In 1909, Gujarati merchant Haridass Asanand, who frequently did business with Al-Matrook and the Al-Kharafis, found himself faced with a claim for Rs 2,200 for rice that he had taken from another Gujarati business associate and forwarded on to the Gulf but had not yet paid for. Court officials in Karachi wrote to the Political Agent in Kuwait asking him to make sure the rice would not make its way to buyers there, at least until the plaintiff's claims could be ascertained. It was there that the confusion began, though. It was unclear whether the rice even belonged to Asanand anymore. The majority of it, officials suspected, had already been paid for: Asanand's practice, like all others in Karachi, was to load the goods onto ships and then send the bills of lading to Bombay. Asanand thus took payment for the goods before merchants in Kuwait received their shipments; by the time the rice arrived in the port, it was already Kuwaiti property. The issue lay in the transaction itself: Asanand had taken the rice on credit, but it was unclear whether he sold it for cash—that is, as a broker—or on consignment, as a commission agent.[33]

Adding to the confusion was the lack of clarity surrounding whether—and how—courts in Karachi might pursue claims that far up the Gulf, where there were British officers but no British courts. When Lalji Lakhmidas filed a claim in the Court of the Judicial Commissioner for Sindh, against the British India Steam Navigation Company, for damage done to a shipment of firewood he sent to Kuwait and Basra, officials in Karachi were at a loss to figure out how they might collect evidence from merchants there. It fell to the Political Agent in Kuwait to interview Lakhmidas's associates, including Al-Matrook and Al-Sager, and remit their testimonies back to the court in Karachi. They repeated the practice for later cases, but the seams in the process—they could hardly even agree on how much witnesses should be compensated for their testimonies—reveal an unresolving tension in Karachi's Gulf trade. Though the transregional trade relied heavily on there being a commercial-legal infrastructure of associations, bills, notes, and credit arrangements, the judicial machinery itself was ill-equipped to pursue legal matters very far downstream.

And yet, trade happened all the time, and in considerable volumes. For every moment of dispute or friction, there were countless times when it worked largely as it was supposed to—when the payments reached the right person more or less on time, when the goods arrived with relatively little damage, and when a kind word between business associates could smooth over a wrinkle in the fabric. Al-Matrook's Gujarati partners did what they could to reach over the linguistic chasm between them, and papered over the distance between Karachi and Basra with letters, cables, and notes. Along with others, they opened up Karachi and its hinterland to merchants in the Gulf, brokering between them and suppliers they did not know. They opened up markets in other directions, too—just down the coast, in Gujarat itself.

. . .

Al-Failakawi and his crew didn't stay long in Karachi; not this time, at least. They spent just one night in town—most likely just enough time to know that the market for dates looked better further down the coast of India, in Gujarat. On the afternoon of October 31, the crew of the *Crooked* weighed anchor and sailed in a southeasterly direction, along the light-brown coast of Sindh, against a backdrop of rocky cliffs. They eventually made their way out on the open water, and for the next two days, as the winds bore down from the north, they crossed the Gulf of Kachchh.

Kachchh was a well-known date market to the nakhodas of the Gulf, who often called at its ports to sell cargoes of dates. *Ruznamahs* from Al-Failakawi's contemporaries are replete with references to ports along its coast. Al-Failakawi even kept a detailed map of the Gulf of Kachchh, annotating ports like Mandvi, Mundra, Bhavnagar, Jamnagar, and many more. He even sketched in a drawing of a lighthouse—"the lighthouse at Toona," he wrote—and marked out the area where the water met Kachchh's salt marshes. Kachchh fielded some tremendously influential merchants around the Indian Ocean, especially in Muscat and Zanzibar.[34] It was also home to its own community of dhow builders and mariners; in other voyages to India, Al-Failakawi would stop at the port of Mandvi to repair any damage done to his dhow and to purchase provisions for the onward voyage.[35]

In the early morning hours of November 2, Al-Failakawi and his crew spotted a protruding obelisk, far off in the horizon—one that gradually took the shape of a "fretted and pinnacled tower" flying a large flag that was visible for many miles from the shore. This was the Dwarkadhish Temple, a Hindu temple more than two thousand years old, where Lord Krishna is thought to have lived out his years. Dhows sailed past the temple, along a coastline "dotted with fortified towns and tree-gilt villages, here glittering in the humid sunshine, there almost hid by dense growth," set against the backdrop of "a range of lofty mountains whose forested crests, unconcealed by even the semblance of a mist, cut in jagged lines the deep blue surface of an Eastern sky."[36]

This was the region they all knew as Kathiawar, the name they gave to the coasts of Gujarat—a region that was distinct from Karachi and its environs. And there was much to distinguish. Whereas Karachi was a bastion of colonial industrial capitalism, Kachhch and Kathiawar were a little more peripheral to that project. These were among the hundreds of Princely States, nominally sovereign precolonial political entities that the Government of India continued to recognize until the mid-twentieth century. Kachchh came under the nominal rule of the Rao, whereas the rest of Kathiawar comprised nearly two hundred Princely States. By the 1820s, the Government of India assumed "paramountcy" over these states: a Political Agent assigned from Bombay would oversee the exercise of jurisdiction by Kathiawar's princes and would ensure that they would cooperate with the wishes of the government.[37]

At noon on November 3, the *Crooked* pulled into the harbor at one such Princely State, Porbandar, which merchants and nakhodas almost always called "Khormiyan," *khor* being a reference to the creek that ran inland from the harbor; they sometimes referred to the port town as Khor Bandar, the "creek port." It was a town well-known to nakhodas; for many, especially those who hoped to make two or more India voyages in a season, Porbandar was their route's terminus. Al-Failakawi's first recorded voyage in his *ruznamah* ended there; he spent a month there before turning back, stopping at Karachi for repairs. Over the next two decades, five of his voyages involved round trips to Karachi, Porbandar, or nearby ports on the Kathiawar coast.

In his notebook, Al-Failakawi's father-in-law, Mansur Al-Khariji, sketched a picture of the port city as seen from the sea. The drawing features a peninsular landmass with a rounded tip pointing westward; directly above it, two waterways snake into the land (in reality there was just one waterway, though with many inlets). The peninsula itself is crowded with buildings, and there appear to be other structures on the northernmost bank of the waterway, most likely signifying the village of Bokhira; stray ship-like objects appear strewn alongside the shore. "This is a picture of the creek at Khormiyan," he notes, "and the town looks like this." The picture bears the date 1949, though it may have been drawn earlier. Just south of the peninsula, he drew two anchors—perhaps an indication as to where one was to lay anchor. In another drawing, he depicted the town from the deck of the dhow: buildings with minarets appear in the foreground, set against a series of hills; a jetty juts into the water, forming a harbor. "This is the appearance of Khormiyan Mountain and the town, as seen from a distance," he wrote in a note at the top of the drawing, adding that those approaching at night should keep an eye out for the lighthouse to the east: "the water is 8 fathoms deep," he noted; "cast down [your anchor] with safety and blessings; the ground will be of mud and sand."[38]

A report on Porbandar written in 1879 noted its long-standing commercial ties with the Gulf and the Red Sea, from which vessels would come to purchase cotton, grain, and Porbandar stone, a limestone widely used in buildings around the region. However, the rise of Bombay as a textile exporter and commercial entrepôt diverted much of Porbandar's trade southward. Meanwhile, port improvements in Karachi, coupled with the development of a railroad line that terminated there, caused much of Porbandar's grain trade to move northward. Though its imports steadily increased from the 1890s onward, and its port facilities improved, the port had little to

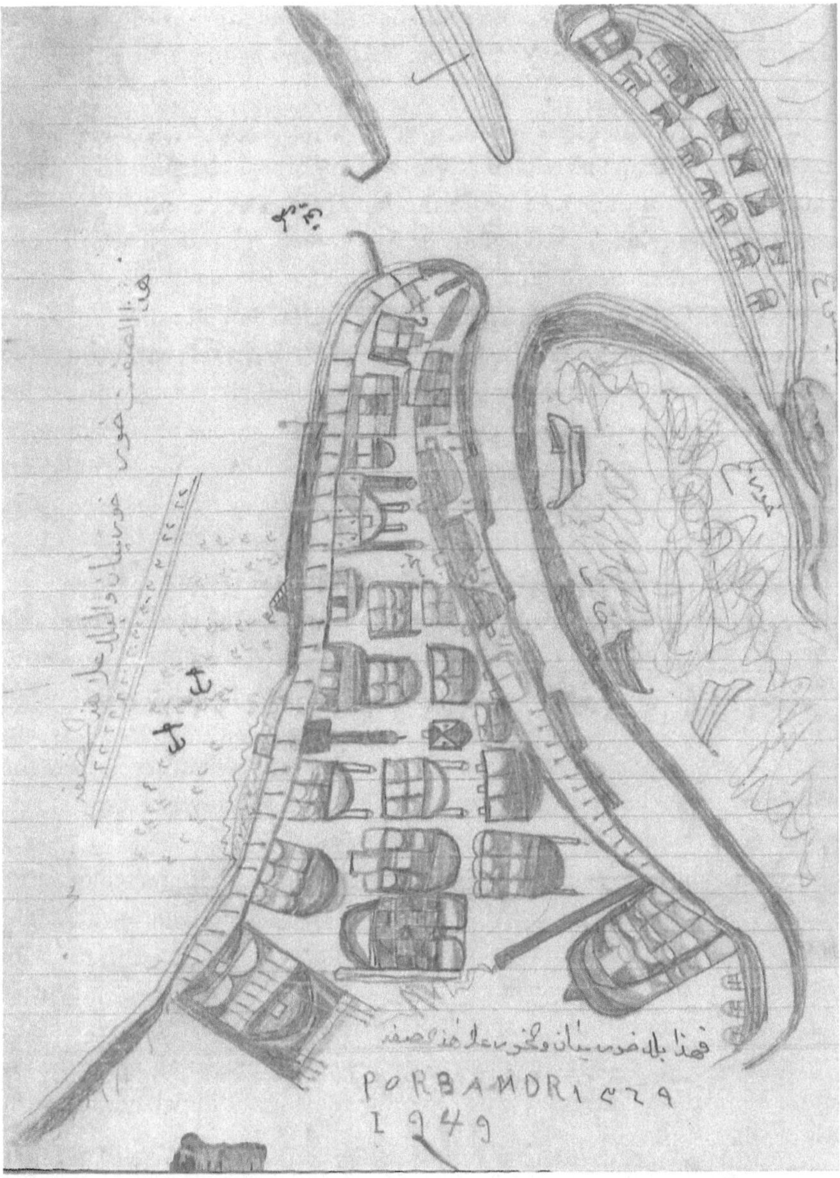

FIGURE 9. Mansur Al-Khariji's sketch of the port of Khormiyan (Porbandar) in 1949. Photo from Al-Khariji's manuscript *Al-Qawāʿid wal-Mīl.*

export beyond grains, and those could be had at better prices and larger volumes at Karachi. Even its trade in building stones had become eclipsed by other ports in Gujarat.[39]

It is more difficult to get a sense of what life was like in the marketplace at Porbandar. Although merchants wrote to one another about prevailing prices there, very few of them ever wrote anything from there. Nasser Al-Kharafi occasionally sent his son ʿAbdulmuhsin down there, and though he had likely gone there himself, he never mentions having done so.[40] Nakhodas visited the port more than merchants did but had little to say. Though many mention it in passing, only the nakhoda Rashid bin ʿAli Al-ʿAsʿousi wrote anything detailed from there. While stopping at the port in April 1918, he wrote to his uncle in Kuwait to let him know that he was in the process of unloading his last six hundred bags before heading to Bombay for repairs. The price of dates that month, he noted, was between Rs 6.25 and 6.50 per bag for good dates and between Rs 5.25 and 5.5 for those of a poorer quality.[41]

Al-Failakawi only spent a few days in Porbandar that November and his logbook makes no mention of unloading any dates. Although it is possible that he failed to make note of having sold dates there, he more likely found the market there oversaturated, causing prices to fall. By nightfall on November 5, he raised his sails and cruised along until noon the next day, when he reached the next closest market, Veravel, and anchored near the shore. He spent the next three days there looking for a buyer before unloading just one hundred *qoṣras* of dates on November 9.

In Veravel, the fog surrounding the activities of Kuwaiti merchants lifts ever so slightly. The merchant Muhammad Thunayyan Al-Ghanim often did business out of there; Veravel was where he carved out a niche in the date business. He started his career as a nakhoda in 1903, at the age of nineteen, ferrying dates for his maternal uncle Hamad Al-Sager. Just four years later, he decided he'd set up a business office in Veravel, first as an agent for his uncle Hamad, and later, in 1923, as a commission agent in partnership with his brother Thunayyan, who traded out of Karachi. He even owned a home there, leased to him by the local ruler, the Babi Nawab, who ruled over the largely autonomous Princely State of Junagarh and was likely eager to attract any kind of trade to his port.[42]

In a series of letters to Al-Matrook in 1921, Al-Ghanim let him know that he dispatched a *dinkiya* loaded with 246 bags of nuts from Veravel to Basra and was in the process of readying an Indian *kotiya* from there as well.[43] Other letters are no more revealing. One Al-ʿAsʿousi nakhoda wrote to his relative in Kuwait in January 1928 to let him know that he had arrived there after unloading some of his date cargo at Porbandar ("Khor Bandar"). "There are 700 *mann* of dates left in our *baghlah*," he wrote, and the going rate was Rs 7 to 7.5 per *mann*; "Allah willing, we will have good fortune." He planned to head directly from Veravel to Calicut.[44] Other nakhodas who wrote longer letters from the port offered very little detail on their activities; they let their associates in other ports know of cargoes sold in Gujarat and of ships that would be arriving in other parts of the Indian coast carrying unsold dates.[45]

Through the limited material, and past the routinized recitation of cargoes and prices, we catch a glimmer of who they were doing business with. In a letter Al-Ghanim wrote from Veravel to a partner in Bombay in 1911, he included a payment from one ʿAbdulkarim bin ʿAbdullatif Al-Maymani—the Memon, a reference to the Gujarati Memon community—and another from an Abdulhussain Juwairi; all were for dates that had been unloaded from dhows visiting Veravel that week.[46] Three years later, he wrote again to Bombay to transfer another payment from Al-Maymani—this time to be added to the account of a Kuwaiti trader who sold the Gujarati a cargo of dates.[47] Juwairi's relationship to the merchants was similarly long-standing: in January 1928, a nakhoda wrote to Kuwait from Veravel saying that he had just sold a cargo of dates to "sons of Juwairi"; the letter was on a sheet with a letterhead bearing the name "Ali Eshak Zaveri, Dates & Dry Dates Merchant & Commission Agent," in English, Arabic, and Gujarati.[48]

Like in Karachi, Gulf merchants around Kathiawar relied on Gujarati merchants to help them dispose of the cargoes of dates that came from Basra. The dates they sold helped them balance their accounts with their Gujarati associates. And like in Karachi, those payments were routed through Bombay, from which Gulf and Gujarati merchants could both draw money and deposit payment orders. Kathiawar was no Karachi, with its steamships, railways, and telegraphs, yet it was no less vital a stop in the itinerary of the dhow trade and constituted an important node in the transregional circuit of commodities, notes, and information. And in Kathiawar, one sees many of the same intercommunal commercial relationships that characterized its busier neighbors.

• • •

On November 10, after nearly four days in Veravel without having unloaded any more dates, Al-Failakawi decided it was time to leave. Before he did so, though, he sailed his dhow in the port city's inner harbor, either in a last-ditch attempt to unload more dates or to get clearance to leave the port. Neither his papers nor those of the merchants operating out of Veravel say anything about it. All we know for certain is that he did not spend much time there: by the evening, the *Crooked* sailed out of the harbor and made its way down the coast. In the morning on November 11, Al-Failakawi noted that they were at Ras Madhwad, a small headland on the southernmost tip of the Kathiawar Peninsula; by noon, they had begun to cross the Gulf of Cambay.[49] There, in the maritime boundary between Gujarat and the Konkan, they sailed past Diu and Surat, both of which stood as remnants of—really, as testaments to—a historical era that had long since passed, one in which merchants in those ports channeled the flow commodities and capital to the coasts of Arabia and East Africa. By Al-Failakawi's time, Gujarat's merchants looked in other directions—to Karachi, from which they controlled much of the movement of grains from Punjab and Sindh outward, and to Bombay, where they grasped at the reins of the Indian Ocean capital market. It was to Bombay that Surat truly lost out—and it was there that the *Crooked* headed as it glided through the Gulf of Cambay and made landfall on the Konkan coast.

INSCRIPTION

Letters

> To the noble, respected, dear brother Fulan bin Fulan: *assalāmu ʿalaykum wa raḥmat Allah wa barakātuhu*. We hope that you and yours are in good health; we and ours are in good health. Greetings to you on the occasion of this happy Eid and blessed new day of the Eid Al-Iftar, may Allah bestow it upon us all in good health, and may He include us and all Muslims among the victors in the many years to come. Eid, on our end, is on date X. Mansur bin Ibrahim.[1]

One of the more puzzling inscriptions in Al-Khariji's notebook is a template for writing a letter bearing Eid greetings. Puzzling, because in a notebook that is otherwise filled with sailing instructions, navigational principles, and models for different instruments, it's unclear why he felt compelled to include instructions for how to write a letter conveying holiday greetings.

But it is maybe not so puzzling at all. Letters formed the lifeblood of the Indian Ocean trade. The tens of thousands of pages of correspondence and letter books that Gulf merchants kept tell us as much. Most were not greetings; they were more functional forms of writing. Through letter-writing, merchants and nakhodas conveyed their most precious commodity to one another: information. They wrote on prevailing market conditions and prices, on goods received, delivered, shipped on, and sold. To send and receive letters was to be a part of a commercial society; to be excluded from the informational circuit, or even to receive letters irregularly, was to be cut out altogether. The mercantile letter—the *khaṭṭ*—was not a medium through which merchants expressed the sort of affect that modern readers associate with letters. It was, at its core, a technology.

My reading of letters as technologies of coordination is nothing new. Historians of the Indian Ocean, and economic historians of the premodern world more generally, have long made the claim that letter-writing served to coordinate action among commercial actors. Some suggested that letter-writing constituted the vehicle through which merchants circulated information on normative behavior among members of their community. Merchants, the story goes, enforced cooperative behavior among their associates by amplifying news of malfeasance within the network and causing recalcitrant agents to suffer reputational damage. Not everyone has bought into that interpretive framework, though—and I've seen nothing in the letters among Gulf merchants that suggests they circulated that sort of information among themselves.[2] When they did discuss the behavior of their associates, it was almost always news of routine activity, not gossip. So then what might one say about letter-writing within this world?

Like their medieval and early modern counterparts, Gulf merchants wrote letters with a strong sense of purpose and adhered to strict genre conventions.[3] The commercial letter articulated and reinforced their accountability toward one another. Writing formed the sinews of a transregional commercial society and nourished the different commercial associations they entered into with one another. Through constant, regular communication, correspondents brought themselves into closer alignment with one another, regularizing expectations and closing the information gap that would otherwise plague any principal-agent relationship; they "rendered the terms of exchange comprehensible to all."[4]

• • •

There was something of a formula to writing a letter. After reading five or six of them, I came to know what to expect of the next five or six hundred. Atop the letter was the standard *basmallah*, the invocation of the name of Allah, followed by a note on where the letter was sent from, where it was going to, and the date (though some writers left the to/from unsaid; it may have been on the envelope, long since discarded). The first line or two of a letter invariably consisted of compliments to the recipient—"after compliments" (*baʿd al-salām*), followed by a string of stock phrases and greetings that wished the recipient good health and fortune. These are easy enough to gloss over, as the more substantive information comes later. However, we might read the inclusion of these greetings in virtually every letter as signaling a broader etiquette in letter-writing, one that structures much of the letter itself. Generic as it was, the form of address established a relationship between

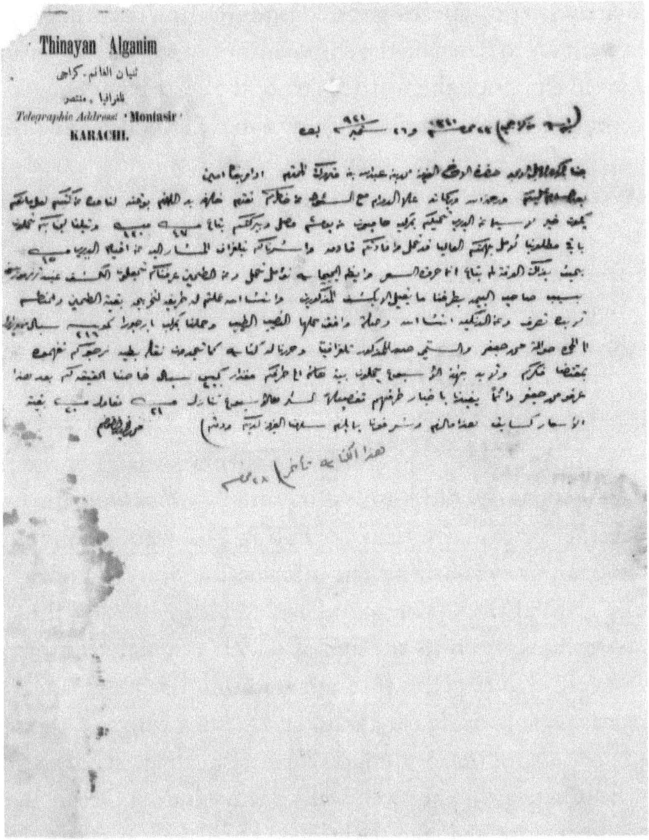

Thinayan Alganim
ثنيان الغانم - كراچي
تلغرافيا . منتصر
Telegraphic Address: 'Montasir'
KARACHI.

FIGURE 10. A letter from Muhammad Thunayyan Al-Ghanim to Muhammad Al-Matrook, 1921. Photo from Aslan Al-Matrook.

the sender and recipient; it implied a bond of friendship and reciprocity between them.

After the front matter came a reference to past communications, especially between regular correspondents. They would specify which letters they had read—of the twelfth of the current month (*jārī*) or the tenth of the last month (*sābiq*)—along with telegraphs and accounts received and would invariably finish with the statement that "we've understood what you've informed us of," or something to that effect. Statements like these anchored the letter, a momentary utterance, in an ongoing conversation. By referring to past communications and stating that their contents were understood, traders threaded together past and present, affirming that the content of their letter rested on a history of transactions between the parties—one of cooperation and sociability.

Immediately after signposting the communication that they received, letter writers went on to describe the shipments of cargo and the money transfers that moved alongside them. This part of the correspondence took up the most space: they let their associates know about nakhodas and other associates who arrived or left, what they managed to sell their goods for, what return cargo they might have arranged for, and the payments the transactions called for. This work that they did on one another's behalf, this iterative exchange of goods and services, constituted the substance of the agency relationship, one anchored in communications received and understood. Through the medium of the letter, they gave cooperation its shape: absent a contract, it was in the mutual back and forth of instructions and news that an institution like "agency" came to be.

Before signing off, merchants informed their associates of the prevailing prices for the goods they did business in, often with commentary on recent trends—whether prices were rising or falling, and whether the supply kept up with demand. They sourced their information broadly, conveying information they learned from correspondents and associates on the ground in different port cities. Even in the age of widely circulated information on markets and commodity prices through mediums like newspapers and price lists, there was no substitute for a man on the spot who had their finger on the immediate pulse of the local marketplace. Epistolary information moved briskly in the nineteenth- and twentieth-century Indian Ocean, in large part due to the steamships that regularly plied the waters between India, East Africa, and the Arabian Peninsula. A merchant in Karachi, for example, could expect incoming mail at regular intervals, with little concern for the vagaries of the monsoon winds and trading seasons. Market information, quick and ready, was not only informative, as it had been in centuries past; it was useful. Merchants could act on it if they got it on time.[5]

As they composed letters in real time, correspondents often hurried to communicate the latest information to one another. We see this in occasional postscripts to their letters, which merchants scribbled below the main text, in the margins, or even above the address line, just before sending them off. In these postscripts, we see them include last-minute instructions: "If you find that flour prices are good in Iraq, then we authorize you to find a buyer, and do not pass up an opportunity," wrote two merchants from Karachi to Al-Matrook in Basra, in a postscript to a letter that principally discussed a shipment of dates between them.[6] Another letter to Al-Matrook, this time from Al-Ghanim, squeezed a few different points and observations into a tight note just below his signature: his upcoming visit to Kuwait and desire to meet with Al-Matrook, and his desire to collect money owed to Al-Matrook's father from Karachi, and the difficulties he might face in doing so because of the impending bankruptcies of several merchants there.[7] In this world of correspondence, the timeliness of information was counted in hours and days, not weeks and months.

Greetings, references to past communications, money and goods received and sent, prevailing market prices, and last-minute information—these comprised the structure of virtually any merchant letter. We see the same pattern across writers and collections from the nineteenth- and twentieth-century Gulf; when you've read one collection, you might as well have read them all. Their Gujarati partners, too, used the same conventions and formula in writing to their Arab partners, drawing themselves into the same epistolary community. The conventions they all used are also remarkably durable across time: one finds strikingly similar genre conventions in letters by Arab, Persian, and Armenian merchants in the Indian Ocean almost two hundred years earlier. Even letters found in the Cairo Geniza, composed nearly a millennium before those by Al-Kharafi, Al-Ghanim, and other Gulf merchants, appear roughly similar in their structure. The merchant letter, the evidence suggests, was about as sticky of a genre as they come.

At least some of what motivates the similarities in letter-writing conventions across space and time is an isomorphism within the genre. Merchants in the Indian Ocean did not learn to write letters independently of one another. They learned at least some of their letter-writing skills in schools, where *khaṭṭ* was a subject they were taught. Elsewhere in the Indian Ocean, there circulated manuals on proper letter-writing etiquette.[8] Together, these allowed for a convergence of norms in letter-writing—a standardization of epistolary forms based on received wisdom on what good letter-writing looked like. Those who did not attend schools or read manuals would have learned it through mimesis: having read enough letters, they would have understood both its form and its function and would have crafted their letters accordingly. Most merchants, we might recall, would have gotten their start by working in another merchant or family member's office. By the time they began writing their own letters, they would have had plenty of exposure to the genre and its conventions.

But there was more to it than isomorphism. The structure served a purpose: by ordering the continuous flow of information on markets, prices, and instructions, merchants reduced the uncertainties inherent in long-distance trade and brought themselves and their associates in closer alignment with one another. The informational ecology of the letter allowed associates to process its contents more easily. One pair of Karachi merchants even went so far as to annotate their longer letters' margins with labels for each paragraph's topic: letters, currency, Australian wheat, money transfers, prices, news on the firm, and updates on the market for dates.[9]

The formulaic structure of the letter also left little room for interpretation, even within the framework of a relationship that could be quite flexible. The use of shared conventions by merchants within and beyond this community provided them with "a decipherable code of expressions and norms that regularized behaviors and expectations."[10] And unlike contracts, which could only capture a single moment in time, letters continually replenished the storehouses of information and mutual expectations and allowed for pivots and adjustments along the way. In bringing this kind of structural and temporal consistency to their communications with one another, merchants and nakhodas could effectively conjure up one another's presence in real time. This was as true in 1924 as it was in 1024, but it was because the Indian Ocean epistolary arena of the twentieth century gyrated more quickly than it had in the past—because the distance between Basra and Bombay or Calicut was less daunting than it might have felt centuries before—that this act of metaphysics felt possible, even plausible.

There was also another technology of communication that merchants had at their disposal in the nineteenth century that they did not have in eras past: the telegraph. For a merchant under time constraints who needed to deliver quick instructions, there was no better solution. "Sell Manchuria mark ETA BLY," wrote an associate in Karachi to Al-Matrook in 1921; "wheat invoices received containing error one thousand in total Mahommerah [*sic*] sale please credit us," wrote another.[11] These were hardly the detailed letters they might have otherwise written to one another, but they communicated orders quickly.

There was some concern as to whether a missive by telegraph rose to the level of a letter, much less whether they could stand in for a real person. One Kuwaiti scholar asked a jurist whether it was possible for news of the commencement and end of Ramadan, usually delivered in person, could be conveyed via telegraph. The jurist, however, responded that in all matters, whether ritual or transactional, telegraph communications could stand in for the absent, just as letters could. They were no less proof of one's intentions, and in some respects even surpassed the bar for evidence, given that they were sent and received instantaneously, and required proof of a sender's identity in ways that even letters did not.[12] In the age of fast communication, letters and cables quickly filled the information gap that had long plagued commercial institutions in the Islamic world and won over what remained of their skeptics.[13]

• • •

"Do not leave us with too few letters," pleaded one merchant to his correspondent in Kuwait, "for a letter is half a meeting" (*al-khaṭṭ niṣf muwājaha*).[14] His concern might have been justified. Letters—correspondence—formed a critical part of an oceanic mercantile infrastructure in a world of fast-moving goods and information and quick decision-making. Meetings were important: they reinforced social bonds, added new dimensions to business relationships, and allowed merchants to establish an agenda together; Gulf merchants made sure to meet with their correspondents whenever they could.

But even an extended in-person meeting was no substitute for regular, predictable correspondence. Even in the age of the telegraph, there was no dispensing with the *khaṭṭ*, a detailed letter that traced past communications, set out expectations, related market information, and conveyed the most recent updates possible. But letters were more than their function. Through regular correspondence, merchants produced and reproduced a transregional commercial society. By adhering to the structure and conventions of letter-writing, they communicated their cultural capital with one another and signaled inclusion in this epistolary community of agents and partners. If a Malabari merchant's letters looked and read like he may as well have been from the Gulf, it was for good reason. And if Al-Khariji included a model for a letter conveying Eid greetings, it was because knowing the proper form mattered.

Was a letter half a meeting? Perhaps. But it may just as well be the opposite. A meeting, one might argue, was half a regular correspondence.

SIX

Bombay

AS THE SUN ROSE on November 14, Al-Failakawi and his crew made landfall on the northern stretch of the Konkan coast; by the afternoon, they could see their next port of call. "We arrived safely at Bombay," the nakhoda wrote, "and we have not seen anything reprehensible in God's bounty."[1]

The crew of the *Crooked* marveled at the sight of Bombay. It was the ultimate metropolis, far outstripping Basra, Karachi, or any other port city on their voyage. The city teemed with life and hummed with activity.[2] The front page of the newspaper that day boasted the city's importance: it overflowed with notices of steamships coming and going to the Gulf, Aden, Singapore, and beyond, of banks opening branches in Bombay from places as far afield as Taiwan and Japan, of law offices, of jewelers, and tailors. Even from where they anchored in the dhow basin, the crew of the *Crooked* could see the marvels that were out on display along the waterfront: the Gateway to India, which had just been completed that year, and the majestic Taj Mahal Palace Hotel, both of which poked their heads out above a noisy crowd of tea stalls, bars, shops, and street hawkers. There was also the imposing Arsenal Castle, the second largest store of weapons in British India, though many more valuable as antiques than as military equipment. Scores of people streamed from one building to another, alongside queues of goods and sailors making their way out of the ships and onto the shore.

Al-Failakawi knew how to guide the *Crooked* into port. Al-Qitami provided his readers with instructions for how to sail into the harbor.[3] Following a series of buoys close to the coast, the dhow made its way into a small basin, sheltered from the sea by a short jetty that curved its arm out into the water, as though to embrace it. By evening, Al-Failakawi dropped anchor in the harbor. The *Crooked* wasn't the only dhow arriving that afternoon: November was the peak of the fall season, and many other dhows had docked in the harbor alongside it.

It was the late afternoon, and Al-Failakawi's business had to wait until the next morning. He moved quickly: barely two days after arriving, Al-Failakawi had cleared customs and began unloading his cargo, noting in his logbook that "we began the morning in good health and blessings in the port of Bombay; we unloaded dates (*nazzalnā tamr*)."[4] He would have felt enormously relieved. Just a few years before, the nakhoda ʿAbdullah bin ʿAbdulrahman Al-ʿAsʿousi wrote to his father in Kuwait to let him know of their arrival and the state of affairs in the city. "Sir, these days, dates have slackened," he wrote in late December 1921.[5] The price for dates had been steadily decreasing for at least a couple of months: letters to Al-Matrook from different Bombay merchants note that the price of *sayer* dates fell from a high of Rs 90 per *mann* on September 21 to Rs 63 by the end of October. *Halawi* dates fared only a little bit better: in late September they sold for Rs 100 per *mann*, and by the end of October they fell to Rs 82.[6] For date traders operating on thin margins, there was too little to be made in Bombay. Al-ʿAsʿousi had to sell the dates he carried, but he made sure to send word to his cousin, who had just arrived in Karachi and was bound for the ports of Kathiawar, letting him know to sell off his dates for the highest price he could fetch there; the market in Bombay was no good that year.[7]

To navigate the vagaries of the date market, merchants in the Gulf and nakhodas relied on business associates there. The merchant ʿAbdullah Al-Fozan, for example, partnered with Gujarati merchants in Bombay to sell dates both locally and abroad.[8] Even amid the slump of the fall of 1921, he and his Gujarati partner Haridass Walji sent three Indian *kotiyas* to Basra to load dates. The first two were to take twenty-one hundred baskets for markets in Bharuch and Bombay, while the third was to load fourteen hundred baskets for markets in coastal Yemen and Somalia. He regularly wrote to Al-Matrook to coordinate the loading and unloading of dhows he sent from Bombay, to better navigate the market's fluctuations. He instructed Al-Matrook to pack specific varieties of the fruit: he wanted *sayer* dates, but dried in an early stage of ripening, rather than fully ripened and packed; he also asked him to avoid packing *ḥalawi* dates, for which demand was very low. "I depend on you in the date business," he told Al-Matrook, "and nobody knows that [business] except he who has bought and sold it in India, especially in a dull market." He ended one of his letters with a note of hope: "everything is upside down," he complained, "but man works and Allah grants him fortune."[9]

By the time Al-Failakawi arrived in Bombay, just over three years had passed since Al-Fozan had written to Basra, and the slump had lifted. The nakhoda found a ready market for dates and was able to dispose of his entire cargo over five days. He probably counted himself especially lucky, as Bombay was not particularly well known as a market for dates. It was an important market for pearls and a critical source of textiles. Above all, though, it was a market for credit. Underpinning the flows of goods and people, of ships and shopkeepers, was a transregional market for credit, and Bombay was its epicenter. Bombay was the Western Indian Ocean's New York City—its monied metropolis, and its financial and industrial center. For nakhodas, these enormous dimensions of the city manifested themselves in small ways: in the merchants they did business with there, and in the notes coming in and checks going out. The sinuous web of debt and credit that Al-Failakawi and his mariners were ensnared in all made their way back to that city.

• • •

If the economic fortunes of much of the Gulf were hitched to India, Bombay was their main anchor. Indeed, the city was developed to serve the interests of merchants. By the late 1800s, it housed dozens of textile mills, a sizable steamship dockyard from which several companies operated, and a railway station into and out of which dozens of trains came and went every day—to say nothing of numerous courts, banks, storage facilities, printers, moneychangers, and more. Despite its distinctly colonial character, the city offered everything a merchant needed to run a business on a local, regional, or even global scale.

The ties to mercantile interests went even further back. King Charles II received Bombay as a dowry from the Portuguese in 1661; shortly afterward, he leased it to the East India Company to govern on his behalf. Company officials soon began building Bombay Castle, which would form the nucleus of the emerging Company settlement. They invested in developing their capacity for governance in Bombay through the establishment of councils, municipalities, and law courts, all of which sought to attract merchants and settlers from other parts of India and the Indian Ocean world. In 1687, as a sign of its commitment to that project, the Company transferred its headquarters from Surat to Bombay.[10]

The Company's move 175 miles southward from Surat to Bombay heralded a transformation far more significant than the transfer of a trading company's offices. The decline of Surat signaled a momentous shift in world history: it stood in for the decline of the Mughals and Safavids and the imperial economies that had given so much life to the port city. As Surat found itself the target of frequent attacks by the Maratha Empire, a Mughal successor state, its merchants, no longer able to access the hinterland like they once did, moved their business to Bombay, where they could count on new business opportunities and the protection of the East India Company. The decline of Surat and the rise of Bombay, then, mapped firmly onto a palpable shift in the political economy of maritime Asia away from the Muslim gunpowder empires and toward the joint stock companies that would come to dominate commerce and politics.[11]

With Bombay firmly established as the major trading port of northwestern India, commercial life in the port city soared to new heights. By the beginning of the nineteenth century, Bombay included merchants from Basra, Persia, Muscat, and Yemen, and attracted settlers from around South Asia. As one writer put it, "Nowhere else probably in the world, not even in Alexandria, are so many and such striking varieties of race, nationality, and religion represented as in Bombay"; in addition to a variety of Indians, there were "many Afghans, Persians, Turks, Malays, Chinese, and Abyssinians . . . Jews and Armenians . . . and many thousand Indo-Portuguese inhabitants"—to say nothing of the Europeans, "a class of the community not strong in numbers, but supreme in social and political power."[12] And after the East India Company dealt the Maratha Empire a series of military blows between 1817 and 1819, there was nothing left to challenge Bombay's hegemony over the trade of northwestern India. The town's population grew from 162,000 in 1826 to well over a half million by the middle of the century; there was no question that its star was on the rise.

Though Bombay's first major windfall came with the opium trade to China during the late eighteenth and early nineteenth centuries, India's principal export in the Western Indian Ocean remained textiles, which chiefly flowed out of the port cities of Gujarat and into the markets of East Africa, the Arabian Peninsula, and the Gulf. Surveys of Gulf ports in the early nineteenth century all point to the enormous popularity of Indian textiles.[13] The rise of industrial manufacturing in Manchester in the first half of the 1800s displaced Indian textile production, and Bombay's merchants shifted to the export of cotton to England. The onset of the American Civil War, which dramatically reduced America's cotton exports, created an opening that Bombay merchants rushed to fill. Together with brokers and moneylenders in the interior, they tried to produce and export as much of the stuff as possible. Indian textiles still enjoyed a regional market, but the new global game was cotton.[14]

As the cotton boom unfolded, officials developed an infrastructure supple enough to handle the increasing quantities of goods flowing through the city. They invested in three steamship docks and ordered work on a railroad system designed to move workers, cotton, and coal (necessary for both the trains themselves but also the steamships) into the port city. Steamship routes radiated out from the city, especially after the opening of the Suez Canal in 1869. From Bombay, one could take passage on steamships up the Gulf, stopping at Karachi and several Arab and Persian ports along the way, or across the Arabian Sea to Aden, and from there up the Red Sea and to the Mediterranean. There were routes from Bombay to Zanzibar, both direct and via Aden or the Seychelles. Steamships sailed in the other direction, too—southward along the coast of India, down to Sri Lanka, and onward to Madras, Calcutta, Rangoon, Singapore, and even the western coast of Australia. Virtually the whole of the Eastern world was accessible from the dockyards of Bombay.

With the growing commercialization of the period and increasing heterogeneity of Bombay's business community came the need for more expansive financial services; merchants, it became clear, needed banks.[15] In the mid-nineteenth century, there was only one bank that enjoyed state sanction and support: the Bank of Bombay, established through a Company charter in 1840, which issued credit, discounted notes, and fueled the commercialization of the city more generally.[16] As rising cotton prices in the 1860s set off a speculative frenzy, many unofficial banks emerged to fuel the ballooning market for shares in railway projects and real estate, virtually all of which enjoyed the imprimatur of Bombay's big merchants—all of whom held seats on the board of the Bank of Bombay as well. Merchants, bankers, investors, and government officials clubbed together to fan the flames of exuberance that engulfed the city.

The good times would not last for long. The end of the Civil War and the resurgence of American cotton precipitated an economic crisis among Bombay merchants, who had poured much of their wealth into fictive ventures promising quick gains. "Those who had strength enough to resist might be numbered on your fingers," wrote one visitor to the city.[17] The losses were enormous: between 1860 and 1898, nearly twenty-one thousand people filed for insolvency. The wave of bankruptcies hit everyone from wholesalers and manufacturers to petty shopkeepers and artisans, men and women alike.[18]

And yet, within this landscape of failure lay the seeds of later success. The infrastructure that had made Bombay such an important player in the global cotton trade (and a central node of imperial governance) was still there: the railways, the steamships, the telegraphs, and the financial institutions and technologies. Merchants unconnected to the cotton trade could still reap the benefits that came with that infrastructure, and those involved in textiles found an enormous opening. In the 1920s, mills in the city still put out more cotton textiles than the 126 factories around the Bombay Presidency.[19] The city's traders were still highly active in the financial sector, too: alongside state banks were at least a dozen private banking establishments, many of which had branches elsewhere in India, Asia, and Europe. As long as there was still commerce, there were generative outlets for capital.[20]

These transformations were as significant for the Gulf as they were for India. The bumpy ascent of industrial capitalism in Bombay in the second half of the nineteenth century marked an inflection point in the history of economic life in the Gulf. The rhythms of industry and banking in India echoed into the Gulf, among merchants and moneylenders, and everyday consumers as well. The region's political fate became more closely hitched to Bombay, too, as the city's administrative contours began to expand well beyond India. British officers stationed around the Persian Gulf, South Arabia, and East Africa reported directly to the Government of Bombay. With an imperial finger in every major port city in the Western Indian Ocean, there was little to challenge Bombay's place as the region's political and economic nerve center.

• • •

Of the Gulf merchants of Bombay, none were quite as prominent as Jassem bin Muhammad Al-Ibrahim. The Arabs there all called him "Shaikh Jassem," in part because his family wielded a tremendous amount of influence in the Gulf: his uncle, ʿAbdullah bin ʿIsa Al-Ibrahim, owned several thousand acres of date plantations along the Shatt Al-ʿArab, and his cousins had married into Kuwait's ruling family. But Jassem had been in India for practically his whole life. He was born in Kuwait in 1869 but moved to Surat soon afterward to join his father, who did business there alongside five other families from Kuwait. Like many others, they eventually moved their business operations to Bombay, if not for its protections, then for its facilities. Jassem's father, Muhammad, and uncle ʿAbdulrahman managed the Indian side of the family business until they died in 1910. Together with his cousin ʿAbdulrahman bin ʿAbdulaziz, Jassem Al-Ibrahim took the reins of the Bombay side of the business.[21]

Gulf merchants' ties to Bombay were in themselves nothing new. Merchants in Basra had been active in the India trade from at least the mid-1700s, trading at Surat and Bombay, among other places. They were particularly active in ferrying horses, which were in high demand among East India Company officials. Gulf traders dominated the horse market in Bombay, shipping stallions from the Arabian Peninsula by dhow and steamer and partnering with Indian merchants to house them in stables and sell them.[22] Reports from as early as the 1820s describe Kuwait as home to a fleet of dhows that sailed from the Gulf to the "Coasts of Sind, Guzerat, and Malabar, and to Bombay."[23] Traders from Muscat, too, frequented Bombay early in the nineteenth century, taking with them cargoes from East Africa, Arabia, and the Gulf.[24]

Bombay emerged as a major trading center for Gulf merchants like Al-Ibrahim. It was as much a part of their world as Basra, Bushehr, or Muscat. In Bombay, they were joined by other merchants from around the Arab world, including the Arabian Peninsula, the Levant, and Egypt—so many that by the early twentieth century there emerged an Arab enclave in the city known as Arab Lane.[25] Nor was everyone a merchant, either: different sultans of Muscat and Zanzibar and rulers of Bahrain had spent varying stretches of exile (imposed or voluntary) in Bombay, along with their wives, children, and members of their entourage. The Arab community of Bombay was a vibrant one, plugged into much broader cultural, intellectual, and political circuits. But they were also locals: their homes were anchors of domesticity in India, and their children were born, raised, and educated there—and they died and were buried there as well.[26]

For Al-Ibrahim and the other Arabs who did business out of the city, the variety of trade goods available in the marketplaces of Bombay was nothing short of stunning. There, they could buy almost every manufactured good under the sun. Above all, the city offered access to all the textiles a merchant could want. Markets around the Indian Ocean craved Indian textiles and yarn: in 1924, when Al-Failakawi docked in the port, more than 528,000 tons of Indian cotton were exported out of Bombay, accounting for more than one-fifth of British India's exports of goods, and an increase of nearly 16 percent from the previous year.[27] The price lists that merchants in the city circulated to associates in the Gulf featured no less than seventeen different kinds

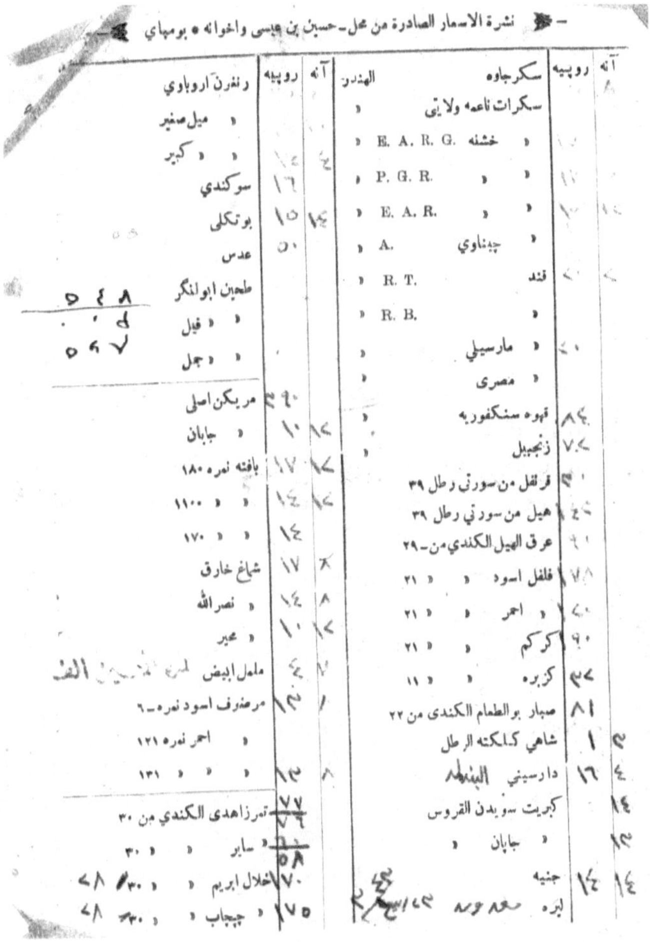

FIGURE 11. An undated price list of goods from the Bombay-based Gulf merchant Hussain bin ʿIsa. Al-ʿAsʿousi Collection, Kuwait.

of textiles: Japanese fabric, American cotton cloth, woven fabrics in different patterns (*Najm Al-Sabaḥ*, *Shams Al-Ḍuḥa*, and *Najm Al-Fajr*, to name just a few), broadcloths and headscarves patterned in different styles (*Khāriq*, *Ibn Nasrallah*, and *Muḥīr*), colored yarns, various silks, and much more—many of which were called by Persian names, indicating possible markets for them.[28] For the textile trader, this variety was what distinguished Bombay from all other ports and what made the journey from the Gulf worthwhile. One Najdi family, the Al-Bassams, made their fortune in the textile trade from Bombay to Arabia.[29] If Karachi supplied the Gulf kitchen, Bombay stocked its clothing chests; the fashion tastes of consumers in the Gulf and Arabian Peninsula formed in conversation with Bombay.

Gulf merchants fed global fashion trends too. Of all the goods they brought in, nothing was quite as valuable as the pearl. Bombay was home to the largest market for pearls in the Western Indian Ocean; virtually all the pearls from the Gulf eventually made their way to its bazaars, where they would then be sold to local jewelers or to buyers from London and Paris, and as far abroad as New York. Gulf pearl traders flocked to the city, and Al-Ibrahim quickly distinguished himself as the crown prince of Bombay's pearl merchant community. His timing worked in his favor. He entered the family business just as the Gulf pearl trade reached a swell: between roughly 1875 and 1905, the value of pearls exported annually from the Gulf rose from an average of around 15,000 British pounds to about 25,000. It was not a steady climb, though; there were enormous fluctuations from one year to another. One year could bring profits twice or three times that of the year before, and the next year's profits might fall to a third of what they had been. Booms and busts took place with regular frequency in the pearl trade, and it fell to merchants like Al-Ibrahim to work out how to manage his portfolio from one year to the next.[30]

To manage the transition from one season to the next, pearl traders like Al-Ibrahim looked to the city's credit market. Bombay was a city of banks: there were more than a dozen that operated in the city. Al-Ibrahim kept accounts open in many of these, including the Bank of Bombay, but also the Alliance Bank of Simla, the Bombay Merchants' Bank, the Indian Specie Bank, the Yokohama Specie Bank, and the Chartered Bank of India, Australia, and China.[31] Those who could tap into Bombay's credit market channeled it into marketplaces around the Gulf and onto the decks of the pearling dhows that set sail every summer. Bombay banks also took Gulf pearls as collateral against the loans they extended. In 1913, The Indian Specie Bank advanced more than $2 million—over 400,000 British pounds!—on Gulf pearls, second only to its loans against silver. So long as pearls continued to command a high price on the market, these merchants could find ready capital in the city's credit markets. Credit from Bombay fueled pearl diving in the Gulf, and pearls from the Gulf fed into credit markets in Bombay—and merchants like Al-Ibrahim stood at the juncture of both worlds, drawing on their access to good catches of pearls to leverage more and more credit from Bombay banks, which they could then use to finance more activity on the pearl banks, then turn back around and drum up even more credit. The pearl merchants of Bombay knew one another well enough that they could even club together, pressuring banks to dole out even more credit.[32]

So it was, then, that in 1911 a group of Gulf pearl merchants led by Jassim Al-Ibrahim came to form Arab Steamers Limited, the first Gulf Arab-owned steamship business in the city, and the only competitor to the British India Steam Navigation Company (BISN) on the Gulf route. The initial group of investors in Arab Steamers was small. The company issued five hundred shares, each valued at Rs 500; most were bought up by Al-Ibrahim himself, along with his cousin, the Bombay pearl merchant 'Abdulrahman bin 'Abdulaziz Al-Ibrahim. Other major shareholders included the Kuwaiti ruler Mubarak Al-Sabah and four of his sons, and three other pearl merchants from Kuwait and Bahrain, along with some of Bombay's Arab merchants. Jassem Al-Ibrahim designated himself the company's CEO and appointed as its managing director Muhammad bin 'Abdulwahhab Al-Mishari, a close associate of Al-Ibrahim's and one of the leading Gulf pearl merchants of Bombay.[33]

The company's early years seem to have gone quite well. One British official in Muscat in 1912 reported that "in consequence of the weekly service maintained by the Arab Steamers the freights to India, etc. were greatly reduced during the year." The cost of shipping dates from Muscat to India fell by over a third, while that for shipping mother of pearl halved.[34] By 1913, the company moved its headquarters from the Arab Lane area to 15 Elphinstone Circle, where it did business alongside the leading firms of the city. Soon after, though, it began to face stiff competition from BISN, which enjoyed access to preferential legislation, rebates, and other subsidies, and could swallow up smaller firms. The onset of the First World War dealt a further blow to Arab Steamers' profit margins, as it faced a tightened market for credit, mounting restrictions, and less passenger traffic than it had in the past. Beginning in 1915, it sold off its ships to a shipping line controlled by BISN, and by 1922 the entire enterprise had been swallowed up by the BISN conglomerate.[35] It was a worthwhile experiment, though, as Arab Steamers' shareholders were paid off handsomely: their shares were valued at three times what they paid for them. Al-Ibrahim and his fellow pearl merchants knew how to thread the needle through Bombay's capital markets even in the most unfavorable of situations.

As he buoyed to the crest of Bombay's mercantile elite, Al-Ibrahim began to embed himself in the changing urban landscape of the city itself. He bought a residence on Nagdevi Street, in the commercial heart of the old city, but as the city expanded, he moved to a new residence—this time, in the ritzy neighborhood of Colaba, built on land that had been recently reclaimed from the Arabian Sea. Al-Ibrahim's new residence was impressive, even by the city's lofty standards; the ground floor alone took up more than eight thousand square feet. Moreover, he had commissioned the Parsi architect Jamshetji Mistri to design the building in a distinctly Saracenic style, and the resulting structure featured all manner of oriental embellishments: horseshoe arches, enormous columns, latticed balconies, and two imposing bulbous domes topped with pinnacles. Though few people could claim to own the kind of magnificent property that Al-Ibrahim did, they would have clearly understood the importance of the nexus between property and credit—and how someone with the kind of access to credit he enjoyed could find themselves in one of the city's wealthiest suburbs. As he and other Gulf merchants wove themselves into the fabric of the city, they took their place among the colonial capitalist elite around the British Empire in the Indian Ocean.

• • •

The Kuwaiti merchant Muhammad Salem Al-Sudairawi might not have been the prince of pearls, but the web of connections he spun from Bombay was no less impressive. Every day, dozens of letters poured into his office from around the Western Indian Ocean and beyond. To call his correspondence network broad would be a gross understatement: he fielded letters from virtually every major port city in Arabia, Persia, and India, and many smaller ones as well. The letters he left behind paint a picture of a man embedded in a broad transregional merchant community; they comprise roughly 15,000 pages of correspondence between him and roughly 750 different individuals, many of whom Al-Failakawi, Al-Matrook, and others knew well.

The majority of his correspondents wrote from the Gulf—Kuwait, of course, but otherwise Basra, Mohammerah, and other towns near the Shatt Al-ʿArab waterway. He also heard from Bahrain, Muscat, the Trucial Coast, and also from Persian ports—Lingeh, Bushehr, and Bandar ʿAbbas. Other frequent correspondents wrote to him from around British India. Al-Sudairawi received regular letters from Karachi but also corresponded with merchants in other nearby port cities: Calicut, Mangalore, and ports around Gujarat. He regularly heard from business partners in Aden, which

FIGURE 12. A group photo of Arabs in Bombay. Muhammad Salem Al-Sudairawi is seated in the front row, third adult from the right. Photo from Muhammad Ghazi Al-Sudairawi.

until 1932 formed part of the Bombay Presidency, but also from Calcutta, Cairo, Mecca, Surabaya, and Singapore.

Virtually all of them wrote about money. As merchants moved goods back and forth between different markets, they debited and credited the value of goods to one another's accounts, and their trading partners regularly wired over ready money so that they might maintain some liquidity and remain solvent. Al-Sudairawi was the central node in this circuit of financial capital: he took deposits and money orders, enacted transfers, issued checks and payment orders, and otherwise coordinated with buyers and sellers on matters of payment. When Gulf traders did business on credit, they would send a payment order to Al-Sudairawi in Bombay, drawing on money they had deposited with him. Al-Sudairawi would then direct banks—and he worked with nearly a dozen of them—to send checks to his correspondents and their associates, who could then look to their banks to honor them.[36] And when rulers in the Gulf wanted to transfer funds to and from Bombay or needed to infuse capital into the commercial arena, they called on his services. The nodes in the money circuit were many, but in the end, virtually all payments were routed through him.

For someone in Al-Sudairawi's position, there was no better place to be than Bombay, a city of banks with branches around the Western Indian Ocean. From his office, he managed a sprawling transregional web of credit. Payment orders on Al-Sudairawi were, in essence, a form of credit—a promise to pay for goods received. The expectation that there would ultimately come a payment, and through clear channels, allowed the financial machinery around the region to keep running. Al-Sudairawi's checks were honored by every bank in the region and were as good as cash. Those with no immediate access to a bank could sign the checks over to their associates, who could in turn deposit them in an account. The notes that circulated between his office and different banks around the Indian Ocean lubricated the wheels of credit within the Gulf merchant community, and among the hundreds of Indian Ocean merchants with whom they kept running accounts. He was, in all respects, a banker to an expanding network of merchants in the Indian Ocean.

Al-Sudairawi was relatively new to his role. His family were recent comers to India: his grandfather, ʿAbdullah, migrated to Kuwait from Central Arabia in 1845. From Kuwait, Al-Sudairawi's father, Salem, took up the trade in horses between Central Arabia and India; he eventually settled in Bombay, setting up a trading office there and hosting the various Gulf merchants who passed through the city. When Mubarak Al-Sabah took the reins as Kuwait's ruler in 1898, he appointed Salem Al-Sudairawi as Kuwait's agent in Bombay. Salem continued his business, but he'd also ensure the supply of critical provisions to Kuwait's marketplace by supplementing what came in by way of private traders. Moreover, he'd facilitate sales and purchases by Kuwaiti merchants visiting the city and would place orders on their behalf.[37]

But Salem was not just there to serve the interests of Shaikh Mubarak and his family; his was a much more broadly conceived role. Through agents positioned at different ports in British India, Shaikh Mubarak aimed to deepen Kuwaiti participation in regional commerce. By extending his presence into port cities like Bombay, he facilitated the circulation of credit and commodities between British India, Basra, and Kuwait. Salem Al-Sudairawi thus formed part of a much broader strategic vision: as rulers like Shaikh Mubarak sought to more fully integrate their marketplaces into those of India and South Arabia, they relied on agents like Al-Sudairawi to make the cash transfers necessary to lubricate the wheels of commerce—including their own.[38] Through figures like Salem Al-Sudairawi, the intersections between public finance and private credit become openly visible.

But Salem's run as Mubarak's agent in Bombay was short-lived. He died in 1906, perishing in one of the series of waves of the bubonic plague that ravaged Bombay at the turn of the century.[39] The work of managing the Bombay agency fell to his eldest son, Muhammad, who at the time was just over twenty years old and who found himself thrust into the center of a sprawling transregional web of transactions—and a vortex of circulating checks and notes.

By the time Al-Failakawi arrived in Bombay, Muhammad Salem Al-Sudairawi had been active for nearly two decades. He knew all the nakhodas who visited Bombay, and virtually everyone they did business with, and often helped them secure onward cargoes.[40] Al-Sudairawi also frequently corresponded with Al-Matrook's business partner in Karachi, Nasser Al-Kharafi, and had also received visits from Nasser's son Abdulmohsen during his travels down the coast of India; he occasionally helped the Al-Kharafis find buyers for dates in Bombay when the Karachi market was slack.[41] He was also in touch with Al-Matrook himself, with whom he seemed to have struck up a friendship. In one letter that he wrote after returning to Bombay from a visit (either to Basra or Kuwait) in January 1922, Al-Sudairawi broke from the usual conventions of business correspondence. Having sensed that Al-Matrook had become preoccupied with money, he offered him reflections on life and wealth. Although he prayed that Al-Matrook would be blessed in his pursuit of wealth, he reminded him that money didn't always bring satisfaction. "Since childhood, [I] have abstained from the material world," he wrote, "and now it means as much to me as an hour's ride in a car." He proffered his friend advice: "Let go of whatever anxiety you have over money," he wrote, "and give yourself time to think about how to nurture your soul."[42]

In part, Al-Sudairawi's advice to Al-Matrook reflected his passion for literature and the arts. He was an avid reader of poetry and regularly hosted gatherings of the Arab literati in Bombay; he even circulated newspapers, journals, and books to fellow readers in the Gulf. His note might have also hinted at how he felt about the transformations he had witnessed over the course of his lifetime. By the time he took over, the nature of the "agency" he inherited had changed. The business of agency in Bombay was much more in the management of the financial capital that pulsed through the transregional marketplace. As the banking infrastructure of the Indian Ocean expanded to include port cities in the Gulf, Al-Sudairawi increasingly found his work involving processing payment orders, issuing checks, and transferring money between different accounts—in oiling the wheels of credit so that Kuwait's merchants could supply the regional marketplace on their own.

The change in the Bombay agency reflected itself in Al-Sudairawi's business address: he initially worked out of the Sitaram Building, which housed the administrative offices of many of Bombay's merchants and manufacturers and gave him easy access to suppliers for the kinds of goods his principals in Kuwait would have asked for. But within a decade, he moved his office to Grant Road, near the city's Arab Lane and closer to many of the city's banks. It also reflected itself in the daily stream of letters into his office, almost all of which asked him to process payment orders and money transfers and issue checks. What little was left of the original job description lay buried beneath mountains of financial paper.

• • •

Working with credit almost by definition meant working with law. The flows of credit that Al-Sudairawi managed from Bombay moved through a legal infrastructure of rights, obligations, expectations, contracts, and legal institutions, all of which allowed merchants to secure their claims against one another. And when it came to law, Bombay did not disappoint. There were many courts that one could go to in Bombay, for everything from small claims to bankruptcies, and even matters of imperial administration. The city housed the Bombay High Court of Judicature, which, during the nineteenth century, came to be the highest court of appeal in the Western Indian Ocean. As the Bombay Presidency expanded to include places like Karachi and Aden, and even (albeit fictively) Zanzibar, legal proceedings in any of those courts could end up being appealed all the way to Bombay. Through Bombay's legal institutions, merchants who resided in the city could make claims on transacting partners far away.[43]

Gulf merchants in Bombay, however, did not limit themselves to the city's judicial machinery. Because of the administrative web that linked Bombay to officials in port cities around the Western Indian Ocean, merchants in the city often mobilized a combination of courts and petitions to press claims against others elsewhere. This was especially true for the Gulf, where there were no staffed British courts until the 1920s, when the first (and only) court was established in Bahrain; and even then, it was difficult to get a Bombay decree executed on the island.[44] In these cases, merchants' claims were managed through actors on the ground: in Kuwait, the Political Agent, who took statements and examined the evidence, and in Bombay, Al-Sudairawi himself, who often facilitated the transfer of money to different claimants and pursued claims on their behalf. In two different cases, we see the workings and limits of this transregional legal machinery, which relied less on bureaucratic rules and procedures than it did on the work of individuals and networks that threaded through different institutions.

In 1910, Muhammad Ebrahim, a Bombay-based seller of enamelware, wrote to the Political Agent in Kuwait to file a claim of Rs 15,000 against a group of shopkeepers there, who had taken goods on credit from him two years earlier and had not yet paid for them. When the Political Agent interviewed the men, they admitted to the debt but claimed they had recently remitted at least some money to the seller through Al-Sudairawi's office. Only one transfer didn't make it through because the payment order that Ebrahim presented to Al-Sudairawi did not bear the debtor's signature and seal. Ebrahim's pursuit of his claims, however, soon ran into much bigger obstacles: one debtor's shop in Kuwait burned, along with all the inventory he had taken from Bombay; another debtor died in Bahrain. And although the qadi in Kuwait was willing to include Ebrahim as a creditor to the estate, he asked him to send in accounts certified by the Shariah Court in Bombay. In the end, to pursue his claims, Ebrahim authorized Al-Sudairawi, who had planned to travel to Kuwait, to settle matters on his behalf.[45] If the writ of Bombay courts could not directly reach the Gulf, the networks that bridged the two places could help facilitate the workings of the law.

The interpersonal nature of these proceedings was never totally shorn of law, though. In 1907, the pearl merchant ʿAbdulrazzaq bin Salim bin Sultan was traveling back to the Gulf from Ceylon with a catch of pearls he had bought as a joint investment with other merchants in Kuwait. On his way from there to Basra, he stopped at Bombay to find a buyer. The market was slack that year, and Bin Sultan needed money to cover his own debts, so he approached Muhammad Al-Mishari, one of the city's preeminent pearl traders and a close associate of Al-Sudairawi and Jassem Al-Ibrahim's. Al-Mishari agreed to loan him Rs 5,000, taking Bin Sultan's pearls as security; he said that he and another pearl merchant would find Bin Sultan a buyer. He could not find one, and the pearls remained unsold.

Months later, Bin Sultan died in Basra, and the merchants who had invested their money in the pearls he bought were looking to recover their property. Al-Mishari demurred: he wanted to sell the pearls to recover the amount he was owed in full before handing over what remained to Bin Sultan's partners in Kuwait. The merchants pleaded their case to Shaikh Mubarak, who then took matters up with the Political Agent. According to both Islamic law and mercantile practice, the losses, he asserted, should be distributed equally between all claimants; "it is not possible that one man should take his right in full and another should be deprived," he wrote. Al-Mishari, for his part, claimed the opposite: because the pearls were mortgaged, "in accordance with religious law and Government law we have every right to dispose of the property by auction and to pay ourselves the money due to us," he retorted. On both sides, claims were grounded in a sharp sense of what "the law" was and a confidence that there existed legal institutions through which they could press their claims. The Gulf Resident was confident of the same: the Kuwaiti merchants, he wrote, ought to appoint an attorney and lodge a claim at the Bombay High Court.

Ultimately, neither Islamic law nor "Government law" prevailed, and nobody lodged any claims in court. Shaikh Mubarak, who was among those who had invested money with Bin Sultan, agreed to appoint a third party—a Bahraini pearl merchant in Bombay—to receive the pearls and sell them, and even assented to prioritize Al-Mishari's claim to their value. All agreed that the proceeds of the sale should go through Al-Sudairawi. The entire matter was resolved through correspondence and was grounded in a desire to keep the wheels of credit turning and to preserve one another's standing. "We paid the money on account of the great respect we had for [Bin Sultan] and to save his honor," declared Al-Mishari. "If we had not paid him, God knows what his creditors would have done with him."[46]

The resolution of the dispute is telling. Though all the parties made claims grounded in law, they actively chose not to litigate. Going to court made sense for those whose ties with one another could be sacrificed—though even Ebrahim, the enamelware seller who pursued litigation, continued to offer his debtors new lines of credit even as he chased after old debts. Among close associates, though, an appeal to law courts would have courted rupture with no certainty regarding the outcome. Instead, in crafting a response to the dispute, Gulf merchants could rely on the dense networks of merchants and rulers that already existed to help circulate credit—which itself was grounded in a deep well of rights, obligations, instruments, and practices that they could appeal to in moments of confusion, all without necessarily having to recourse to courts. As the principal conduit into Bombay's credit market, Al-Sudairawi helped mediate these transregional processes. His work, the cases make clear, was as much legal as it was financial.

But sometimes the complexity of the organizational forms that Gulf merchants entered into made going to court unavoidable. Just a few years after

Al-Mishari's pearl dispute, in 1915, Arab Steamers—whose managing director was Al-Mishari himself—found itself in the Bombay High Court. It had agreed to ship twenty-five hundred bales of cotton to ports in Italy on behalf of an Indian firm in April 1915, just weeks before the Italians entered the First World War. The British government announced a prohibition on trade with Italy, and in October 1915, Arab Steamers found itself sued by its erstwhile Italian clients for money owed and for not fulfilling its end of the bargain. To represent the company, Al-Mishari and the board hired one of Bombay's leading law firms, Messrs. Little & Co., who argued that Arab Steamers was only unable to carry out their end of the agreement because of government restrictions. Moreover, they argued, Arab Steamers ought to be construed as a "common carrier"—a company that carried goods for any and every customer, along specific routes—and should thus be exempted from the provisions of the Indian Contract Act, under which the company would have been obliged to pay for nonfulfillment of the original agreement. The judge ultimately disagreed.[47]

Compared to the issues at stake in the Ebrahim and Al-Mishari cases, the Arab Steamers case rested on somewhat esoteric points of law—unsurprising, perhaps, given that it was heard in the High Court and pleaded by professional layers. The thirteen-page report from the High Court, replete with references to Indian and English case law, seems a world apart from the more muted claims about law and obligation that were articulated to the British Political Agent and within the network. And yet, it underscores the point that Gulf merchants rested their commercial work on legally grounded understandings of what they owed to others and what was owed to them. That they, or even their lawyers, could claim they were "common carriers" tells us a little about their legal consciousness. As they drew on the organizational forms available to them in Bombay, they learned to speak in different legal registers, amplified and translated through lawyers and in legal forums. But it was all law talk all the same, and those who could massage claims through a network could just as easily pursue them in court. The arenas of credit and law did not bifurcate, they converged, generating multiple scales of claim-making that sought to manage the movement of goods and money into and out of Bombay.

• • •

Al-Failakawi was ready to leave Bombay within a week of arriving, but he had to put off his departure. Shortly after unloading the last of his cargo of dates, he noticed that his dhow needed repairs. A brief note appears in his logbook five days later, on November 26: it reads much like his other notes from Bombay, except that this time the nakhoda noted that they "began the morning in good health and blessings in the *jaddāf* at Bombay." The *jaddāf* referred to the shipyard where dhows underwent repairs, when structural damage to the vessel required the assistance of more experienced shipwrights and better tools than they had on board the dhow.

The *Crooked* was out of the *jaddāf* three days later, but Al-Failakawi did not leave Bombay for another week. He could hardly sail back to the Gulf with an empty hold; the dhow would likely lose its keel and would be much more difficult to steer. He could have carried ballast cargo, which was readily available, but would have lost out on the opportunity to earn more freightage. And so, he spent the next two weeks trying to secure a cargo—preferably a bulky, high-freightage cargo from the city itself that he could take back to the Gulf with enough time left over to squeeze in a second round-trip voyage to India in the early winter. But failing that, he would settle for a cargo that he could carry between ports in India so that he might make his current voyage worthwhile for both him and his crew. He spent his days visiting merchants, waiting for them to telegraph their partners elsewhere, and patiently waiting for the replies to come, as he and other nakhodas in the port exchanged leads on potential cargoes to carry.

He eventually landed upon something to carry onward, though he never tells us what. In the early morning of December 6, more than three weeks after docking his dhow in the harbor, Al-Failakawi ordered his crew to weigh the anchors and raise the sails. His logbook entry for the day read: "In the name of Allah, most gracious, most merciful, on the date Saturday, 8 Jumada Al-Awwal, 6 English Nūvember"—he seems to have gotten so used to writing November that he forgot that they were now six days into a new month—"we entered upon the morning in good health and blessings in the port of Bombay, and we set sail." As the ship left port, he added his daily prayer: "Oh Allah, ease things upon us and upon all Muslims, by the power of the Prince of Messengers, Amen oh Master, in safety oh Master, protect us and all Muslims."[48] In the light of the morning, the *Crooked* left Bombay and followed the currents heading south.

INSCRIPTION

Transfers

AL-FAILAKAWI KNEW A THING or two about how money moved. Inscribed into the notebook that Al-Khariji had given him were two templates for different kinds of money transfers, evidently copied from real instruments.

The Form for a Kambiyāla Ḥawāla

> We would like a *ḥawāla* upon you, Muhammad Thunayyan, for Rs 10,000 by check (*ṣakk*) to Hajj Taher bin Jassim, and have it be from our account so that there is no confusion. Signed, Hamad bin ʿAbdullah. 22 Muharram 1341.

The Form for a Ḥawāla *for a Known Period*

> To his excellency, Yousef bin Marzouq, long may he remain, Amen: greetings. If Khalifa bin Shaheen arrives, please hand him Rs 20,000 by check (*ṣakk*) due in one month, and have it be on our account; we send our greetings again. Signed, Hamad bin ʿAbdullah. 22 Muharram 1341.

The two inscriptions are short, but their brevity belies their importance and ubiquity. From Africa to Southeast Asia and beyond, commercial actors made use of notes like these to move money, through their commercial partners. Among Indian merchants, these were called *hundis*; others referred to them as *ṣakks* and *hawalas*, or transfers. Al-Matrook's archive teems with different forms of these money transfers—roughly four hundred in total, both handwritten and printed. They join a long line of instruments from around the Indo-Islamic world aimed at moving money, principally through different commercial associates.

For historians of capitalism in Europe and the Americas, these instruments might seem familiar, and for good reason. The *hawalas* and *hundis* of the Indian Ocean are in many ways analogous to the bill of exchange. Anchored in the markets of early modern Europe, the bill of exchange has come to epitomize the history of commercial capitalism in the Western world, allowing for the movement of money and the spread of commercial activity across space. The acts of negotiability and discounting that the bill of exchange required rested on a much broader institutional framework: banks, exchange fairs, and financial houses specializing in discounting; the historian Fernand Braudel went so far as to say that the widespread use of the bill of exchange generated these institutions. And as European settlers began to colonize the Americas, bills fueled the early Atlantic economy.[1]

Like their Atlantic cousins, the *hundis* and *hawalas* of the Indian Ocean traced the outlines of a broader infrastructure of credit and capitalism in the region. They intersected with Western banks, but also other institutions that were not obviously financial, like temples.[2] *Hundis* and *hawalas* moved through networks of merchants, brokers, and retailers, forging line items in the account ledgers that linked the metropolitan centers of financial capital to producers, shopkeepers, and laborers in more distant credit markets. They lubricated the wheels of production and commerce and allowed traders and statesmen to move money over long distances.[3]

It wasn't all nearly so straightforward, though. With every transfer, with every change of hands, came questions about what these instruments were, what they were supposed to do, and what they meant for the people who did business in them. Among Gulf merchants in the Indian Ocean, these boiled down to a simple question: Were these instruments money?

• • •

Upon his return home from years of study in nearby towns and cities, the Kuwaiti scholar ʿAbdullah Al-Duḥayyan found himself confronted with a wide range of different inquiries, many of which touched on matters economic—of production, commerce, and consumption.[4] People wanted to know whether men wear silver or cloth embroidered in gold, and whether divers who were out in the pearl banks were obligated to observe the fast during the month of Ramadan.[5] And they wanted clarity on money—on banknotes and money transfers, and what they owed on them. The world of modern capitalism posed serious moral challenges, and they looked to scholars like him for answers. Al-Duḥayyan, in turn, reached out to his mentor in Damascus, the Hanbali jurist ʿAbdulqadir Ibn Badran Al-Dumi, who later compiled his responses to Al-Duḥayyan's questions into a short treatise, which he called *Al-ʿUqud Al-Yāqūtiyya fī Jayd Al-Asʾila Al-Kuwaytiyya*, or *The Ruby Necklaces in the Excellence of the Kuwaiti Questions.*

Writing to Ibn Badran, Al-Duḥayyan posed three questions about some of the new monetary technologies and practices his constituents had been asking about. First and foremost, he asked, were banknotes—a relatively new form of money in the Gulf—considered to be hard currency, and could the obligatory alms (*zakat*) be collected on wealth in banknotes? Second, was transferring money by way of post permissible? And finally, could a sale agreement that had been conducted in Ottoman currency be completed in Indian rupees?[6] Though his questions suggested that these forms of money enjoyed widespread use in the Gulf economy, there was clearly not a lot of consensus on what obligations they might generate.

For Ibn Badran, at the core of the matter lay a fundamental question worth considering out loud: Did banknotes and postal transfers constitute money, or did they simply represent value in a different form? Were they forms of currency—did they have value in and of themselves, as forms of corporeal property (*ʿayn*)—or were they really only paper instruments that ultimately emerged from a relationship of debt (*dayn*)? These were useful questions to consider in the abstract, but the questions also had real-life implications: whether *zakat* could be collected on them and whether (and on what terms) their bearer could transfer them to another party.

In his long answer, the jurist declared his stance almost immediately: banknotes and postal transfers could not be considered to be money as such. In his view, they were lacking money's most essential characteristic: intrinsic value. The person holding a banknote or postal transfer had to call upon the issuer to pay its value in currency, which was simply not the case when it came to gold, silver, copper, or other metal currencies. "Metal currency does not require payment of its value," he wrote, "nor is it backed by another currency." It didn't matter to him what the specific metal content of the currency was. Rather, it was the fact that everyone agreed these were the species of metals that enjoyed value, a genus whose species naturally had more in common with one another than they did with paper.[7] By contrast, the paper itself was mostly worthless: it only had value in that it denoted a preexisting transfer of value from one person to another. An insolvent person who had issued these notes held no money as such, he wrote; his creditors would have to go to others to recover the value of their loans to him, and usually at a discounted rate. This was simply not the case with metal currency.[8]

The lack of intrinsic value in the instruments didn't render them useless, though. They were in fact quite useful—but as guarantees of debts incurred, for an item of value that had been sold before it physically changed hands. They were representations of value rather than forms of value in and of themselves. These instruments, Ibn Badran reasoned, were forms of the *ṣakk*: instruments that guaranteed the transfer of value, whether as money or goods, from one person to another. Their passage from one person to another did not constitute a sale, nor could it; it was simply a transfer of a guarantee for a debt, and the debt itself was where the value was nested.[9]

To help his colleague better grasp the nature of the paper instruments, Ibn Badran took him through an extended detour through the hallways of the history of capitalism. The history of banknotes, he suggested (drawing on the writings of an Egyptian jurist) was a distinctly European one. "The first banking associations emerged in the Italian Republics of the Middle Ages, like Venice and Genoa," he asserted, "and in Spain, they emerged in Barcelona, and then in the Free Cities of Germany"—by which he meant the towns that comprised the Hanseatic League. From there, he wrote, they arrived in Holland "driven by its urban development and mercantile expansion, and its confidence in the transactions of bankers." He listed dates (using the Hijri calendar) for the establishment of different banks throughout Europe—in Venice, Barcelona, Amsterdam, Hamburg, Nuremberg, and Stockholm, all over five hundred years beginning in the late 1100s. As the banks expanded the scope of their activities and began lending to both individuals and governments, a market emerged for shares in those banks. The governments, for their part, would issue their own debt certificates, which gave the holder a claim to debt on the government itself. These, Ibn Badran stated, were banknotes, and people began to transact in them because of their convenience: "A person could carry in his hand or pocket the equivalent of many *qinṭārs* [a unit of weight for precious metals]."[10]

Ibn Badran was cognizant of the tensions built into the instrument. He understood that people's transactions in banknotes were directly correlated to their confidence that the government would, at some point down the line, be able to pay their bearers the value of the note. This, he recognized, created problems: if a government was delayed in being able to meet its obligations—or worse yet, if it defaulted—the value of the paper dropped precipitously and people refused to accept it. The lesson, to him, was clear: "It is invalid for a government or bank to issue these instruments unless the value of the paper was present," even though at times governments would loan out more than they have in ready cash in their coffers.[11]

Alongside his discussion of the history of Western banking, Ibn Badran offered another, parallel genealogy of the *ṣakk*—one grounded in the history of the Islamic world. He began with the eighth-century jurist Malik ibn Anas, whose legal compendium *Al-Muwaṭṭa* took on the *ṣakk*: these, Malik wrote, were issued under the Umayyad Caliph Marwan Ibn Al-Hakam as guarantees for foodstuff, and people almost immediately began to buy and sell them in a secondary market. Upon learning of this, the caliph ordered his soldiers to confiscate the *ṣakks* and return them to their original owners. The moral dangers present in paper instruments, Ibn Badran suggested, were present from the earliest days of Islam. He then traced commentaries on the episode through several genres of legal writing: it was picked up by Muslim bin Al-Hajjaj in his authoritative ninth-century compilation of hadiths, and then commented upon by the thirteenth-century Shafiʿi scholar Al-Nawawi, both of whom agreed that *ṣakks* were only problematic when they came to be traded on the secondary market—that is, when they became a commodity in and of themselves.[12] The clear difference between the food and the paper highlighted his point, that *ṣakks* were written guarantees for property that had been bought before it could be physically handed over; they were representations of items of value, and could not be treated as having value.

At its core, Ibn Badran's discussion reflected the slippage between fact and fiction that characterized money more generally, but especially paper money. For thinkers around the world, money always raised the question of representation: was it inherently valuable, or did it simply stand in for something of value, as a token? If communities converged on a consensus that gold and silver enjoyed intrinsic value, it was not always clear that other forms of currency did as well. At different junctures in world history, thinkers raised questions surrounding the use of copper and other coins. For the most part, people accepted these as part of the normal course of business; it was often during moments of crisis or downturn that people would then call into question their routine use of monetary tokens that bore no intrinsic value.[13]

Around the Islamic world in the early twentieth century, though, the question of how to handle new forms of money felt more pressing than it had in centuries past. Ibn Badran's answer to Al-Duḥayyan's question anchored itself in a longer history of thinking about finance and markets for monetary paper, but also reflected contemporary concerns—he himself referred to scholars in India, Java, and Cairo, all of whom grappled with identical issues, and who turned up different answers.[14] The pearl merchant Muqbil bin ʿAbdulrahman Al-Dhakir, who spent a lot of time in Bombay and knew

Al-Sudairawi well, asked similar questions of the jurist Rashid Rida, editor of the enormously popular journal *Al-Manar*, in 1902. Staking out one position among many, Rida declared that the banknotes were currency, and their usage was permissible and entailed the same obligations as the use of gold or silver.[15] The questions asked of ʿAbdullah Al-Duḥayyan in Kuwait thus connected that port city to broader ripples throughout the Islamic world in the time of capitalism.

• • •

When Al-Failakawi and Al-Khariji inscribed the *hawalas* into their notebooks, they knew the broader contours of the world they were tapping into. They knew of Al-Duḥayyan, of Al-Dhakir, of Rashid Rida, and of the questions surrounding paper and value that they grappled with. How could they not have? These were literate men who voraciously consumed books, especially on matters of religion, and who were in contact with other literate men who read widely. The questions people asked of a local scholar—to say nothing of a jurist of international repute—would not have gone by unnoticed. What these pieces of paper could do, what they stood in for—and fundamentally, where value came from—may have been questions they asked as well.

The circulation of money and transfers of obligations from one account ledger and bank to another moved alongside the dhow, passing across its deck. These were the genres of the bazaar economy—chief among them the *hawala* and *ṣakk*, but also the many other forms and instruments that formed the paper infrastructure of circulation in the Indian Ocean. Through their inscriptions, nakhodas anchored these instruments in the dhow's itineraries, and wrote the dhow into the circulations of money, people, and goods that conjured up capitalism itself.

SEVEN

Malabar

THE *CROOKED* WAS BARELY a day out of Bombay, but the appearance of the coast had changed so much that it felt like it had sailed to a different part of the Indian Ocean altogether. Gone were the desert stretches of the Makran coast and the rocky outcroppings that flanked Bombay. The southward coasts were far more dramatic: verdant hills towered above the coastline as the Western Ghats, draped in their lush greenery, tumbled into the sandy beaches that marked the outer edges of the Arabian Sea.

It was a relatively smooth stretch of the voyage, too. "On these routes, nothing will ever overburden you, and nothing will take you too far away from the coast," Al-Qitami reassured his readers; "it will be an even-keel journey." The navigator was right: Al-Failakawi could not have hoped for better conditions. As he sailed down the Konkan coast, the winds blew southward, and the currents ran in a south and southeasterly direction. The *Crooked* glided effortlessly past one landmark after another, and Al-Failakawi carefully noted them, checking them against his notes and Al-Qitami's manual. First came Malwan, a coastal town bound by three creeks, which the *Crooked* passed in the afternoon on December 8; Al-Qitami wrote that "you can see [the creeks] whether it is night or day." Then came Goa, which nakhodas knew from its lighthouse, which sat atop a large hill.[1] Al-Failakawi sailed past Goa and then Karwar, and by December 10 he could spot Netrani Island (which nakhodas called *Natloh*). Within a day, he could see Mount Eli, which nakhodas and medieval geographers alike called *Jabal Al-Hayli*; it rose nearly one thousand feet high and signaled to nakhodas they were close to Mangalore.

When Al-Failakawi spotted "Kaki Island" (most likely Velliyamkallu, a two-acre rocky island) he knew that he was approaching his destination. "If you want to go from Kaki Island to Calicut," wrote Al-Qitami, "it is but one line—take it from between the setting *ʿagrab* and *al-ḥimārayn* [i.e., between southeast and south by southeast] and be wary of the headland at Koyilandy [*Ras Kolandi*]." Al-Failakawi also knew to keep an eye on the changing water depths: Al-Qitami instructed his readers that the water depth would be "five to six fathoms [*bāʿ*] before it plunges deeper; across from Calicut, it will be ten fathoms."[2] At 11:00 p.m. on December 13, the nakhoda pulled his dhow into the harbor: "We arrived in Calicut," he wrote in his logbook. "May Allah ease things upon us and upon all Muslims, by the power of the Prince of Messengers; Amen."[3]

The next morning, as Al-Failakawi looked out from the deck, he could see the dockyards at Calicut buzzing with activity. Busy sawmills flanked the Chaliyar River, and timber merchants eagerly awaited the arrival of fresh supplies of wood that came downriver from the forests of the interior. Piles of ready-cut lumber lined the waterfront; the entire place was awash in the earthy, almost leathery smell of teak and the acrid smell of sawdust. Beyond the beach extended a thick stretch of tall palm trees, occasionally punctuated by white buildings with gabled, ribbed roofs, laid with the distinctive red clay tiles of the region. Alongside the *Crooked* were other vessels—some from Kuwait, but just as many from other ports in the Gulf, from India, and from Yemen. All of them eagerly anticipated a good cargo to carry back: a good bit of lumber, some tamarinds, and sacks of coffee and spices.

Al-Failakawi had been to Calicut before. His *ruznamah* had, too: a note on the first page of the book declared he bought it in Calicut two years before, in March 1923, for one rupee. It was something of a homecoming for the *Crooked* as well, which at one point in its journey would have taken the form of those very piles of lumber on the beach. But that was but one node in a much broader circuit through which nature made its way into the marketplace. As nakhodas, merchants, brokers, and shipbuilders moved between the Malabar coast and the Gulf, they fashioned a circuit that connected the forests of Malabar to the shipyards of the Gulf, and to marketplaces around the Western Indian Ocean.

. . .

The merchant Yousef Al-Sager often spent spring mornings watching the dhows that pulled into the harbor. He eagerly anticipated the arrival of ships from the Gulf, and specifically from his home port of Kuwait, for they brought news of family and friends, orders from abroad, and the familiar company he often lacked in Calicut. He had his home built specifically to receive travelers from abroad. The house was built in the Malabar style, with a wide, slanting red-tile roof, white exterior, and deep brown wooden frames for the door and window. It rose two stories high: the first floor included a large reception room and home office, while the second floor included a large room that housed visiting merchants and nakhodas. A balcony jutted out from the second floor, giving Al-Sager an unobstructed view of the Arabian Sea; during this time of the year, he'd walk out every morning and, eye pressed to a telescope, scan the horizon for incoming dhows.

It was Al-Sager's job to anticipate company. Sometime around 1920, he moved to Calicut to set up a permanent business office there. He was the only Kuwaiti trader who permanently resided in Calicut. All orders for that port city's goods—spices, tamarinds, and most importantly, lumber—went through him, and nakhodas visiting the port did their business through him. He was an agent to anyone who needed one: he placed orders, arranged cargoes, stored goods, and brokered between local producers of goods and buyers from around the Indian Ocean—including Al-Matrook in Basra, with whom Al-Sager corresponded. And though he was central to Gulf commerce in the area, he missed his home dearly. He would frequently travel to Kuwait to see family and friends, and he married and had children there, but he spent most of the year in Calicut. And when he wrote to friends and colleagues in Kuwait, he complained of his loneliness there. "We wish to see the people of Kuwait," he lamented, "but Allah ordained that we were to be in Malabar."[4]

It was not just that Allah ordained it; Al-Sager chose to thrust himself into the oceanic trade by way of Calicut. He knew the Indian Ocean trade all too well; his family was steeped in it. His great-grandfather, also called Yousef, was one of the first family members to engage in that trade. A nakhoda and merchant, he ferried goods from the Gulf to Calcutta and would frequently stop in Malabar along the way. He died when his ship foundered off Kothi beach in Calicut. Al-Sager's grandfather, ʿAbdullah bin Yousef, entered the same business, and at some point decided to migrate to East Africa, which was then still under the rule of the Al-Busaʿidi dynasty; he never returned.[5] Given the family history, Yousef's father Sager and uncle Hamad perhaps unsurprisingly chose more sedentary careers, sending ships out of Basra and Kuwait to engage in the Indian Ocean trade. But when it came time for Yousef to strike out on his own, he took the more adventurous path.

Al-Sager was only the latest to join the long-standing Arabic-speaking diasporic merchant community in the region. Malabar, which Al-Sager and other Gulf merchants called *Al-Naybār*, was for centuries well-known among Arab and Persian merchants for its spices.[6] By the mid-1100s, merchants were already calling at different ports in the region to purchase cinnamon, pepper, ginger, and cardamom, all of which grew in abundance in the hills of the Western Ghats. Calicut was widely known as a regional emporium: one Muslim traveler wrote that Calicut was "equal to Hormuz in its mercantile population from every land and region, and the availability of rarities of all sorts from . . . the Land below the Winds [Southeast Asia], Abyssinia, and Zanj."[7]

By the twelfth century, too, Muslim merchants were able to take advantage of the emergence of coastal states in Malabar so that they could carve out a place for their own trade and community.[8] A few hundred years later, Muslim merchants in the region began to mark their community's presence with mosques—though instead of minarets, they built "turret-like edifices," which allowed them to visually blend more seamlessly into the Hindu-dominated port city landscape. As the Muslim community in the region grew, Malabar emerged as both a destination for and a source of Muslim scholars in the Indian Ocean, who took up positions as preachers, community leaders, jurists, and qadis.[9]

Though the Portuguese arrival in India disrupted Malabar's trade with Arabia, it was only momentary. The region's merchants were frequently able to circumvent the Portuguese; they rerouted trade through friendlier ports, paying taxes and protection costs to the Portuguese only when they had to. Trade continued, even if under a slightly different carapace: as late as the eighteenth century, ships, people, and goods continued to flow between the Persian Gulf, Malabar, and Bengal.[10] Malabari merchants continued to supply the Ottomans in the Red Sea with shipbuilding timber, too, despite the animosity that ran between the Portuguese and their Muslim imperial rivals.[11]

As the Portuguese position in the Western Indian Ocean waned, local rulers on the Malabar coast once again reasserted themselves, seeking out renewed alliances with other powers. Between the mid-1600s and late 1700s, the Zamorins of Calicut courted both the English East India Company and Dutch VOC, and ultimately persuaded the latter to help them dislodge the Portuguese from the coast altogether. And in the mid-1700s, when the ruler of the Kingdom of Mysore, Haidar Ali, invaded the Malabar coast and briefly occupied it, he and his son Tipu Sultan sought to strengthen the ties between southwestern India and the Gulf, signing commercial treaties with the Sultan of Muscat and sending embassies to the Shah of Persia; Tipu Sultan even appointed a commercial agent in Muscat in 1785.[12] Even after he lost his newly acquired coastal possessions in a series of battles with the Company in the 1790s, the connection remained: the sultan of Oman and Zanzibar, Saʿid bin Sultan Al-Busaʿidi, commissioned several vessels from Calicut in the early nineteenth century.

The Calicut that Al-Sager and Al-Failakawi knew, however, was less a mart for spices and ships—though it continued to put out fine vessels—and more of a source for shipbuilding materials like lumber and coir rope, all of which made their way to shipyards in the Gulf. For despite the Gulf's vibrant shipbuilding industry, it produced none of the materials with which dhows were built. For those, Gulf traders looked to Malabar, tracking Calicut's transformation from shipyard to exporter of raw materials over the course of the eighteenth and nineteenth centuries. When the British Empire lost its American colonies with the Revolutionary War of 1776, England also lost its principal supplier of timber. British shipbuilding, like Britain's imperial administration writ large, swung east; Bombay emerged as the principal shipbuilding center in British India, and Company officials descended upon the forests of Malabar to meet Bombay's shipbuilding needs. Calicut gradually became a satellite to Bombay, feeding its shipyards with raw materials from the forested interior.

Very quickly, however, it became clear that Bombay's shipbuilding demands, coupled with the regional market for lumber, would soon exhaust Malabar's forests. Officials in India attempted to regulate the trade in lumber, alternatively trying to channel it through their own licensed merchants or government-run shipyards, where the bulk of the supply would be reserved for the construction of British ships. The policy ultimately failed, largely due to the hold that Malabar's merchants had on the timber trade. Officials also experimented with teak plantations and other regeneration techniques as they attempted to balance the needs of an export-oriented market with the desire to conserve and sustain the region's forests. Those twin impulses—export and conservation—would shape much of the forestry policy during the nineteenth and early twentieth centuries.[13]

At the same time, British officials recognized that lumber exports could end up enlarging the capacities of those who threatened Britain's imperial ambitions in the Indian Ocean. In the early nineteenth century, they sought to ensure that shipbuilding supplies would not end up in the hands of imperial rivals like the French, with whom they vied for supremacy around the region. And in the wake of the Gulf maritime treaty of 1820, they also sought to restrict lumber sales to Gulf merchants, whom they suspected of supplying timber to the Qawasim of Ras Al-Khaimah—though, as in the past, Gulf merchants circumvented the regulations by routing their supply networks through other ports.[14]

The movement of shipbuilding materials from Malabar to the Gulf continued throughout the nineteenth century and may have even expanded alongside Gulf trade in the Indian Ocean. Traders from Kuwait would sail to Malabar and carry back shiploads of lumber and coir, which they would sell around the Gulf. Merchants in Kuwait and Basra wrote to their associates visiting Calicut and Kochi in the 1870s and 1880s, too, asking them to bring back different varieties and cuts of lumber. One asked for a tall piece of lumber to be used for a mast; he wanted it to be free of knots and was willing to pay whatever the asking price might be.[15] As the Gulf dhow trade itself expanded, so, too, did the demand for shipbuilding materials.

To procure shipbuilding materials from Malabar, traders from the Gulf worked through local merchants; Al-Sager was but one of many traders looking out across the water. Among them were members of the Barami family, who migrated there from Hadramawt in the late eighteenth century and worked in the lumber trade. Many of them managed workshops in the shipyards at Beypore.[16] Gulf merchants also did business through the Malabari merchant Kunji Ahmad and his son Ahmad, both of whom were brokers.[17] Even Al-Sudairawi in Bombay had a business partner in Calicut: Ahmad Abdulqadir, a timber merchant and commission agent to whom he regularly sent orders for tamarinds and shipbuilding materials—coir, timber, and even anchors—which he had him ship directly to business associates in Basra and Mohammerah. He'd also forward him dates from those places to sell in Calicut.

Access to shipbuilding materials became especially important in the Gulf in the context of the dhow shipping booms that took place during the two world wars. As Europe descended into war, the British government requisitioned privately operated steamships for the war effort. Shipping moguls like the British India Steam Navigation Company were left with a minimal fleet, and the weekly steamer service between India and the Gulf had all but stopped, leaving a yawning chasm in regional transport that dhow traders were eager to fill. Shipbuilding boomed, and the demand for timber was

FIGURE 13. The merchant Yousef Al-Sager (seated front row, center) pictured with business associates and British officials in Calicut.

ravenous. Any merchant who put in the time to secure shipbuilding materials could make a fortune off the commissions alone.

Though he may have at first thought of himself as a world away from the Gulf, over time, Al-Sager became firmly established on the Malabar business scene. He had a home and office, and sent out letters on letterhead bearing his name, address—"Beach Road, Calicut"—and even a telegraph address, "ALSAGAR." One photo from Calicut shows Al-Sager sitting alongside three British officials; flanking them are roughly twenty Arab and South Asian-looking men, many of whom were finely dressed. Some of these were likely Al-Sager's employees, but most would have been other timber merchants and brokers—members of Calicut's business community. That Al-Sager was front and center attests to just how embedded in the Calicut business world he had become.

Al-Sager gradually anchored himself in Calicut society as well. Eventually, he took a wife in Calicut—the daughter of a Bahraini merchant who had long resided in the port city. She gave birth to a daughter, Layla, in 1947, and the three stayed in Calicut together until 1961, when they were joined by her half brother, Yacoub; her mother remained in Calicut. Layla remembers that theirs was one of only a few Arab families in the city, but that they were thoroughly integrated into the local social fabric. "We ate Indian food," she told one interviewer. "In fact, we lived like Indians. Their situation was our situation."[18]

• • •

Just outside Al-Sager's office lay freshly cut piles of timber of varying shapes and sizes: long, straight trunks, thicker nubs, Y-shaped, wishbone-like cuts, and many more. All were marked for the ports of the Gulf, waiting to be loaded onto the dhows that would carry them away to markets abroad. Their journey began much further inland. Upriver from Calicut, in the forests and teak plantations, independent proprietors waited to hear from brokers who would communicate orders from merchants on the coast. They marked specific trees for cutting, moving between timber of the "first sort"—older teak, from trees between sixty and one hundred years old, roughly forty feet in length—and the second and third sorts of timber, which were younger and produced shorter logs. As they felled the trees, they replaced them: they planted, germinated, transplanted, and thinned teak in the forests to create room for supplies to grow.

After cutting down the trees, the timber merchants in the interior had to send them down to the coast. Until 1927, when railways connected the coast to the interior, standard practice was to float the logs down the Chaliyar River, which snaked down the mountains of the Western Ghats and through the countryside before emptying into the Arabian Sea. Timber merchants hired the owners of draft animals—buffaloes and bullocks, but also elephants—to transport the felled trunks to the riverside. There, transport workers tied cuts of timber together, often with bamboo to increase their buoyancy. They then floated them downriver, following them along the way to ensure a smooth journey—no lost cargo, no damage, and no logjams. The workers moved the timber down the Chaliyar River until they reached its mouth, not far from Al-Sager's home office on Beach Road.[19]

Coastal merchants like Al-Sager knew how to distinguish between the kinds of timber that made it to the coast. What to many would have looked like indistinct logs piled up around Calicut's business offices were, to the trained eye, different varieties of wood, suited for different purposes. Most merchants bought large planks of teak (*sāj*), which shipwrights used to build a dhow's keel, the stem, and sternpost—that is, much of its outer hull.

Shipwrights often made use of other cuts of lumber—smaller planks (called *chakī*) for smaller boats, squared planks (called *sakatlī*) for building cabins, or short, round cuts of lumber called *kōl*, from which they fashioned the support beams that dhows rested against during the summer months. A buyer would need to know what kind of teak they needed, how they wanted it cut, and for what purpose. They would make no concessions on the teak itself, though: it was easily workable, and lasted a very long time, particularly under the brutal heat of the Gulf summer; there was no other wood that could match it.

Teak may have been king, but buyers also sought out other varieties of wood for shipbuilding. Ben teak, which Gulf merchants called *mantī*, was a slightly less durable variety of teak, and dhow builders would often use it for parts of the dhow that were not exposed to the sun's rays; they would also use it to fashion the helm. They also used Indian laurel, called *jangalī*, to form the dhow's keel, and for parts of the upper and lower deck as well. Shipwrights knew it to be a hardy wood that resisted shock and breakage, especially when submerged under water. Another popular variety that Al-Sager sold was *pali*, which Gulf merchants called *fīnī*; shipwrights preferred this variety when fashioning the topmost plank of the dhow's hull. And then there was *punna*, called *fann*, which grew upward of ninety feet tall and was strong, rendering it ideal for masts and yards. There were many other varieties as well, all suited to different purposes.[20]

It was Al-Sager's job to find his buyers the product they needed, and to translate orders for shipbuilding materials into terms legible to his suppliers inland. The first step was clarifying the different varieties of lumber that were on offer. It was not enough to know that one wanted teak or Indian laurel; there were different grades that people could buy: did they want government (*serkāli*) grade teak, or Calicut (*Kalikūtī*) teak—and was it of the first, second, or third grade? There were similar varieties and grades for virtually every kind of plank that went into the dhow, and each commanded a particular price on the marketplace. Through his careful coordination of information, Al-Sager could match buyers in Kuwait or Aden to specialized lumber yards around Calicut.

So, too, with coir rope, of which there were large, heaping coils that lined the floors and shelves of rope sellers' shops around town. Coir rope was integral to the dhow building; the ship's rigging—the cables that tied the sails to the yards, stabilized the masts, and connected the helm to the rudder, and much more—was all made from coir rope. To produce coir, rope makers would harvest coconuts and then separate the nut from its fibrous hull. They then cured the hull by soaking it in water for several days before thrashing it to separate the fibers from the pith and outer skin. They would then lay the fibers out to dry, and women workers spun them into yarn, twine, and ultimately rope.[21]

Like lumber, there were different varieties of coir—different species and different grades, all of varying strengths and length and thickness, all with corresponding prices. Al-Sager's price list enumerated no fewer than eight different varieties of coir rope for sale, and most came in two or three different grades. Altogether, his buyers could choose from fifteen kinds of coir rope, ranging from the most expensive—in 1928, it was "*Isti'māla* No. 1"—to the cheapest, "*Shawkat* No. 3" or "*Pattani* No. 3," which he sold for nearly half the price. Different grades of coir were suited for different purposes: the strongest was necessary for the ship's rigging, but those who needed rope for fishing or for lashing together bales of goods could do just as well with lower-grade varieties.[22]

As the goods made their way to merchants like Al-Sager, he measured them using two different units. The more standard unit for lumber was the *korja*, or score—a bundle of twenty planks. The *korja* was the standard unit for most of the precut wood he sold. Teak and coir, though, were sold by the *khandi*, a unit of weight that one nineteenth-century British official described as being roughly equivalent to 560 pounds. *Khandi* could be a unit of volume, too, he pointed out: "In Malabar, there is also a *khandi* for timber about 2 feet 4 inches square."[23] As commodities of standardized and measurable quantity and quality, lumber and coir could make their way through the communities of merchants and nakhodas, and into marketplaces and shipyards around the Western Indian Ocean.

By the mid-1920s, Al-Sager began circulating printed price lists among his family members and associates, in which he listed the prevailing prices for the goods they most frequently ordered: lumber and coir rope, but also pepper, ginger, cinnamon, coffee, tamarinds, and various oils—all in the different grades and measures they were available in. He made sure to include the going price for dates, too, in case anyone wanted to sell a cargo in Calicut.[24] When Al-Sager communicated information to his buyers and sellers in the language of grades and standards, he rendered the terms of exchange mutually intelligible to all, opening up the Malabari marketplace to a broad audience of buyers. He accumulated specialized knowledge of the region's products and compressed it onto a list that rendered it legible to his correspondents.

In the process, he turned the gears that made the Indian forest market-ready—transforming a tree into planks of specific types and grades of lumber, aimed at specific purposes and markets, all through the efforts of loggers, brokers, and merchants. Through his communication to his customers, he rendered the forests of Malabar commodifiable.

• • •

"My lord and uncle ʿAbdulrahman bin Hussain . . . we have arrived safely in Calicut, and we have seen nothing displeasing from God's blessings," wrote the nakhoda Hussain bin Rashid Al-ʿAsʿousi to his relative in Kuwait in January 1924, just a year before Al-Failakawi docked at the same port. The Al-ʿAsʿousis were regulars in Calicut: at least one member of the family visited the port every year. The person to whom Hussain addressed his letter, ʿAbdulrahman bin Hussain Al-ʿAsʿousi, was once a nakhoda who used to travel to Calicut to do business on behalf of himself and others, well before Al-Sager ever arrived. Though he stopped sailing in 1910, he continued to travel between Basra, Karachi, Bombay, and Malabar by steamship, and prepared orders for dhows captained by his sons, nephews, and cousins. He would write to his family members but also Gulf merchants elsewhere, including Al-Sudairawi in Bombay, and would report the prices of lumber, coir, and spices. His sons ʿAbdullah and Hussain carried on the same work, along with his cousins Rashid bin ʿAli and Khalid bin ʿAbdulaziz, all ʿAsʿousis.[25]

The ʿAsʿousis blurred the easy distinctions between nakhodas and merchants. They owned and captained their own ships, arranged for their own cargoes, and hired their own family members to transport them. Through their travels back and forth between the Gulf and the Malabar coast, they became intimate friends with Al-Sager too. When they'd visit Calicut, they'd stay with him, writing to family members in Kuwait on Al-Sager's stationery.[26] And when they were away, Al-Sager would write to them, letting them know of lumber prices in Calicut; he'd end his letters to the family by conveying his greetings to his friends—to "our father ʿAbdulrahman, our brother Rashid, and all the boys," he would say.[27] Like many others, the Al-ʿAsʿousis relied on Al-Sager's services: he would communicate to them prices on his end, and arranged for cargoes to be loaded onto their dhows, often on credit.

ʿAbdulrahman Al-ʿAsʿousi and his son ʿAbdullah built up extensive knowledge of the lumber trade. Their letters to one another brim with orders for various species of lumber, for different purposes, and they often specified how they wanted the wood to be cut. One letter related all the purchases of lumber one member of the family had made in Calicut that year: different varieties, cut at different lengths and thicknesses, measured by the *tisō* (an Indian measure equivalent to an inch) and at times for clearly designated purposes ("for the deck" or "for the water tank"). Sometimes they requested specific cuts: the *ank* or *ang*, for example, a thick, narrow board located above the deck and nailed to the inner surface of the frame, allowing water to flow outboard.[28] Their lumber talk came from their buyers; it bore the imprint of the shipbuilders with whom they worked. The process of transforming timber into planks, their writings suggest, unfolded as much in the shipyards of Kuwait as it did in the sawmills of Calicut, and actors like the ʿAsʿousis stitched the two worlds together. As these local measures and terms for purchasing and working lumber circulated between the coasts of the Gulf and Malabar, they forged the transregional pathways of commodification.

But the ʿAsʿousis' portfolio involved much more than just lumber. They traded in a variety of Malabar goods, chief among them tamarinds and coffee. Both were highly sought after in the Gulf and Central Arabia. The Malabar tamarind, or brindle berry (Gulf merchants called it *ṣbār*), was distinct from the varieties available elsewhere: it was bulbous in shape with vertical lobes, unlike its reddish-brown, pod-like cousin. By the time it made its way onto the market, it was dried—shriveled, black, and prune-like. And it was very sour—which is why consumers in the Gulf sought it out. They'd use it as a souring agent in different curries, particularly fish curries, but also in various chutneys; it was also known to treat inflammation and aid digestion. If Karachi was a port from which they would buy the basic necessities of life in the Gulf—rice, wheat, pulses, sugar, and tea—Malabar was where they would pick up some of the flavoring agents that shaped the Gulf palate.

The ʿAsʿousis and other merchants would purchase large quantities of tamarinds from Calicut and Mangalore and tracked its changing prices carefully. In the spring of 1922, they all wrote to one another in search of affordable supplies of the dried fruit: stores were low in Kuwait and the price was steadily rising. In Malabar, supplies were disrupted "because of the movement that is happening in the jungle that is keeping goods from reaching the town," wrote one merchant to Al-Matrook in Basra—an offhand reference to the Mappila rebellion, the latest in a string of agrarian revolts in the region that emerged largely in response to growing colonial regulations on land tenure.[29] Amid the unrest on the Malabar coast, the ʿAsʿousis' usual suppliers were no help: "We wanted tamarinds, but they are too expensive these days," grumbled ʿAbdullah bin ʿAbdulrahman Al-ʿAsʿousi to his father in Kuwait. "This year is not like it normally is."[30]

In their search for supplies of tamarinds, members of the family traveled into the Malabari interior, to Palakkad, where they sought to purchase tamarinds directly from the source. There, they found themselves in direct competition with other merchants for the highest grade of tamarinds—the seedless "Number 1," as they called it. "Every time anything came, the people of Madras would buy it for a higher price, and the market price is too high for us," complained Hussain to his brother ʿAbdullah, who was on the coast. Going further inland—to "the jungle," as he called it—offered no respite from the competition: "There are more buyers here than in Pallakad!" he exclaimed. They managed to get their hands on some second-grade tamarinds ("Number 2," which contained seeds), but too few to meet demand in the Gulf.[31]

The bigger concern was coffee, which was in short supply. Writing to his father amid the disruptions caused by the unrest in the Malabari interior, ʿAbdullah bin ʿAbdulrahman Al-ʿAsʿousi suggested that he hold on to whatever little coffee he had rather than trying to sell it: "if you have two bags' worth, keep it for our home in Kuwait," he wrote; "it would be better than buying any in Mangalore."[32] The coffee they had on hand was valuable because it was in high demand; buyers in the Gulf consumed it with relish. When Gulf merchants bought coffee, there was no shortage of outlets for it: they could sell it directly in their home market, reexport it to smaller ports in the region, sell it directly in their home markets, or sell it to buyers coming in from the interior of Arabia, where coffee consumption formed a pillar of daily sociability.[33]

To procure coffee beans, the ʿAsʿousis and other nakhodas and merchants regularly visited Mangalore, which was the principal outlet for the produce of the lush coffee plantations that sprawled along the highlands and valleys of the Western Ghats. Mangalore chiefly sourced the coffee of Mysore, which is thought to have been transplanted from Yemen by a pilgrim returning from there.[34] Growers experimented with different strains and curing methods; the ʿAsʿousis themselves frequently distinguished between green "plantation" coffee, which always commanded a high price, and "red" coffee. By the end of the First World War, the port city had established itself as an active market for coffee, marked by regular auctions. One coffee estate manager recalled later that "in the earlier days local buying of Coffee outside European exporters was almost entirely done by the Arabs, who came to the Coast in large numbers from the Persian Gulf."[35]

But "the Arabs" did not buy these goods directly. Seasonal traders like the ʿAsʿousis relied on the services of a community of local brokers. In Mangalore, the ʿAsʿousis purchased much of their coffee through Mahmood bin Mohideen, a prominent Malabari broker whom they came to know over the course of their visits there. Mohideen had some experience with Kuwaiti merchants and nakhodas: he corresponded with Al-Sudairawi in the early 1910s and frequently arranged to sell the cargoes of dates that nakhodas brought to Mangalore. Merchants and nakhodas kept open accounts with him, and he would help them transfer money throughout Malabar, through his own network, when they needed it.[36] He also wrote to his Arab correspondents in their own language, and over the years he learned to communicate in a distinctly Gulf idiom: his letters mimicked virtually all the conventions of letter-writing among Gulf merchants, even if in an unsteady hand.[37]

It was by way of this network of merchants and brokers, and through the circuitry of letters, lists, and credit notes they moved between one another, that the Indian countryside entered the marketplace. Through brokers like Mohideen, who learned to speak to their Arab buyers on their own terms, Gulf merchants learned to make distinctions between "Number 1" and "Number 2" tamarinds and "Green" and "Red" coffee—how one might process them differently, and what uses they might be put to. The process of commodification took place as much through the chains of communication and knowledge circulation about the features of different goods and the measures by which to grasp them as they did through the disembedding of the object from its natural world through finance and structured labor. As they passed into the marketplace, directed by merchants and brokers on the coast, pods on trees in the valleys of the Ghats became sacks of coffee and dried tamarinds, all piled into the holds of dhows like the *Crooked*.

• • •

We do not know what Al-Failakawi came to Calicut to load. All he lets us know was that he and his crew "awoke in good health and spirit in the port of Calicut," over and over again until he lost interest in noting where he was and instead gave only the date. In the month he spent there, he most likely ended up loading a mix of different Malabar goods—lumber, tamarinds, coffee, and various spices. He and his first mate watched over the sailors as they loaded the lumber into the dhow's hold, giving orders to stack the lumber and maintain the distribution of sacks of other goods carefully so that the *Crooked* wouldn't list on its way back to the Gulf.[38]

As Al-Failakawi loaded the *Crooked* with lumber, he reinscribed a commodity circuit that saw natural growth in the forests of Malabar converted into stacks of lumber, taken to the shipyards of the Gulf only to be reassembled into dhows that would sail to Malabar to pick up more stacks of lumber. He and his dhow, together with the brokers and merchants he worked with, helped form the infrastructure of circulation that animated the process of commodification. Through letters, price lists, standards, measures, and shorthand, and through the vehicle of the dhow itself, they propelled the products of Malabar from the first nature of the forests to the second nature of the market.

By January 10, it was time to set sail again, homeward bound: "We departed at night, in pursuit of our livelihoods [*ʿalā bāb Allāh*, as he put it] alongside the *boom* of Ahmad Salim." He would return to Calicut many times again—on every voyage for the next few seasons, and then many times beyond that. The journey out from ports like Calicut could take nakhodas in any number of directions. They could go a little further south, to Cochin, though for most Kuwaiti nakhodas, Calicut was as far south as they would go. Most would have loaded the cargoes they were after and would head on to the next market—perhaps back up the Indian coast, across the water to Muscat, Persia, and maybe to Basra, or perhaps elsewhere altogether, crossing the open sea to Yemen, to Somalia, or the Swahili coast, often alongside others. They would write of the other dhows that were also sailing out of the port: "With us here is the *boom* of Bin ʿAsfour, who sails tomorrow with the Bishara and Al-Kharafi *booms*," wrote Hussain Al-ʿAsʿousi to his father in 1924.[39] Another year, he named different dhows carrying cargoes to Bahrain, to Kathiawar, and to Basra.[40]

It was not an ideal time for the *Crooked* to leave Calicut; the winds were unfavorable. The dhow barely crawled up the coast, facing winds that alternately blew at it from the sea, or worse yet, came directly from the north. Al-Failakawi often had to anchor the ship overnight just so it wouldn't lose what little distance it had gained that day. It took him nearly two weeks to cover a distance that, a month earlier, he had sailed in just three days. When he reached Goa on January 24, he decided that it was no longer worth trying to push forward, and docked at the port.

Another nakhoda faced the same issue. Rashid bin ʿAli Al-ʿAsʿousi wrote to a relative that he had arrived in Goa, but "we took eighteen days to reach here from Calicut because the winds were either too weak or blew too strongly from the north." His dhow had taken in some water, too, and so he had to dock in Goa for repairs. Al-Failakawi likely also took the time to write to his employers, letting them know what had happened to him since he left Calicut. Goa was a good place for that: ships passed through all the time, and one could send letters with them or make use of the facilities there to send a telegram back home. It was also a good place to find a fellow vessel to sail out onto the open sea with. "We are leaving in the company of a *kotiya* belonging to Muslims from Jamnagar," wrote Rashid bin ʿAli to his cousin in Kuwait; "they are our convoy [*sinyār*], and they are good people; they are carrying timber and other goods, from Calicut to Qatar."[41]

Seven days into the *Crooked*'s stay in Goa, the winds finally turned favorable. It was time to sail out.

INSCRIPTION

Conversions

> A principle for knowing how to calculate lumber: you take the radius (*rub' al-dawrī*) and multiply it by itself, and then you look to the length, and what the result is, and then do the same for each [piece of timber] from the whole, and in each 576 is a *kandi* [*khandi*]. For example, if the radius is 8, we multiply it by 8 and get 64, and if the length is 12, we multiply that by 64 to get 768, and each 576 is a *kandi*, and only Allah knows what is correct.[1]

The note would have been easy to miss. Al-Failakawi—or, more likely, Al-Khariji, who included a slightly longer version in his own notebook—wrote it near the bottom of a page that was otherwise crammed with notes on navigation. But as the note tells us, before leaving the Malabar coast, nakhodas had to know how to measure what they carried and how to take the cargo of lumber they loaded onto the dhow and convert it into a known unit of measurement, the *khandi*.

The note would have been easy to ignore, but for the fact that there were many others. Al-Failakawi's notebook includes notes on converting between the Maria Theresa Thaler (the *riyāl*) and the Indian rupee, on how to get from annas to rupees, on how to calculate the value of a *qoṣra* of dates, and how to convert between a range of different currencies and measures, all by way of arithmetic. Al-Khariji's contained several of these as well: a note on converting between different measures of length—from the planetary to the length of a camel's hair—and another on converting between different units of weight: the *raṭl*, the *mann*, and the *ḥakk*, including their regional variations. None of these were exact. Instead, they functioned as rules of thumb, meant to guide a nakhoda as he made his way around an oceanic marketplace marked by a dazzling variety of goods, weights, measures, and currencies.

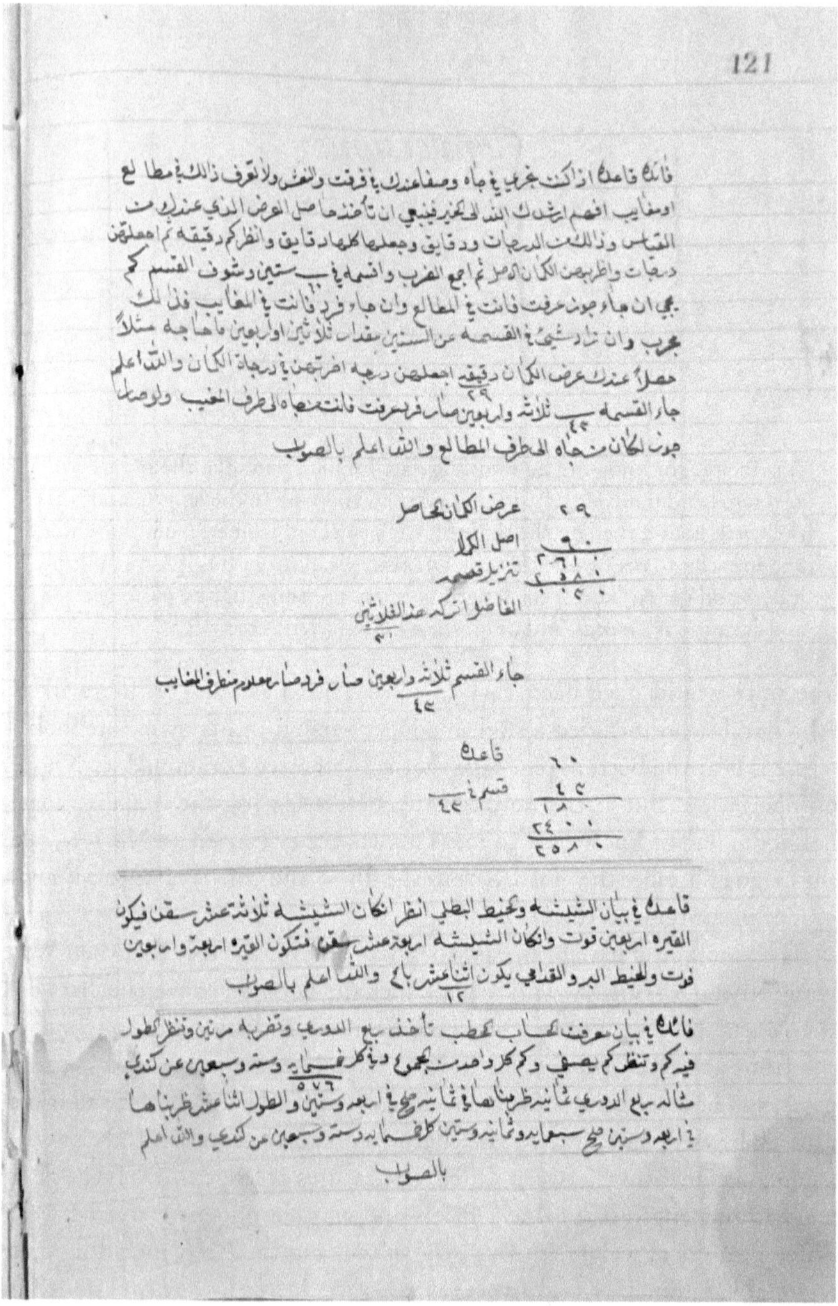

FIGURE 14. Some of Al-Failakawi's copied principles for sailing and market activity, with principles on converting timber quantities at the bottom. Al-Failakawi Collection, Kuwait.

But the notes signal something larger than a nakhoda feeling his way through different marketplaces. The lessons on conversion that Al-Khariji tried to impart onto Al-Failakawi taught him to translate between different markets, to attain commensurability between different commodities and currencies. In guiding Al-Failakawi through these conversions, he laid out a guide to moving between the different marketplaces the *Crooked* would invariably find itself passing through over the course of the season. And in laying out how to determine the *khandi* value of a cut of lumber, he signaled the processes of abstraction that commercial actors like him engaged in—a process by which wood that grew out in the forests of Malabar came to be converted into a market value that could be entered into account ledgers, all through simple arithmetic.

· · ·

In the history of economic life, there is no escaping math. It was the framework through which different economic actors increasingly came to make sense of the world around them. I am not speaking here of the regressions that economic historians like to run on historical data. Rather, I'm referring to another, far more mundane kind of arithmetic: that of the marketplace, wherein actors took the flows of goods and precious metals and sought to convert them into units that were measurable and accountable. Some of the earliest evidence we have of human civilization points to a tendency to represent these processes through abstract symbols, especially in the realm of taxation and public finance. And the important watersheds in economic history—the commercial revolutions and industrial revolutions of world history, irrespective of where and when we might locate them—all involved a process of thinking mathematically about nature and human beings. "The West's distinctive intellectual accomplishment," writes the historian Alfred Crosby, "was to bring mathematics and measurement together and to hold them to the task of making sense of a perceivable reality, which Westerners, in a flying leap of faith, assumed was temporally and spatially uniform and therefore susceptible to such examination."[2] Others have been less celebratory, pointing to how factory owners and planters used quantification to effectively dehumanize their laborers, reducing them only to their outputs.[3] The process of commodification that students of capitalism point to as the defining feature of its history all happened by way of numbers.

So, too, with the natural world. Long ago, the environmental historian William Cronon suggested that industrialization in nineteenth-century

America involved a process by which humans subjugated "first nature," the nature of the forests and prairies, to a built infrastructure that facilitated the circulation of commodities that were marketable and consumable, one that became so deeply embedded in the way things were that it was indistinguishable from nature itself—"second nature," as he so eloquently put it. He was pointing to a process of commodification, by which natural goods came to be standardized and could more easily travel through a marketplace. Accompanying this was a process of abstraction, through which natural products, with all of their idiosyncrasies, came to be flattened into measurable, quantifiable features that could be grouped under broad categories and more easily make their way through circulars, price lists, stock exchanges, and company books. It was, at its core, the process by which nature was rendered portable and salable.[4]

. . .

Al-Failakawi was familiar with the process of abstraction in service of commodification. He would have seen merchants and nakhodas engage in exactly that sort of thing in the pearl trade. To arrive at a precise measure of their pearls' weight, merchants used a variety of tools: sieves for sorting them by size, magnifying glasses, precise weights, and calibrated hand scales. But they also relied on a manual that helped them convert the observed features of a pearl—its size and weight—to a market measure called the *chau*, a unit of size and weight akin to something like a carat. Row after row and column after column, the authors of these *chau* manuals guided pearl merchants down the path from the specificities and idiosyncrasies of nature to the abstractions of the market.[5] We've already met the author of one such manual—the nakhoda ʿIsa Al-Qitami, who had penned the *Dalil Al-Muḥtār fī ʿIlm Al-Biḥār.* Alongside his writings on navigation, Al-Qitami authored a pocket-sized *chau* manual that he somewhat aptly named *Al-Khāliṣ Min Kulli ʿAyb fī Waḍʿ Al-Jayb*, or *The Perfect Pocketbook*, which he published in 1924—the very year of Al-Failakawi's voyage.[6]

By the time Al-Qitami published his *chau* manual, many similar manuals were circulating around the region, the majority of which were printed in Bombay in the 1910s and 1920s, in the heady years of the pearl trade. The genre itself was an old one, as evidenced by a 1730 *chau* manual, written in Armenian. And although they all comprised the same genre, *chau* books served several different immediate purposes: some involved different classes of weights, while others highlighted different subdivisions of the *chau*. Not all the twentieth-century *chau* books were in Arabic, either; there exist similar manuals in Gujarati.[7]

In theory, pearl merchants had no need for manuals. The *chau* was a known quantity, and there was a known process for arriving at it. Merchants first moved the pearls slowly down a cascade of sieves, sorting the pearls into different batches roughly grouped by size. Pearls that were too large to move through the first sieve, along with smaller pearls of exceptional roundness, luster, and color, were set aside for individual valuation; the rest were collected together to be weighed and valued in packets. The standard unit of weight used was the *mithqal*—nominally 4.5 grams, though it could vary from one market to the next.[8] The *mithqal* was hardly ever used to weigh individual pearls, though; it was simply too heavy (though it could be used for large batches of lesser-quality pearls). In most cases, merchants drew on subdivisions of the *mithqal*: the *rati* (1⁄24 of a *mithqal*), the *anna* (1⁄16 of a *rati*), and in some cases, the *ḥabba* (1⁄66 of a *mithqal*).[9]

Once a merchant sorted and weighed his pearls, all that was left to do was to determine the *chau* value of a single pearl or a packet. One British official in 1886 observed that there were essentially two methods for determining a pearl or packet's *chau* value, an Indian method and an Arab one; both produced the same result. In the Indian method, one multiplied the square of the *ratis* weight of the pearl by 55⁄96 to get the *chau* value of a single pearl. The same method could be applied to a packet of pearls, though one would have to then divide that result by the number of pearls in the packet. In the Arab system, one multiplied the square of the *ḥabba* weight of a pearl by 30⁄4 to get the value in *dokras* (i.e., 1⁄100s of a *chau*), and then added to it (as a correction) 1⁄100.[10] Put more succinctly, the Arab formula was $ḥabba^2 \times 30/4 \times 1.01 = chau$.[11]

The *chau* manuals offered merchants a shortcut through this complex, time-consuming arithmetic. Once a merchant had determined the weight of the pearl or packet, all he had to do was look it up in the manual to find out what the corresponding *chau* value would be. Because the *chau* value was calculated on the square of the weight, it increased for individual pearls of a large size and weight and thus accounted for the rarity of a particular pearl, even as it tried to flatten distinctions between pearls through a process of abstraction. And at its core, this is what the *chau* manual sought to achieve: to take a natural object with all of its idiosyncrasies and to transform it into an abstraction, a standard, for which value was as objective as possible.

Though Al-Qitami might not have understood it as such, what he guided his readers through was a process of commodification by which a pearl, whose worth combined the vagaries of nature with the bodily labor of the mariners who had to dive for months on end to fish it, was essentially reduced to its value as an exchange object. The *chau* stood in as a proxy for all these processes, flattening them into a single abstract unit that could later be given a market value. It rendered pearls comparable and interchangeable as market goods, collapsing a wide variety of attributes onto a single scale.

. . .

Much of the discussion surrounding abstraction in service of commodification has proceeded from the premise of exactitude—that humans have at least sought to arrive at increasingly precise instruments for measuring goods and calculating value. But this doesn't seem to be the case with Al-Khariji's and Al-Failakawi's notes, many of which seem strangely imprecise. There could be no mathematical laws for converting between currencies, for example—nor even for arriving at measures like the *khandi*, given the variations in what a cut of lumber might look like.

But what if exactitude was not necessarily the goal of measurement? Al-Khariji's and Al-Failakawi's notes suggest that there might be other metrological orientations: rules of thumb, or rough estimates, which they used to guide themselves through these processes of commodification. This would be no surprise to historians of science, who have suggested that attempts to introduce standards always came up against other ways of knowing and measuring, and that these different metrological standpoints could easily exist alongside one another; the transitions from one to the other were hardly ever frictionless.[12] Among nakhodas, too, more precise tools of estimation could just as easily exist alongside rougher estimates. In Al-Failakawi's and Al-Khariji's writings, the two were intertwined: the nakhodas drew on the tools of mathematics to arrive at abstractions that were less precise though no less useful than those we might find in more fleshed-out manuals on matters of the market.

We might not make too much of a distinction, either, between more precise measures like the *chau* and the rough estimates we see in Al-Failakawi's notebook. Even as it attempted to abstract a natural object into a market variable, the *chau* left much unsaid. Its market value was subject to negotiations between buyers, sellers, and intermediaries. Like the rules of thumb that Al-Failakawi drew on, the *chau* constituted a starting point for market processes, not an end. It may have given merchants and nakhodas something to anchor their conversations in, but there was plenty left to talk about.

Whether in the *chau* book or the notebook, conversions allowed actors like Al-Failakawi to realize commensurabilities across different commodities and marketplaces. In this respect, conversion constituted a form of translation, allowing practitioners to speak across different market registers and to move from one marketplace to another with fluency. As such, it enabled the sorts of circulations that actors like Al-Failakawi were principally interested in. By stripping goods of their idiosyncrasies and reducing them only to their observed features, and then taking those and reducing them further into an abstraction that they could communicate across marketplaces, merchants lent the commodity its most salient feature in a regional marketplace: portability.

Perhaps this was what Al-Failakawi and Al-Khariji were after when they inscribed the note on how to arrive at a *khandi* into their notebooks. Moving goods from one market to another involved more than the acts of loading and unloading. For logs of teak and other kinds of wood to move from the "first nature" of the Malabar forests into the arena of "second nature"—in the hold of the *Crooked*, and from there to the markets of the Gulf—they first had to pass through the arithmetic of the marketplace, with all its abstractions. But before they could do that, they had to first pass through the nakhoda's hands, and the rough and ready estimates he arrived at by way of his notebook.

EIGHT

Crossings

AL-FAILAKAWI WEIGHED ANCHOR and had his dhow's sails hoisted on the last day of January 1925; he noted that he left in the company of three dhows. This was no coincidence: the return home was the most difficult stage of Al-Failakawi's voyage—really, for any Gulf nakhoda sailing to India—as it involved crossing a large stretch of the open sea. That morning, Al-Failakawi and the other dhows sailed out, but had to turn back; "the wind," he wrote, "is weak."[1] As they waited for the conditions to improve, they headed back to a spot further up the coast from Goa, anchoring for the night at Malwan. From the deck of his dhow, the nakhoda looked out across the vast expanse of water that he was about to cross on his voyage back—what he and other mariners from the Gulf called *Al-Ghibba*, literally a deep gulp of water, but more commonly understood as referring to any deep body of water. It was not inaccurate: parts of the Arabian Sea are more than five thousand meters deep. By contrast, the coastlines that Al-Failakawi had been sailing along for most of his voyage were barely fifty meters deep; even the deepest parts of the Persian Gulf were only about twice that. For the homebound nakhoda, crossing the *Ghibba* meant losing the comforts of a shallow draft but also dispensing with the geographical orientation that coastal landmarks furnished.

Nakhodas very rarely crossed the open sea on their own; they preferred to sail in a convoy, which they called a *sinyār*, so that one might guide the other, and for strength in numbers if they found themselves accosted by an unfriendly vessel. Al-Failakawi sailed out "with the Al-Qitami *boom* and the Al-Qitami *baghla*, as well as the *boom* of ʿAbdulrasul." The *baghla* belonged to ʿAbdulwahhab ʿAbdulaziz Al-Qitami, who would have been captaining his own ship for roughly seven years by then.[2] The identity of the captain of the *boom* is unclear; later in the crossing, Al-Failakawi mentions having taken items from "the ship of ʿIsa." If that ship was the same *boom*, its nakhoda would have been none other than ʿIsa Al-Qitami, famed captain and navigator, and author of two guides to sailing the Indian Ocean. The third nakhoda, for whom we only get a first name, is harder to identify.

Lurking beneath the perils of the open-sea crossing were its ambiguities. Shorn of the usual landmarks, nakhodas who crossed the *Ghibba* had to grasp at whatever tools they had at their disposal to locate themselves upon the seascape: instruments, combined with texts in both Arabic and English, mathematical formulas, and anything else they could mobilize to approximate their place. But even if the wayfinding tools they grasped at could help them plot a course in physical space, that was only part of the picture; they also had to know where they were in legal space. Whether on shore or at sea, where one fell in the jurisdictional fabric of the Western Indian Ocean mattered—and as Al-Failakawi and his companions knew, this was often not straightforward. Movement across the Indian Ocean was less movement across a sheet than a patchwork quilt, and when it came to delineating and navigating these spaces, neither sextant nor sovereignty could furnish the kind of certainty one might hope for.

The departure from Malvan had to be put off for yet another day; in his entry for February 1, Al-Failakawi noted that "the wind is blowing in from the sea and the ship's bow is being pulled to the northwest." And so, on February 2, 1925, the nakhodas would try their luck again. They weighed anchor once more; "a weak wind is blowing from the south," he wrote. The conditions were not ideal for sailing, but the nakhodas may have felt that they could not justify waiting much longer. With every day that passed, they'd have fewer provisions for the crossing and less time to make it back home. At least the wind blew from the south, which would have facilitated the northward movement of the ship.

As he pointed his dhow out to the open waters of the Arabian Sea, Al-Failakawi would have taken one look back at the coastline before jotting down his prayer: "O Allah, ease our journey and those of our Muslim brothers, by the power of the Prince of Messengers; Amen, O master of the worlds, O merciful one."[3]

• • •

When a dhow sailed out to sea, the logbook entries often looked very different. When nakhodas sailed along the coast, they were dry and terse. By contrast, logbook entries from on the open water were long and detailed but spoke in what feels like an altogether different language. Al-Failakawi's first entry from the crossing reads:

> In the name of Allah, the gracious and merciful, the date is Monday the 7th of Rajab, the 2nd of February English, 177 Nawruz. We awoke in good health. When it came time to take measurements, we measured 56 degrees and 35 minutes, and when we subtracted from one another what remained were 33 degrees and 25 minutes. We had the English *mīl* [declination of the sun] for February at 16 degrees and 52 minutes, and the sextant read 6 degrees, from which we subtracted the *mīl*, and we determined the latitude (with the assistance and strength of Allah) to be 16 degrees and 27 minutes, and took last evening's latitude of 15 degrees and 62 minutes and subtracted them from one another to get a difference of 35 minutes. We entered it into the *nūri* and we found the departure [*ṭūl*] to be 21 degrees and 6 minutes and the *masāj* is 41, and on the rhumb it is 31, and what remained was . . . and a weak wind is blowing from the south. May Allah ease things upon us and upon all Muslims, by the power of the Prince of Messengers; Amen, O Master of the World, O Gracious one.

What we see in entries like these are nakhodas laboring to determine their position at sea. There were none of the landmarks that nakhodas normally relied on nor any other fixed indicators of the dhow's position. There was only the open sea and the tools that the nakhoda had at his disposal to approximate his place upon it.

Among nakhodas of Al-Khariji and Al-Failakawi's generation, calculating longitude had become part of the standard repertoire. But there is no evidence that either one of the nakhodas made use of a chronometer. Like many others, they approximated longitude by way of math. Through combinations of arithmetic and trigonometry, they determined their position by calculating the known variables to determine the unknown variables. There were at least a couple of ways nakhodas could begin the process of measuring their position; all began with determining latitude.

As the sun began to ascend into the sky on February 2, right around 9:00 a.m., Al-Failakawi walked out onto the open deck, took out his sextant, and aimed it at the morning sun as the wind blew hard from the north.[4] Moving his sextant's arm back and forth, he calculated the sun's angle relative to the horizon: 56 degrees and 35 minutes. This then went through the regular adjustments: the declination of the sun and the instrument error, which he calculated at 6 minutes. When all measurements were taken, he determined that the dhow's true latitude that day was 16 degrees and 27 minutes.[5]

Calculating latitude was only the first step in the process. It could tell him how far north or south he had traveled, which was useful information when sailing along the coast. However, on the open sea, what he needed to know was how far east or west his ship had gone. To approximate this, he needed to know the difference in latitude from the previous day's reading—in this case, 35 minutes—and the distance the ship traveled (called *masāj*). To determine *masāj*, captains often used a chip log (called *bāṭili*), which consisted of a reel of knotted rope attached to a weighted board. The nakhoda would feed the rope into the water for a set period—one nakhoda measured it by the time it took to recite the Quranic verses of *Sūrat Al-Ikhlāṣ*—and would then measure the length that the rope extended to determine the ship's speed and distance, quite literally in knots.[6]

With the difference in latitude and *masāj* known, there were a couple of ways nakhodas could figure out the distance they covered going east to west or vice versa (i.e., its departure). If the ship were imagined as sailing on a plane, then one should think of the differences in latitude (translated into nautical miles; each minute would count for one nautical mile) forming one leg, and the *masāj* forming the hypotenuse. With a knowledge of trigonometry, a nakhoda could easily determine the ship's departure. However, most nakhodas (including Al-Failakawi himself) preferred to rely instead on mathematical formulas they had committed to heart, or in Al-Failakawi's case, to pen and paper. In the first few pages of his logbook, Al-Failakawi laid out five different formulas for determining latitude, longitude, and *masāj*, each from the other. Each of these relied on a constant-value assigned to the stars that stood in for the cardinal and intercardinal points on a compass. These charted out the constant-values for the stars in each formula, along with the process for determining one variable from another.[7]

The notebooks that Al-Failakawi carried on his dhow were full of navigational formulas like these—mathematical shortcuts that helped him determine his position based on the limited information he had at his disposal. The notes were written in Al-Khariji's neat, careful handwriting, most likely somewhere on land. Whether by latitude, longitude, or *masāj*, the bulk of a nakhoda's work in reckoning his dhow's location happened in math. From his perch on the dhow's poop deck, maps unfurled, Al-Failakawi added, subtracted, multiplied, and divided his way toward an approximation of his position out on the open sea. Nature, the historian Richard White once wrote, can be learned through labor. Al-Failakawi might have agreed; theoretical knowledge enhanced practical knowledge, and a nakhoda could ultimately only really learn by doing. The vast expanse of the Western Indian Ocean was rendered legible by the sun, the stars, the sextant, the chip log, and pen and paper—and of course, the guidance of Allah and his messenger.

There were other guides too. Many of the nakhoda's notes on navigation, it seems, were copied from ʿIsa Al-Qitami's manual on navigation, *Dalīl Al-Muḥtār.* Navigational knowledge, the notes suggest to us, did not only move from master to apprentice through training; it circulated between the deck of the dhow, the logbook, and the printed manual through a process of learning, testing, copying, and excerpting. It was preserved in writing and reanimated at different moments in the voyage.

Al-Failakawi drew from other texts as well. As he was wrapping up his arithmetic, he reached into his chest and pulled out a well-thumbed copy of *Norie's Nautical Tables*, a nautical almanac that gave mariners accurate information on the declination of the sun and other celestial bodies for every hour of every day of the year. Nakhodas around the Gulf were familiar with Norie; "we entered it into the Norie" is a phrase that appeared across logbooks. Al-Qitami himself incorporated the book into his navigational repertoire in his discussion of wayfinding via sextant.[8] The tables gave nakhodas a shortcut through the trigonometric calculations they needed to correct their readings and determine their dhows' true latitude and departure.

The book's author was perhaps an unlikely protagonist in the intellectual histories of Arab navigation. Born in London in 1772, John William Norie established himself as a maker of British Admiralty charts—several of which ended up in Al-Failakawi's possession—and writer on navigation. He began his career working with another publisher of nautical charts, instruments, and books, and took over the business in 1813, renaming it J. W. Norie & Co. The firm went through several iterations and rounds of renaming after Norie's death in 1843 but continued to publish all manner of nautical texts: pilots, sailing directions, charts, and, most famous among the nakhodas of the Gulf, the nautical almanac *A Complete Set of Nautical Tables*, which was later renamed to *Norie's Nautical Tables*—or, in the Gulf nautical vernacular, *Al-Nūri.*[9]

The choice to use *Norie's Nautical Tables* was not an obvious one, and in some ways seems counterintuitive. This was a text produced for the British Admiralty, not nakhodas in the Gulf. Moreover, it was a text produced in service of an expanding British Empire. Though the earliest attempts at charting the waterways through which European vessels arrived on the shores of Africa, Asia, and the Americas took place in logbooks, by the late 1700s these came to be understood as insufficiently objective—and for the imperial projects that were unfolding around the globe, insufficiently scientific. By the time Norie arrived on the scene, European empires began to reconceive oceans as spaces teeming with commercial and imperial significance. Printed charts, instruments for nautical surveying, and the publication of books on sailing directions were among the tools of imperial expansion; knowing the oceans was seen as an essential prerequisite to ruling the waves.[10]

The extent to which nakhodas drew on the text reveals much about how their science of navigation had become increasingly bound up in an imperial epistemological framework. The notions of space and time that came along with imperial expansion by sea drew in other maritime actors and other ways of knowing the ocean. Nakhodas actively took these technologies of imperial epistemology and domesticated them. They produced their own texts like Norie's, too, such as the *Al-Natīja Al-Kuwaytiyya*, an Arabic nautical almanac authored in 1933 by two Kuwaiti nakhodas, Muhammad bin ʿAsfour and Hussain bin ʿAbdulrahman Al-ʿAsʿousi. Unlike the received wisdom on Ahmad Ibn Majid's writings, which historians like to think of as being unsullied by European contact, the navigational thinking of nakhodas like Al-Failakawi, Al-Khariji, Al-Qitami, and many others like them was generated through active engagement with European empires. Theirs was an entangled nautical tradition—a product of the times in which they lived, in which European notions of space and time emerged as hegemonic, but also a reflection of the simple fact that the world in which they sailed and did business bore the indelible imprint of European imperial expansion. Even as they continued a tradition that hearkened back to the time of Ibn Majid, nakhodas' changing writings on navigation reflected the changing boundaries of knowledge—but also the imperatives—of their own time.

But even their usage of texts like Norie's could only bring them a little closer to precision. Through the tools at their disposal—sextants, chip logs, mathematical formulas, British nautical tables, translated almanacs, and more—nakhodas could only grasp their way along the open sea. Theirs was not an exact science, if ever there was one: it was a way of approximating their place upon the seascape: of rendering the vast expanse of water legible and of carving out pathways along it. Through their tools, they tried to envision where they were within the broader watery geography of the Western Indian Ocean and worked to figure out how to move through it.

Others were crossing the Arabian Sea as well. Just a few days into his open-sea voyage, Al-Failakawi noted that after taking measurements in the morning, the crew "saw a ship coming from Aden, going to Bombay." Eight days later, they saw another ship heading from Aden to Karachi.[11] The crew of the *Crooked* would have been forgiven if they felt the slightest bit of unease as they passed and hailed the two ships. For in addition to navigational tools and texts, there were other ways of determining one's place on a seascape—ones that often had a damaging bluntness to their imprecision. For moments in which ships encountered one another at sea, there was often law. And it is by way of the British colony of Aden, from which the two ships sailed, that we can see the possibilities—and more importantly, the grasping limits—of imperial jurisdiction across the Arabian Sea.

• • •

The English captain Philip Howard Colomb arrived in Aden in the early fall of 1868. He came by steamship from Southampton by way of the Mediterranean—around the Iberian coast, through the Straits of Gibraltar, and into Alexandria. He then journeyed by train to Cairo, and then on to Suez, where engineers were digging the canal that would ultimately connect the Mediterranean to the Red Sea. From there, he took a steamer to Aden. Colomb was a naval officer who, at the age of thirty-seven, was sent to the Indian Ocean to captain a naval cruiser, HMS *Dryad*, one of a handful of ships that were commissioned to patrol the Western Indian Ocean in search of slave traders. It was at Aden, where the Red Sea and the Arabian Sea meet, that Colomb would first board the *Dryad* and begin his "slave-catching" tour of the Western Indian Ocean.

Nothing exercised British imperial anxieties in the Western Indian Ocean quite like the slave trade did. The archives teem with material generated by encounters between British naval cruisers and dhows suspected of involvement in the slave trade, deliberations surrounding which seemed to occupy much of the attention of British consuls around the region. In 1861, the British government decided that in addition to the Resident at Aden's normal duties, he would be authorized to convene a vice-admiralty court with the power to adjudicate cases related to the slave trade, inasmuch as they fell within the scope of the treaties that the government had entered into. A few years later, they did the same in Zanzibar, where the consul himself could authorize the search and seizure of suspected dhows.[12] An apparatus aimed at surveilling the movement of dhows around the Western Indian Ocean slowly came into being.

FIGURE 15. British steam pinnace HMS *London* chasing a slave dhow, printed in the *Illustrated London News*, December 1881.

The captain marveled at the ease with which he could navigate the bureaucracy. "No process can be more simple or less formal than that of adjudicating in the Aden Prize Court on cases of undoubted slave trading," he wrote. Because of the multiple duties the Resident at Aden performed, "practically, he is the whole court—vice-admiral, judge, and registrar," and "the court is held generally in one of the rooms of the registry." Even the processes—the examination of the vessel and witnesses, the discharging of the enslaved passengers to the "slave island" in the harbor, and the issuing of a decree of condemnation, were all "very simple."[13]

Colomb quickly realized, though, that the simplicity of the process belied a substantive complexity. The Western Indian Ocean was highly variegated as a jurisdictional space; British naval officers like him had to be mindful of the "where" and the "when" of slave trade suppression, not just simply of the how. His power was constrained by treaty, and within prescribed limits—usually the territorial waters of a ruler, which he noted were "waters within three miles of any shore whose owner is strong enough or important enough to assert his jurisdiction." Moreover, between May and the end of December, the Sultan of Zanzibar's subjects were free to traffic slaves "from any one port of his dominions to any other."

In practice, the coastlines of the Western Indian Ocean that Colomb passed through comprised a patchwork of different jurisdictions over different shorelines and waterways, and the sovereignty of the Busaʿidi sultans waxed and waned depending on where the *Dryad* happened to be. The territorial waters of the sultan extended from Zanzibar to Lamu, then disappeared altogether until one reached the southeastern tip of the Arabian Peninsula. "At, or around, Ras-el-Hadd begin the dominions of the Sultan of Oman, whose power and will to claim jurisdiction increase up to Muscat, his capital," wrote Colomb as he tried to chart out the boundaries of his own authority. "They then decrease again towards Cape Mussendom, long before arrival at which point petty tribes erect 'fortlets' and would no doubt claim territorial waters—if they dared." On the Persian side of the Gulf, "somewhere between [Lingeh] and Bunder Abbas we come, not upon Omani territory, but upon Persian territory farmed by Oman."[14]

Colomb's writings capture a much more pervasive sense of ambiguity surrounding dhows and whether they might legitimately be seized and destroyed. The moments in which British naval officers could confidently assert that a dhow was carrying slave cargo were few and far between. Most of the cases were shrouded in ambiguity: questions of whether the people on board the dhow were enslaved or free, and whether the dhow even fell under the jurisdiction of the court—what officials called the "character of the vessel"—often remained unanswered. This was particularly true when it came to reading the documents on board the dhow, which one consul noted "generally tend to determine the character of the vessel."[15] The problem, of course, was that none of the officers could themselves read the Arabic script. George Sullivan, another naval officer who published his memoirs on chasing and seizing dhows during the mid-nineteenth century, admitted in his reflections on one particular encounter that the papers they found "might have been, for all we knew, Bills of Sale for the niggers on board, or warrants for their execution; or, more probably, directions as to where our boat was, how to avoid it, or to cut the throats of every Englishman if they could get the chance." Unable to read the papers, he was forced to let the dhow go.[16]

In Colomb's case, the matter of dhow papers was only slightly alleviated by the presence of an interpreter on board the *Dryad*—a Comorian named Saleh bin Moosa. When Colomb would stop and board a suspicious dhow, Bin Moosa would be the one tasked with reading the dhow's papers. This was less straightforward than it might seem: "The literary knowledge of [Bin Moosa] is not extensive—I believe him to be acquainted with some of the

Arabic characters when plainly written," wrote Colomb; "I doubt whether he can readily detect the difference between Swahili, written in Arabic characters ... and pure Arabic itself."[17] Colomb's experience with Bin Moosa was hardly unique. The interpreters who British naval officers hired to overcome the linguistic barrier were of limited use: either their Arabic or Swahili were too colloquial, or their English was not up to snuff. And those who possessed both languages would simply not take the job. "The position of an interpreter, partaking largely as it does of that of a spy and informant, is not favorably regarded in native society," wrote a consul; those who were qualified "could easily find employment without having any necessity to go sea."[18]

The difficulties of translating from dhow to cruiser were realized on more than one occasion when Bin Moosa seemed to have difficulty reading the documents that nakhodas produced. When one nakhoda from Sur produced a safe-conduct pass issued to him by the sultan of Zanzibar, Colomb observed that "it was a terrible trial to Moosa, whose finger wandered with great uncertainty over the Arab characters; by dint of much spelling out, however, he managed to gather that the clearance stated the vessel was fishing trader to and from Zanzibar." His attempts at communicating orally were no more successful: Colomb described an exchange between Bin Moosa and a nakhoda sailing out of Aden in which the two exchanged single-word grunts with one another. Bin Moosa declared that the dhow's name was *Haaf*, though the dhow's papers, issued at Aden, identified it as the *Summah*.[19]

The enduring confusion surrounding seized dhows could sometimes have enormous consequences, at least for the nakhodas and dhow owners. While sailing in the Persian Gulf, the *Dryad* crew spotted a dhow with two Africans on deck; upon seeing Colomb's ship, the dhow made for the coast. The *Dryad* gave chase and finally caught up with the nakhoda, who could not account for his behavior, nor for the two Africans on board the ship. Though Colomb tried to allow the nakhoda to explain himself, the "unfortunate Arab will not get himself out of the hobble he is in ... and the dhow must die, either for her actual sins, or for the evidence which is so presented as to show her sinful." It was Colomb's first time destroying a suspected dhow; "I cannot say that I regarded it with any feeling of satisfaction," he wrote, likening it to the first time he ever shot a rabbit.[20]

Though he expressed his unease about the proceedings and would do so again on different occasions, Colomb acknowledged that it was very much a product of the circumstances the *Dryad* found itself in. The *Dryad*'s crew had to board suspected dhows and sail them, sometimes many hundreds of miles, or tow them to the nearest port of adjudication, and would have to leave their normal patrol duties behind while adjudicating the case. There was something of an escape clause, though: if a vessel was not deemed to be sufficiently seaworthy as to be sent in for adjudication, the captain was authorized to destroy it. The warped incentives that the clause created were immediately clear to Colomb: "What is the necessary result of the conditions described and of this clause in the instructions?" he wrote. "It is that the latter becomes the rule. Every detained vessel, unless the capture be made almost within sight of the port of adjudication 'appears to be unfit to proceed' there, is formally surveyed, formally reported unfit, and very informally scuttled or burnt."[21]

By the time of Al-Failakawi's voyage, this was no longer the case. Of course, dhows were routinely harassed, especially when they were suspected of carrying contraband—especially arms. However, nakhodas had recourse to the British Political Agent in Kuwait, who would investigate their complaints and file for compensation for any damage done to their dhow or cargo. For Colomb, whose mandate for the suppression of the slave trade felt more urgent, due process was a luxury he could not afford.[22] Whether along the coasts or on the high seas, "the captain of the ship is judge, jury, and executioner"—though one whose ability to grasp his surroundings through law was never quite as precise as he might have liked.

. . .

For nakhodas crossing the sea, Aden was never too far off the horizon; Al-Failakawi and his colleagues frequently sailed there. It enjoyed an outsized role in the Indian Ocean trade and global commerce writ large. The opening of the Suez Canal in 1869 transformed the port city from a growing colonial outpost of still-relative obscurity to a central node in a global shipping network. Just five years after the canal's opening, one official observed that the trade there had nearly tripled—and that was only for business transacted within the settlement itself. "Were it possible to show the value of the goods transshipped at this port either to Europe and America, to the westward; or Natal, Réunion, Seychelles, Zanzibar, Muscat, Persian Gulf, British India, and China, to the eastward," he wrote, "it would be made apparent how commercially important Aden has become."[23]

Aden was no ordinary settlement for the British. It formed a critical node in an expanding imperial regime in the Western Indian Ocean. From Aden,

British officials could direct an ever-broadening stream of commercial and military vessels making their way up and down the Red Sea and could coordinate the movement of information and personnel between the Gulf, India, and an expanding consular presence in East Africa. Indeed, so critical was Aden to the administration of a growing British Indian empire that officials designated it as a district of the Bombay Presidency—though, as we will see, one with a more ambiguous status than most others.[24]

Merchants around the Western Indian Ocean flocked to this British imperial apparatus at Aden, arranging themselves around the port's changing infrastructure. Among the more energetic of these was Aden's Parsi business community, which channeled goods and credit from India to the port city, and reinforced its commercial connections with Bombay, their other metropolitan anchor. Some, like the firm of Cowasji Dinshaw Adenwalla, grew to regional and international prominence; Dinshaw's firm ran a regular steamer service that dominated the routes that linked Aden to Bombay, Karachi, Mombasa, and Zanzibar.[25] There were others as well: Persians, Gulf Arabs, Hadhramis, Ethiopians, Somalis, and traders from the Swahili coast, all of whom flocked to the rapidly changing port city to expand the scale and scope of their businesses.

Kuwait's merchants maintained a presence in Aden too. The firm of Muhammad ʿAbdullah Hasanali & Brothers did extensive business on behalf of the Al-Sager brothers and other Basra-based merchants. The Hasanalis were importers of different goods: one registry lists them as importers of Italian goods, among others, into Aden.[26] The firm also imported Basra dates through partners like the Al-Sagers and would sell them on their behalf in the Aden marketplace. After selling the dates, they would remit the profits to Basra by way of Bombay. The firm's secretary would write to Al-Sudairawi in Bombay and include *hawalas*, checks, and banknotes, which he would then ask the Bombay banker to deposit into their principals' accounts. When the Hasanalis wrote to the Al-Sagers, they would include accounts that detailed proceeds from goods sold, but also the money they loaned out and collected for the Al-Sagers and others.[27]

Later, the Al-Sagers would send one of their own employees to Aden to act as their agent. Khaled ʿAbdullatif Al-Hamad initially worked for the Al-Sagers in Basra as a letter-writer, but within a couple of years, they decided that he was better suited to act as their agent in Aden. He had been there before with his father, who worked for another Basra family, and he was still young and energetic. Khaled consented: he went to Kuwait, where he applied to the British consul for travel papers, then arranged for his brother Ahmad to take his place in the Al-Sager office. As the Al-Sager agent in Aden, Al-Hamad was responsible for everything. "I was the cook, the writer; everything!" he remembered later in life. He would coordinate with his principals in Basra on the shipping of dates to Aden and would oversee their sale. From Aden, he would purchase cargoes of coffee, which arrived in the port on caravans from the interior, and sometimes by way of dhow; other cargoes of coffee would come to him from Ethiopia and the northern coasts of Somalia. Al-Hamad would arrange for these to be shipped to Kuwait, Basra, Bahrain, and other port cities, based on instructions from the Al-Sagers.[28] He also worked with the Hasanalis and other agents that the Al-Sagers appointed, especially when it came to collecting outstanding debts in the Adeni marketplace. And on at least one occasion, when he was away from Aden, Al-Hamad appointed Hamad Al-Sager as his own agent, to pursue debt claims on his behalf in Aden's courts.[29]

As the population of Aden ballooned, so did the legal questions that its administrators had to grapple with. If one of Aden's perceived virtues as a place of commerce stemmed from its deep connection with British India, and with Bombay in particular, it was also the source of much juridical confusion. The precise legal relationship between Aden and Bombay was never set out plainly. The 1864 legislation that had established the Court of the Resident at Aden also subjected it to the "superintendence" of the Bombay High Court, but few officials on the ground had any sense of what that term meant, and the Letters Patent (effectively, a charter for the settlement) to which the 1864 legislation referred offered little guidance. The Letters made it clear that only the Bombay High Court could try "European British subjects" for crimes punishable by death and that it could issue general rules to regulate the proceedings of the Resident's court, but they left the Bombay High Court's power to review decisions handed down at Aden highly ambiguous. If the advocate-general at Bombay determined that there was an "error in the decision of a point or points of law decided by the Resident, or that a point of law decided by the said Resident should be further considered," the Letters stated, then the High Court had the authority to review the case in question and pass judgment and sentence.[30]

But what did any of it amount to? The vague language left the matter in the grasping hands of litigants, lawyers, and judges. In a 1903 case, an Indian residing in Aden sought to appeal the Resident's decision regarding property that he owned. The Bombay High Court affirmed his right to appeal, the majority of the judges arguing that appellate jurisdiction might be understood as a form of superintendence.[31] The case ultimately made its way to the Privy Council, the highest court of appeal in the British Empire, which struck down the Bombay court's decision. Superintendence, judges declared, differed substantively from jurisdiction. The High Court enjoyed "the power to make general rules for regulating the practice and proceedings of the Court of the Resident," and could oversee and alter the paperwork issued by the Aden court, and could even alter the rules by which the court was run, but did not necessarily enjoy appellate jurisdiction.[32]

That hardly settled the matter. A few years later, the court clarified that Bombay had appellate jurisdiction, but only when the Resident at Aden submitted to it; otherwise, he was only to be "guided by the spirit and principles of the laws in force in the Presidency of Bombay."[33] Despite the guidance, the legal geographies of the oceanic imperial space between Western India and South Arabia were often difficult to delineate. It remained unclear where Bombay ended and Aden began, at least as far as the law was concerned.

Nor did the confusion settle when Aden was cleaved off from Bombay, first to be administered by the Colonial Office and then as a "Crown Colony," directly administered from London, during the 1930s. The connection to India was a dense one that was difficult to unravel. For several years, litigants in Aden could continue to appeal court decisions to the Bombay High Court, even though the Government of Bombay itself no longer enjoyed jurisdiction over the port city. Even after 1936, when the British government established the Aden Supreme Court, civil cases valued at more than Rs 5,000 or criminal cases could still be appealed to Bombay.[34] More generally, the specter of British Indian law loomed large in Adeni judicial proceedings. Cases heard in the Supreme Court teemed with citations of Indian legislation and cases from around India; throughout, there was a distinct lack of clarity as to whether Aden might still be best understood as a juridical extension of British India, even as it was twice removed from the Indian administrative orbit.[35] Judges there also grappled with questions surrounding the place of Islamic law, Jewish law, Somali custom, and other normative orders in Aden—and throughout, they looked to Indian legislation to furnish guidance on the place of other legal orders within the colonial legal administration.[36]

What Aden's place was within a burgeoning imperial legal geography in the Western Indian Ocean was never fully clear—not to litigants, nor to judges and administrators, nor even to the members of the Privy Council, whose job it was to bring legal order to the British Empire writ large. The jurisdictional quilt that they tried to produce in the Western Indian Ocean required endless restitching. Ultimately, they produced as many loose threads as they brought together.

• • •

There was little to occupy mariners as they sailed across the open sea. For the whole of the stretch of the crossing from India to the Arabian Peninsula, there is nothing in Al-Failakawi's logbook beyond the record of sextant measurements and calculations, save the occasional observation. He barely even recorded wind conditions unless they had changed; the wind was either strong, weak, or not worth commenting upon at all. Meanwhile, his crew members occupied themselves with daily routines. If they weren't manning different masts and halyards, they were mending the sails and rigging. Long stretches of work would be punctuated by three daily meals and a coffee break, and set to the clock of the five daily prayers: work began after the dawn prayer, and after the nighttime prayer the sailors would sleep in shifts while others kept watch.[37] Whether at night or during the day, the nakhoda and his crew kept their eyes on the horizon, watching in eager anticipation of a glimpse of land.

Had the *Crooked* been moving toward Aden, Al-Failakawi would have known what to keep an eye out for. Among the sketches of coastline that his teacher Al-Khariji kept in his notebook was a curiously rounded drawing of Perim Island, which he called by its Arabic name, Mayyun—"the island that is near the Bab [El-Mandeb] straits," just ninety miles west of the port of Aden itself. The nakhoda marked out the customs house (*sirkar*), a lighthouse, an airstrip, a wharf that he labeled "Mayyun Company" (presumably belonging to the Perim Company), and a handful of structures; among them were two British flags.[38] He kept the drawing nestled along sketches of many different landmarks that he relied upon to help him identify his place on the coast. On the same page are drawings of different hills and mountains in the Gulf, in India, and along the coast of Aden itself.

The nakhoda's choice to include a drawing of Perim is illuminative: the island's place in the imperial jurisdictional seascape was perhaps even more shrouded in ambiguity than Aden's. When one Indian working on Perim Island was charged with murder there in 1885 and was convicted by the Resident at Aden, he sought to appeal the decision to the Bombay High

Court. There, judges expressed confusion as to whether Perim formed a part of Aden, and whether it might be included within the Bombay Presidency. After much back and forth in a series of decisions, judges held that there was nothing to substantiate the notion that Perim Island fell under the jurisdiction of the Aden Resident, though they also suggested that cases arising in Perim might be transferred to other courts in the Presidency, including the Bombay High Court itself.[39] In the jurisdictional quilt of the Western Indian Ocean, Perim was a square apart from even Aden itself. Perhaps that's why Al-Khariji felt the need to include not one, but two British flags in his picture.

But there were many possible Perims out there, in that vast seascape that Al-Failakawi and his crew traversed—many ports, islands, stretches of coast, and courts whose places in the legal landscape were never fully clear to anyone, places that were always slightly beyond the reach of the law, whose jurisdiction was never as precise as any of its practitioners wanted it to be. Whether in Aden, whose relationship to the Bombay High Court never emerged from the shroud of ambiguity, or in places like Karachi, whose path to appeal was clear but its ability to enforce decisions outside of the city was not, the arms of the law were invariably shorter than anyone might have liked. Much like Al-Failakawi's navigational instruments and texts, law could only act as a rough guide through the watery worlds of the Western Indian Ocean.

But part of the reason why people clung to rough guides was that oftentimes, they worked. On the morning of February 16, two weeks into their voyage across the Arabian Sea, Al-Failakawi looked out across the ship's bow, then back down at his *ruznamah*, and wrote: "We saw the coast . . . thanks to Allah for our safe arrival."[40]

INSCRIPTION

Maps

THE FIRST THING I SAW was the maps. Yellowed and rolled up, they formed the topmost layer in the Malabar teak chest that sat in the Al-Failakawi family home. ʿAbdulrahman Al-Failakawi and I didn't immediately unroll the maps; he set them aside so that I could see the nakhoda's logbooks first, and so that he might show off to me some of the instruments his father took on his voyages: his sextant, a spyglass telescope, and a set of parallel rulers. As I'd later learn, he had good reason to put off showing me the maps. They were in a fragile condition: unfurling the maps placed an enormous amount of stress on the canvas paper, which cracked every time we tried to flatten out the large sheets. I was only able to take a few pictures of each map with my camera before it tried to roll itself back up into a tight scroll or before the paper began to crack under pressure.

I didn't know what to expect from the maps, but I certainly could not have expected what I saw as we unfurled each of the scrolls. The maps—fifteen in total, each roughly a meter long by a meter and a half wide—were mostly British Admiralty charts of different stretches of coastline around the Western Indian Ocean; there was also one French map of the Arabian Sea, produced by the French Navy. The collection included maps from the mid- to late nineteenth century that had been corrected in the first half of the twentieth century, but also up-to-date maps from the 1910s and 1930s. These on their own were not especially surprising: Alan Villiers wrote about how he had seen "uncorrected out-of-date Admiralty charts" on a Kuwaiti dhow off the coast of East Africa—one of his many jeremiads on the decline of the Arab art of navigation in the time since Ahmad ibn Majid. The fact that Al-Failakawi made use of those very Admiralty charts would have perhaps confirmed that Villiers was right—that Arab navigation had become

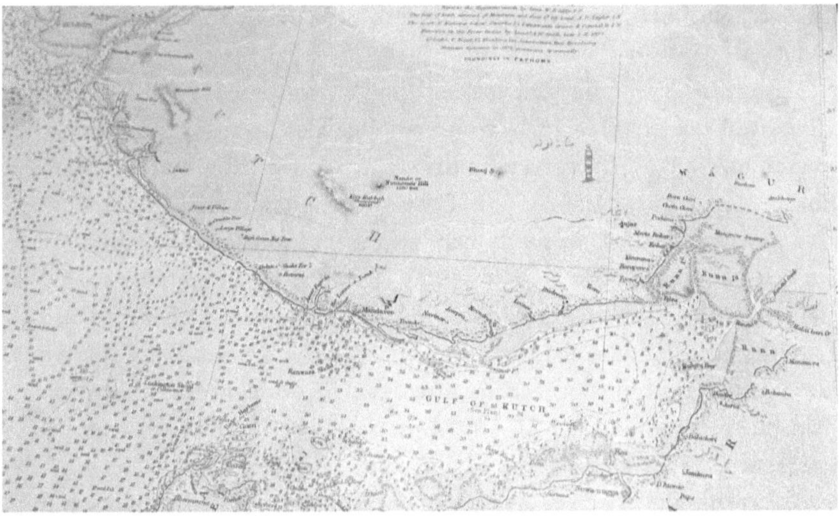

FIGURE 16. Al-Failakawi's annotated map of the Gulf of Kachchh. Al-Failakawi Collection, Kuwait.

derivative, that there was nothing to see, even less to write on, and much to lament.

Except that none of that is true.

As I unfurled the maps, I immediately noticed that many of them had been carefully annotated. Port cities had been translated from English (or French) to Arabic, and at times added different landmarks—a lighthouse on the coast of Gujarat, for example. In different places on the maps, too, the nakhoda wrote out different notes, describing when and where the maps entered his possession. The story, it seemed, was not how it had been told. Maps were not indications of the nakhodas' long fall from grace; rather, they were evidence of the nakhodas' hunger for information, of their desire to fill out their repertoires, and of the agility with which they seized upon the technologies of imperial cartography and incorporated them into their own worlds of knowledge.

• • •

A map, any good historian will tell you, is never a straightforward or neutral representation of reality. Many of us will have learned some version of this in our high school history classrooms, when we were told that the standard projection of the world, the Mercator projection, inflated the size of

Europe and North America, and shrunk most of Latin America, Africa, and Asia.

Historians of empire have taken things a few steps further. Maps, they argue, were the technologies by which empires came to know the regions they sought to control. By surveying different parts of the world, empires—European, Islamic, Asian, and otherwise—rendered them legible, conquerable, and taxable. In some cases, maps were aspirational: they sought to stake out claims to particular areas, often by color-coding an empire's territorial claims vis-à-vis others—the flesh-pink hue that often coded the British Empire's jurisdictional claims, for instance. At other times, maps could conjure up the very kinds of territorial realities they sought to project. In this field, the medium—the map—is the message; it is the tool through which states come to see, know, conquer, and rule.[1]

The Admiralty charts in Al-Failakawi's possession lend themselves well to this sort of analysis. The charts were printed by the Admiralty's Hydrographic Office, for use by the Royal Navy. The process of charting the coastlines of the Indian Ocean, as elsewhere in the world, was bound up in a process of imperial expansion, even as it often relied on native informants.[2] To take just one example: one of the maps of the East African coast in Al-Failakawi's possession was produced by Captain William Fitzwilliam Owen during his voyages around the region. Owen's voyage was, on its face, a scientific one: he was to survey the coasts of the Western Indian Ocean and supply information on "the number and character of the natives, their occupations, modes of subsistence, &c the nature of their soil, and also the productions of the surrounding country."[3] But Owen's scientific voyage also included audiences with rulers in different port cities, plans to suppress the slave trade out of East Africa, and the establishment of a short-lived protectorate in Mombasa. The plan to survey and chart the East African coast was inseparable from an ambition, latent though it may have been, to know, regulate, and rule it. Owen's map may have been a representation of a reality, but it was also an attempt to project power onto a part of the Indian Ocean world where the British Empire had not yet fully established itself.

But maps are also canvases that can be inscribed upon, material objects that can be rolled up and stored away, and market goods that have lives; they were bought, sold, and handed off from one person to another. The maps were not in Al-Failakawi's possession because they were monuments to the expansion of the British Empire in the Indian Ocean. They were tools. He bought them from different vendors and from other nakhodas so that he might use them. To pluck the phrasing of one social scientist totally out of its context, people can encode objects with significance—they can make maps stand in for imperial ambitions—but it is by following the movement of those objects that we can see the human and social contexts in which they existed.[4]

It wasn't always the case that maps moved from one person to another. Before the nineteenth century, maps were too valuable to circulate freely. Even with the advent of engraving and printing, reproducing maps was expensive, and mariners mostly accessed maps and charts through institutions of learning. When maps and charts formed part of the process of settlement and colonization overseas, they were produced by and for particular organs of the imperial bureaucracy; they rarely circulated, even among the elite.[5]

In England, where most of Al-Failakawi's maps came from, mapmakers were appointed as hydrographers to the Admiralty and produced charts specifically for that institution; the first was Alexander Dalrymple, who was appointed to the position in 1795 after having served as hydrographer to the East India Company. It was under Dalrymple's tenure that the Admiralty adapted the printing press to chart-making: it wasn't until around 1800 that the Hydrographic Office first began printing charts, and they only began offering them for sale to the public in 1821. The emergence of lithographic printing during the second half of the nineteenth century did little to change things. The Admiralty continued to print charts from copper plates until the twentieth century, even as it occasionally used lithographic technology.[6]

Outside of the Admiralty, there were a handful of private mapmakers and map sellers; John William Norie, the author of *Norie's Nautical Tables*, was one such seller. Still, they were relatively expensive and could not keep up with the changing needs of merchant vessels. By the beginning of the twentieth century, the mapmaking firm that Norie founded had merged with three others, forming the company of Imray, Laurie, Norie, and Wilson—the only private publisher of Admiralty charts and "blueback charts" (known for the rough, blue-colored manila paper they were printed on) in the United Kingdom.[7] Al-Failakawi owned one of these blueback charts, a map of the Shatt Al-ʿArab waterway produced after a survey conducted in the summer of 1924.

By the time Al-Failakawi began sailing, Admiralty charts could be found in the marketplaces of major port cities around the Western Indian Ocean. The expansion of map printing, and the frequency with which maps were corrected, meant that older maps made their way into the hands of other buyers, including Gulf nakhodas. A note on the back of one of Al-Failakawi's maps suggests that he bought it in Mutrah, the marketplace that served

Muscat. The map was a chart of the coasts of Oman, from Muscat down to Ras Madraka, past the island of Masirah; it was where the nakhoda had made landfall many times after the open-sea crossing from India. Alan Villiers also observed that nakhodas bought "uncorrected out-of-date Admiralty charts" from Bombay.[8] Another nakhoda from Kuwait purchased maps on his visit to Mombasa in 1944, for 33.5 shillings.[9]

Alongside this market for uncorrected maps existed a robust secondary market, in which nakhodas would sell used maps to one another. These accounted for the bulk of the charts in Al-Failakawi's possession. Of the fifteen maps the nakhoda owned, nine came from individuals that he named. Four of these—maps of Cape Comorin, the Gulf of Kachhch, and East Africa—were purchased from Ahmad bin Hussain Ka-Ibrahim sometime during the 1904–5 sailing season. The buyer was most likely not Al-Failakawi himself, but Al-Khariji, whose neat handwriting states that "he bought this map [*nāyila*] of Ahmad bin Hussain Ka-Ibrahim" for Rs 3 per map. One of the maps of East Africa—a chart originally produced by Captain William Fitzwilliam Owen—was bought in Bombay; the other, a map of the Gulf of Aden, was also bought from Ka-Ibrahim, for Rs 10 and 7 annas.

The other six maps were bought by Al-Failakawi himself, who wrote his name on them. Three of the maps of the coasts of the Gulf and of Makran were bought in 1936–37, the French map of the Arabian Sea in 1939–40, and four maps—of the Arabian Sea, Red Sea, and East Africa—were bought during the 1943 sailing season. None of them indicate how much he bought them for, but some do list names: a Mohammad Sa'ad Palat in Basra, who sold Al-Failakawi a map of coastal Oman in 1936; Hajji 'Abdullah Hamza, who sold him the French map sometime in 1939 or 1940; and Mohammad Ibrahim Karam, who sold him maps of the Arabian Sea and East Africa in 1943; we can only guess that all of these were nakhodas. As they sailed around the coasts of the Western Indian Ocean, then, nakhodas would find people who would sell them the maps they no longer needed.

The movement of maps from the Hydrographic Office to Naval officers, and from them to secondhand markets around the Western Indian Ocean, suggests that imperial cartographic knowledge was not as bounded of a phenomenon as historians might think it is. Maps and charts could be products of a history of imperial knowledge production—of the process by which empires came to "know" and ultimately rule the shores of the Western Indian Ocean. But they were also objects that changed hands, that moved from the deck of one ship to another. Imperial knowledge thus circulated far beyond the structures of imperial authority. It seeped into the maritime marketplace and made its way into the hands of the very people the empire sought to surveil.

• • •

But the maps that circulated among nakhodas were not exactly the same as those that the Hydrographic Office produced. Nakhodas were not passive consumers of Admiralty charts; they actively engaged with them, inscribing upon them the information that they thought useful. Villiers noticed this as well. He described one Admiralty chart of the Indian Ocean as being "decorated with Arabic script giving landmarks, distances, and other information of importance to the Arab mariner"—though the observation did not prompt him toward a more nuanced understanding of the changing contours of cartographic knowledge among Arab nakhodas.[10]

Al-Failakawi's maps included many such inscriptions, at least some of which were in his own hand. Another nakhoda's annotated maps survive as well—the nakhoda Sa'id bin Salamah, who was a rough contemporary of Al-Failakawi's, having sailed during the 1930s and until the mid-twentieth century. We might consider the two sets of maps together, since they're not very different from one another, at least in terms of how the nakhodas engaged with them.

When the nakhodas purchased maps, they inscribed onto them the names of the different port cities they visited. Some of these are immediately recognizable to the modern land-based reader: Aden, Bombay, Basra, Calicut, Dubai, Sharjah, Lamu, and more. But the bulk of them marked out lesser-known ports, towns, and harbors—ones that have since been renamed, absorbed into larger developments nearby, or completely erased altogether. The inscriptions that pepper the littorals of the Western Indian Ocean correspond to places that the nakhodas referred to in their logbooks but are difficult to locate on a map: *Jazirat Natlōh* for Netrani Island, for example. Some referred back to names that were firmly embedded in an Arabic geographical lexicon—like *Khormiyān* for Porbandar, *Bukīn* for Madagascar, and *Al-Jazira Al-Khaḍra* for Pemba. Al-Failakawi's maps included at least one coastal marker, too: a drawing of a lighthouse on the Gulf of Kachhch, at the harbor at Toona; he included the note "the lighthouse at Toona."

On the face of it, this was a straightforward act of translation: they simply annotated the maps with the Arabic names of places that were already there in English. But the act of translation is an illuminative one in its own right: it is an act of encoding as much as it is one of decoding. In annotating the maps, the nakhodas transposed one spatial imagination onto another. They layered on the geography that they lived and their own activity space: a web of mostly secondary and tertiary port cities that cut across continents, jurisdictions, and empires. This was the Indian Ocean world as they knew it, separated though it may have been into different coastlines spotlighted on different charts. As the nakhodas accumulated charts of the Gulf, India, East Africa, and South Arabia, along with more detailed maps of places like Zanzibar, Gujarat, and Basra, they stitched together the world they saw and traversed, marking out the places and landmarks that rendered it legible to them. What we see on the maps, then, are two different ways of making the Indian Ocean world "knowable"—one layered atop of another, rather than one displacing the other.

But there are more than two layers to the maps. Several of them include annotations that predate the nakhodas'—notes in English that were not in the nakhoda's hand. Some added information: lines marking out sailing distances across the Gulf, for example, or a note on shoals near Ras Al-Hadd. Others involved corrections to the charts, and occasional notes on the corrections (one includes a note of caution, that the map was not corrected with the use of navigational aids). All these point to the past lives the charts had before entering the nakhoda's possession—the hands they passed through, and the different palimpsests of cartographic knowledge that others layered onto them. Even the Arabic annotations on the maps are in an inconsistent hand: perhaps a product of the moment of inscription, or maybe a clue pointing to yet another stop in the map's journey from the Hydrographic Office to Al-Failakawi's Malabar teak chest.

. . .

Al-Failakawi's maps had seen better days. Many were falling apart—and indeed, quite a few had already fallen apart at some point in the past, when the nakhoda decided to patch them together with scrap bits of paper: clippings from newspaper articles during the Second World War or strips of paper with his arithmetic calculations on them. But rather than lament the sorry state of the artifact, we might use these patches to think about these charts as tools that once had robust lives. At their inception, they might have been the tools of imperial expansion—of knowing, assessing, and conquering—that historians have long associated with maps. But they very quickly became material objects with life trajectories, passing through the hands of different historical actors: cartographers at the Hydrographic Office, naval officers, shopkeepers, and communities of nakhodas—all of whom left their imprints on them.

As artifacts that circulated between nakhodas, maps carried with them the many layers of cartographic knowledge that had sedimented onto them over the course of their lifetimes. When the maps changed hands, a sedimented knowledge of geographies, routes, and landmarks, of places and how to spot them, passed from one nakhoda to another. The circulation and recirculation of spatial knowledge between nakhodas conjured up a particular understanding of the ocean space, one that foregrounded the lived geographies that the dhows and their crews moved within, and that rendered them legible, and ultimately navigable. If Al-Failakawi's maps are worn and crumbling, it is because they carry the weight of history.

NINE

Muscat

AFTER TWO WEEKS OF SAILING on the open sea, the *Crooked* found itself opposite Ras Jibsh, a headland just west of Ras Al-Hadd—a little south of where most nakhodas, including Al-Failakawi himself, normally would have made landfall.[1] From there, Al-Failakawi and his crew crept along the coast of Oman; they lost sight of land, but only for a day, and when the winds picked back up on February 19, they anchored for a night across from the village of Abu Qumaylah. "We saw a dhow," Al-Failakawi wrote, "but it did not pass us, and we could not identify it." It was the first of many dhows that he'd encounter, many of which he knew very well.

For the next week, the *Crooked* crawled along, using its jib sail to tack against heavy winds bearing down from the northeast. One night, the winds were so strong that it was forced to anchor off the village of Al-Ashkhara, where other dhows had also sought safety. There, Al-Failakawi wrote, they "took supplies from the dhow of ʿIsa and others"—though what, he never says. Then, suddenly, the winds calmed. Freed from the oppressive headwinds, the *Crooked* could follow the coastal currents until it rounded Ras Al-Hadd. Al-Failakawi continued to hug the coast, sailing against the prevailing currents, propelled by occasional gusts of wind blowing from the south.

As the *Crooked* sailed along the Omani coast, Al-Failakawi would have noticed how it changed. Past Ras Al-Hadd, his view of the coast became increasingly dominated by the imposing Hajar Mountains, which, peaking at nearly ten thousand feet, formed one of the highest mountain ranges on the Arabian Peninsula. The highest peaks were far off in the distance, but dark, rocky outcrops along the coast thrust their gravely fingers out into the sea. The *Crooked* maneuvered around the rocky coast for a few days, and on March 3, it rounded a promontory and found itself in a still harbor. "We arrived at the port of Muscat," wrote Al-Failakawi. "Praise to Allah, the master of the universe."[2]

Nakhodas that stopped at Muscat rarely spent time there. Most preferred to anchor in Mutrah, the city's main harbor and principal commercial district, and land the longboats on the beach near the fish market. By the waterfront, Omanis, Persians, Indians, and Baloch maneuvered along the fish carcasses, dogs, goats, and human waste that peppered the shore. "Mutrah was the dirtiest place we were ever in, and the stench of the beach was frightful," complained Villiers; "the beach itself is used as a road, park, junk yard, market, fish shed, landing stage, dog house, and public lavatory by all Oman and half Baluchistan."[3]

Despite its foul stench, the shore buzzed with activity. Dhows unloaded sacks of rice, grains, sugar, tea, tobacco, charcoal, and building materials from India. As they hummed their chanties, sailors stacked soft mattresses and tins of ghee, fish oil, and kerosene from Persia, alongside piles of mangrove poles from East Africa that they heaped onto the beach, laying out with precision the dimensions of many of the houses in the Gulf.[4] Merchants and shopkeepers streamed into and out of the marketplace, shouting out to stevedores and carriers to carry the goods into the stores and warehouses that lined Mutrah's narrow alleyways.

But the coffeehouses were what attracted nakhodas the most—not for their wares, though few Kuwaiti nakhodas could resist the clove- and cardamon-flavored coffee and the famed sweetmeats of Oman. They came for the news: Villiers observed that "these coffee shops are like exchanges and clubs and most of the sea business is done in them."[5] There, they would drink tea and coffee and take in news of markets all around. The ifs, whens, and wheres of the Indian Ocean marketplace all passed from one nakhoda to another as tea glasses clinked against their saucers and patrons called loudly for boys to refresh the coal on their tobacco pipes. The coffeehouse acted as an information clearing house. A nakhoda on his way back from India, Yemen, or East Africa could stop in Mutrah and learn of where he might go next. He would also hear news from around the region, as fellow nakhodas traded stories of political, religious, and social developments in the places they had just visited. Still others would read aloud from Arabic newspapers carried in local stores, sent in from cities around the Middle East, South Asia, and East Africa.

From these coffeehouses, nakhodas and their crews tapped into the intellectual currents that ran through the Indian Ocean world. In Muscat, as in different port cities, they consumed news and editorials from around the

FIGURE 17. The harbor at Mutrah in 1939, with sailors and smaller dhows in the foreground. Photo from National Maritime Museum, Greenwich, UK.

region and the broader globe, alongside pamphlets, treatises, and all kinds of literature. At times, these could speak to biding interests of theirs—in poetry, ethics, or religious obligations. But in times of political change, like the imperial reconfigurations that characterized much of the nineteenth and twentieth centuries, the transregional intelligentsia that produced these writings pushed their readers to imagine different futures—of Muslim unity and reform or pan-Arab solidarity. Amid the geopolitical convulsions of the early twentieth century, writers and readers around the Indian Ocean forged the contours of a transoceanic public sphere.

• • •

Virtually every Kuwaiti nakhoda who spent time in Mutrah and Muscat knew Yousef bin Ahmad Al-Zawawi. He was among the leading merchants of that port city, and his family had been there for at least a century by the time of Al-Failakawi's visit; he also had cousins up and down the Gulf. Al-Zawawi's grandfather had migrated to Muscat from the Hejaz in the early 1800s, during the reign of Saʿid bin Sultan. The family, which included merchants and scholars alike, had stayed in Muscat through the worst of times.

The port city had suffered a long, torturous decline since its heyday in the time of Saʿid bin Sultan, when the riches of Zanzibar and East Africa buoyed its coffers. After Saʿid's death in 1856 and the partition of his dominions, his son Thuwaini, who was awarded the throne at Muscat, found himself confronted with a restive population of tribes and family members—including his own son Salim, who murdered him in February 1866. Two years later, Salim was overthrown by his brother-in-law and distant relative, ʿAzzan bin Qais, who sought to reestablish the Ibadi imamate in Muscat. ʿAzzan himself was killed just three years later—this time by another one of Saʿid's sons, Turki, who led a military campaign to recapture the coast, fueled by the financial support of his brothers in Zanzibar. By the time he managed to establish himself as sultan in Muscat in 1871, more than five years of damage had ravaged the port city.[6]

After the dust settled in Muscat's political arena, new opportunities emerged for merchants who stayed. By the last quarter of the nineteenth century, the establishment of a regular steamship service between India and the Gulf ensured, among other things, that there would be a regular flow of goods, money, bills, and paperwork between Bombay, Karachi, and Muscat. Though the regularization of steamship traffic cut into the dhows' share in trunk routes, nakhodas still managed to carve out a role for themselves in the new geographies of regional trade. Merchants like Al-Zawawi benefited from the steamer service, too, and used it to position themselves advantageously in the circuits of money and goods that ran between India and the Gulf. He managed to embed himself in Muscat's political community as well and was a close friend and adviser to Sultan Faisal bin Turki. When the sultan received an invitation to celebrate the coronation of King Edward VII in 1903, he sent his son and future successor, Taimur, in his place, and asked Al-Zawawi to accompany him. His son ʿAbdul-Munʿim joined Taimur's entourage years later, when the young sultan went on a world tour in 1937.[7]

Al-Zawawi also acted as Mubarak Al-Sabah's agent in Muscat and kept an extensive correspondence with his counterpart Muhammad Al-Sudairawi in Bombay. None of their exchanges betray any of his political activities, though. Like so many of the Bombay banker's correspondents, Al-Zawawi's missives revolved principally around his financial activities. In letter after letter, Al-Zawawi sent Al-Sudairawi *hawalas* to add to different accounts. The bulk involved members of Oman's Gujarati merchant community, who looked to Al-Zawawi to manage payments to the Bombay merchants they did business with; he'd even forward *hawalas* from Gujaratis in Zanzibar to Bombay for fulfillment.[8] He sent Al-Sudairawi *hawalas* from merchants in Kuwait and Bahrain and asked him to credit his account with the cost of the goods they purchased from him. It was all business, though there were rare moments in which he broke genre conventions to convey more personal messages: news of the death of his son Salem in January 1916, and then of his father just a month later.[9]

But Al-Zawawi's relationship with the Bombay Arab community was perhaps more multidimensional than his letters suggest. Through Al-Sudairawi and other contacts in Bombay, Al-Zawawi became embedded in a transregional network of thinkers and writers. Bombay had long established itself regional center for print culture: by the early nineteenth century, the city was already producing books, treatises, and tracts in English, Persian, Urdu, and Arabic.[10] When nakhodas, merchants, and writers from the Gulf wanted to publish their writings in the early twentieth century, they took them to Bombay. There, Al-Sudairawi often purchased printed materials on behalf of buyers. In addition to regularly sending newspapers to Kuwait, in 1912 he wrote to the publishers of *Al-Manar*, the popular Cairo-based journal, and ordered two sets of poetry: one by the Persian poet Hafiz, and the other a compendium of poems by the Arab poet Ahmad Shawqi. A year later, he ordered two maps, one of Asia and the other of the world. People in Kuwait looked to Al-Sudairawi to send them books from Bombay too. In 1913, Shaikh Mubarak's son Nasser wrote to him asking him to purchase the novel *Al-Naḍrāt* by the Egyptian author Mustafa Lutfi Al-Manfaluti, along with an eight-volume set of Pierre Alexis Ponson du Terrail's *Rocambole* series.[11] Drawing on different texts and writers that moved through Bombay, Al-Sudairawi was able to use his network—colleagues like Al-Zawawi—to channel circuits of literature and political thought in the Arab world writ large into the Gulf, where they found receptive audiences.

Among the readers in the Gulf was Al-Failakawi himself, who was an avid consumer of books. Over the course of his travels, he collected dozens of books. The majority of these were religious in nature: books on the correct performance of prayers, the hajj, and different rituals; books on Muslim ethics; and a book of scenes from Heaven and Hell. A substantial part of his collection, though, was literary: stories of the early Muslim community, collections of parables from the Muslim literary tradition, books of poetry, and a multivolume history of Islam. Though some of these books came from later periods in his life, they suggest a picture of Al-Failakawi as a reader who took advantage of a far-flung market in printed and lithographed texts to nurture his intellectual curiosities and sustain his religious and ethical proclivities.

Markets were only one source of books, though; some came as gifts, too, through relationships between writers and readers. The writer ʿAbdulmasih Al-Antaki, who edited the journal *Al-ʿUmrān*, met Al-Zawawi in Cairo; there, the Muscat merchant gifted him a manuscript copy of Ibn Maʿsum Al-Medini's seventeenth-century travelogue, which detailed, in both poetry and prose, the author's travels from the Hejaz to India.[12] Antaki, for his part, did the same: writing to Al-Sudairawi in 1911, Al-Antaki referred to a shipment of books he had sent to Bombay that Al-Sudairawi had attempted to pay for—apparently many times. "I have repeatedly told you that it was a small gift to you," Al-Antaki wrote; "I am thus unable to assign it a value, and a gift cannot be returned." In any case, he wrote, "your graciousness is greater."[13]

Together with Al-Antaki, Al-Sudairawi, and others, Al-Zawawi formed part of an interregional intelligentsia—one that sustained intellectual centers throughout the Middle East and South Asia and oriented them toward new audiences and supporters. He established himself as the main point of contact for the Salafi scholar Rashid Rida during the latter's trip to India. Al-Zawawi was not in Bombay for Rida's visit; that honor fell to the Bombay-based Arabs. However, when Rida departed from Bombay, he stopped in Muscat after a brief stay in Karachi; there, he was received by Al-Sudairawi, who was visiting the city, and Al-Zawawi, who put up the scholar in a room "with modern furnishings." He also invited Rida to give a sermon in a mosque the family had endowed—"one of the cleanest and most magnificent mosques in the city," Rida wrote, complete with "water supplied through lead pipes." The scholar delivered a speech that moved his audience to tears.[14]

Al-Zawawi's activeness in intellectual and political circles brought him to the attention of British officials in the region. A British listing of individuals active in politics around the region in 1919 describes Al-Zawawi as "a strong Muhammadan [whose] family has great influence in Musqat and is probably responsible for the undercurrent of pan-Islamic feeling which exists there." British officials also suspected the Al-Zawawis of harboring pro-Ottoman sympathies and fomenting resentment toward the Sharif of Mecca, most likely because he did not think him powerful enough to lead an independent Muslim state after the Ottoman Empire.[15]

Whatever his feelings about the Sharif may have been, what is clear is that Al-Zawawi actively thought about the changing map of the Islamic world, especially in the context of the political upheavals of the early twentieth century. His engagements with Rashid Rida, Al-Antaki, and Al-Sudairawi were interwoven into a circuit of printed materials—books and newspapers—that animated new intellectual networks that emerged in the early twentieth century. From Cairo to Delhi and beyond, Muslims grappled with the question of how to address challenges posed by political reconfigurations that were taking place amid the First World War—a matter that took on greater urgency after the abolition of the Ottoman sultanate and dismantling of the caliphate.[16]

Seen from this perspective, Al-Zawawi's treatment of Rida and friendship with Al-Antaki was intentional. These were figures whose writings commanded the attention of a broad audience of readers around the Arab and Islamic world. In Al-Antaki, the rulers of Kuwait and Mohammerah, Shaikh Mubarak and Shaikh Khazʿal, saw a figure whose writings could bolster their own political claims in a changing world. In 1907, he wrote a tract in praise of Mubarak; a year later, he did the same for Khazʿal and then published an account of his travels through Kuwait and Mohammerah that brimmed with praise for the two rulers. In 1911, Al-Antaki also ghostwrote a tract on political ethics on behalf of Shaikh Khazʿal, to which he gave the obsequious title *Al-Riyāḍ Al-Khazʿaliyyah fī Al-Siyāsa Al-Insāniyya*, or *The Khazʿalite Gardens in Humanitarian Politics*.[17] Rulers who could align the outlooks of influential literary figures with their own, and who harnessed the power of print, could stake a plausible claim to the region's political future—and people like Al-Zawawi and Al-Sudairawi could help sustain those relationships.

If Al-Zawawi's role in channeling currents of pan-Islamic thought into the Arabian Peninsula touched a nerve among British officials in the region, it is because they had long associated Muslim movements with anticolonial sentiments. From at least the early nineteenth century on, officials in British India suspected Muslims of drumming up resistance to British rule in the subcontinent. Indian Muslim pilgrims to Mecca in the 1810s found the city to be a hub of emerging anticolonialist thought, and their movement back and forth between Arabia and India generated concerns that they might spread their subversive ideologies. The Indian Rebellion of 1857 itself was thought to be animated by anti-British ideologies expressed among Muslim scholars. And in the decades that followed it, British officialdom found itself gripped by a fear of "Wahhabi fanaticism" that could fan the flames of anticolonialism. By Al-Zawawi's time, these manifested most urgently in concerns surrounding the Khilafat movement, which opposed British involvement in the dismemberment of the Ottoman Empire and the removal of the Ottoman sultan in the wake of the First World War.[18]

None of Al-Zawawi's antipathies seemed to be directed toward the British themselves, though; the entry notes that the Gulf Resident, Sir Percy Cox, "has always found him [Al-Zawawi] useful in political matters."[19] Around the Indian Ocean, Muslims like him found different ways to engage with European empires, not all of which were, strictly speaking, anticolonial. Drawing on printing presses, steamships, telegraphs, and other technologies of empire, scholars forged intellectual currents that thought through new forms of Muslim sovereignty and belonging in the age of European colonialism and a time of political reconfiguration around the Islamic world. Pan-Islamic sentiment did not necessarily incline Muslims toward anticolonialism.[20] If anything, Al-Zawawi's connections to imperial centers like Bombay and Calcutta only reinforced his position within a transregional Arab community that stretched between the Mediterranean and Indian Ocean. They plugged him into a broader circuit, one in which books, scholars, and ideas moving between India and the Mediterranean through channels of steam and print became much more readily accessible to him, and to a broader community of readers.

• • •

Late February wasn't an especially busy time in Muscat. Dhows coming back from India would have only just begun to make their way back; they usually reached Oman sometime in March. Others would have been catching up to the Zanzibar-bound fleet, which would have already made it to the Swahili coast by February. At least some would have begun loading mangrove poles from the Rufiji delta, just south of Zanzibar. The Rufiji mangrove pole trade was a delicately timed business: a nakhoda coming to East Africa had to be sure to give himself enough time to load his dhow and sail back before the summer pearl dive began. He also needed to give himself time to sell the poles to buyers in different Gulf ports, where they were in demand as roofing material. Zanzibar-bound nakhodas who got a late start—and if they were still in Muscat in February, they were very late—would have to content themselves with poorer-quality mangrove poles they bought from sellers in closer ports like Lamu.

At least some of the mariners Al-Failakawi spent time with in Muscat had seen the Rufiji delta trade firsthand and would have spent their time in the coffeehouse telling stories of the Swahili coast to a rapt audience of local patrons, tea boys, and other nakhodas. The mangrove pole business formed the latest frontier in a Gulf-East African dhow trade that had once brimmed with a wide range of commodities. With the establishment of British protectorates in Zanzibar and the East African coast during the 1890s, alongside a German colony in Tanganyika and an Italian one in Somalia, came infrastructural projects aimed at remaking the coastal economy and entrenching European claims to the region. By the turn of the century, the building of the Uganda Railway, which connected Mombasa to towns deep into the interior, was well underway. The port facilities in Mombasa and Zanzibar also changed to accommodate the new modes of transportation: officials built modern harbors for the steamship traffic, while dhows were shunted off to older wharves. To anyone visiting, the message was clear: East Africa's commercial future lay in rail and steam, not sail.[21]

Within this burgeoning colonial economy, nakhodas carved out an opportunity in the mangrove pole trade. It was a different kind of business for them. The trade in dates, timber, coffee, and other commodities between the Gulf and India was firmly in the hands of merchants. Nakhodas only made money through the freightage fees they charged to ferry different goods around, and whatever else they managed to trade on their own accounts. The mangrove pole trade, however, was theirs alone: whatever they earned selling them to merchants and home builders went right to the captain and crew's accounts.

There were a few different routes that nakhodas took to East Africa. Most nakhodas sailing down the Gulf from Muscat would round Ras Al-Hadd and sail westward along the South Arabian coast until they reached Mukalla. From there, they would cross the Gulf of Aden and make landfall at Ras Hafun, the easternmost point in Africa. When Al-Failakawi sailed to East Africa in 1943, though, he crossed from the Malabar coast, where he had been for nearly a month. In Mangalore, he loaded a cargo of terracotta roofing tiles, which were in high demand among home builders in East Africa, and sailed out into the Arabian Sea.[22] He and his crew spent the next two and half weeks crossing the open water, making landfall on the coast of Somalia. For the next nine days, they sailed southward, reaching Mombasa on February 10.[23]

Al-Failakawi never made it as far as Zanzibar, at least not in 1943. While he was sailing toward Mombasa, his ship struck one of the many coral reefs along the coast, badly damaging his rudder's head. His crew members tried to repair the damage themselves, but they were unable to. Instead of sailing down to the Rufiji delta, Al-Failakawi spent two months in Mombasa, waiting for repairs on his dhow to finish. He would have also had to deal with a burgeoning bureaucracy. In the wake of the *Muscat Dhows* ruling in 1905, officials in East

Africa wove a dense web of red tape: fees that nakhodas had to pay, rules surrounding sick or injured seamen, and a whole battery of regulations on ship licenses, manifests, and other bureaucratic minutia.[24] Meanwhile, he found buyers for the cargo of tiles that he and his crew had purchased in Mangalore, and may have even found a companion to spend his nights with.[25]

Other dhows would have kept going to Zanzibar, and from there to the Rufiji delta. In 1940, Hussain bin ʿAbdulrahman Al-ʿAsʿousi arrived in East Africa following very much the same route that Al-Failakawi did. He had spent the first half of December 1939 in Calicut before sailing up to Mangalore. On December 24, he sailed out with two other dhows, one Kuwaiti and another from the Iranian port of Kung, and spent the next two and a half weeks crossing the Arabian Sea, making landfall on the Somali coast on January 9. He and his companions reached Mombasa a week later and spent nearly a month there before sailing to Zanzibar on February 18. Two days later, he sailed down to the Rufiji delta.[26]

The journey to the delta was short. The northeasterly monsoon winds combined with the southward-moving current meant it only took a day and a half for Al-ʿAsʿousi to sail from Zanzibar to the customs house at Kwale, a low island by the northern end of the delta. There, "a Swahili customs clerk enters the ship inwards, collects their dues, and watches that they do not smuggle too much."[27] Dhows' visits to the Rufiji delta were mediated through many layers of colonial administration, each of which, like this one, peeled apart from the other. In the 1870s, the sultan of Zanzibar built a station in the Rufiji, levying a 10 percent toll on poles being exported. A decade later, he leased the entire delta to the German East Africa Company as a concession, which it turn leased much of the it to private firms, each of which held individual rights to stretches of the forest.[28]

When the British government took over Tanganyika from the Germans after the First World War, it continued the practice of leasing it as a concession to one private company after another, roughly every five years—not, as in Malabar, as a way of balancing exports with conservation, but as a means to offload the cost of making the delta export-ready. The concessionaire was supposed to invest in infrastructure in the delta and streamline the process of cutting and marketing poles. In practice, however, they seldom did that.[29] Between a colonial government eager to pass off forest administration to a private enterprise and concessionaires who were uninterested in doing more than marking up the price of mangrove poles, there emerged a visible gap through which nakhodas happily sailed their ships.

Al-ʿAsʿousi reached the mouth of the Kiasi Creek, at the southernmost edges of the delta region, on February 25, 1940, anchoring just off the town of Dima. At Dima, Al-ʿAsʿousi likely met with a labor recruiter and began the process of hiring the crew of pole cutters. The work of cutting poles was difficult: one had to wrestle eighteen-foot poles out of the tidal mud and hack away at them. To do this work, they relied on communities of seasonal pole cutters that settled in small villages around the delta. From Dima, Al-ʿAsʿousi assembled the crew of cutters that he'd be working with for the next two months, in the inner reaches of the Kiasi creek. As they worked to supply the dhow with poles, the cutters lived and ate on board the dhow, alongside the mariners. Villiers, who was in the Rufiji delta just the year before, wrote that the dhow he sailed with took on eighty Swahili men and children during their time in the area.[30] Swahili labor, and local knowledge of a challenging coastal ecology, made possible the export trade—not the half-hearted attempts of concessionaires.

Al-ʿAsʿousi's *ruznamah* tells us little about the month he spent in the delta, loading poles; he only writes that on March 2, roughly a week after reaching Dima, they "began loading."[31] His contemporary, the nakhoda Mufleh Al-Falah, gives a better sense of the process. He had spent the season at "Khor Boniyan," most likely Kiomboni Creek. At different times, Al-Falah recorded payments made to a *sipahi*, the forest guard whose job it was to ensure that all the poles being loaded were accounted for. The guard received a payment of 5 shillings (presumably a fee) but also gifts of *halwa*—sweetmeats that Al-Falah had picked up in Zanzibar. He also records having gifted the *chimamizi* (the labor headman) two measures of cloth. Over the course of three weeks, Al-Falah records having officially loaded 208 *korjas*, or score, of poles, writing that "we imprinted them *yondo*," marking them with the official axe. Those were only what he recorded as the marked, official cargoes. Villiers described how the nakhoda he sailed with "loaded, and paid for, 150 score poles . . . yet we had loaded another 150 score before we left that place." The Swahili crews would hide entire stores of cut poles and secretly transport them to the dhows at night. Al-Falah's gifts to the guard may have been to encourage him to look the other way; his gifts to the headman were likely a way of ingratiating himself with the entrepreneurs who operated in the shadows of the forest bureaucracy. In Villiers's words, "it was like a racket in America, as well run and as unscrupulous."[32]

His dhow loaded to the brim with poles that were most likely only half paid for, Al-ʿAsʿousi sailed out of Dima by May 5, in the thick of the rainy season; his *ruznamah* entries recall again and again that "the *suhayli* was full, and it rained." And when it was his turn to leave the delta, Al-Falah noted that he sailed out with a dhow belonging to "Bin Gurnah": Salim Bin Gurnah, one of Zanzibar's dhow agents. Together with other Zanzibaris of Arab extraction—often Hadramis—Bin Gurnah formed a community of brokers who would receive the nakhodas and their dhows, arrange for a place for them to stay in Zanzibar, and help them secure orders of return cargoes. The dhow agents were also tasked with ensuring the maintenance of public order among dhows visiting Zanzibar, as part of a growing bureaucracy aimed at regulating dhow movement in East Africa.[33] For nakhodas making their way down the East African coast in the time of European colonialism, movement was punctuated with an increasingly invasive set of rules and forms: customs documents, permits, quarantine regulations, and more. And as they would learn, things were no better for those living in Zanzibar.

• • •

"There was a war and clash [*al-ḥarb wal-ḍarb*] between the Christian British and Omani Arabs," wrote Mansur Al-Khariji into his notebook. On February 7, 1936, "five Arabs were killed, and six or seven more were injured; and on the Christian side two were killed, [one of whom was] an inspector named Qamar Al-Din, a Punjabi; and a slave; and one askari; and Shihiri coffee seller." The cause, he wrote, was "the copra coconut." "This," he wrote, "was what we witnessed with our own eyes."[34]

The clash Al-Khariji witnessed was sparked by the newly introduced Adulteration of Produce Decree, which sought to regulate the cash crops brought to market. The decree was part of a battery of regulations introduced by the British government aimed at rehabilitating the market for Zanzibari produce, whose prices had cratered during the global depression. However, it ended up targeting recently arrived Arabs from Oman (called the "Manga" Arabs), many of whom were small-scale entrepreneurs who made a living by buying copra from planters and selling it in the market. The decree had only been applied to copra the week before the riots, and Manga Arab copra sellers found themselves at the mercy of the inspecting staff, many of whom were non-Arab Zanzibaris and mainlanders. On February 7, forty or fifty of the copra dealers armed themselves and made their way to Malindi, where incoming dhows docked, and clashed with the Zanzibar police.[35] Al-Khariji was likely in the area, with his dhow crew, as the riot unfolded.

But copra was only the most proximate cause. The tensions that led to the clash had a much longer history to them. Over the nineteenth century, Arabs from Oman and elsewhere in the Arabian Peninsula began migrating to Zanzibar and elsewhere on the East African coast. Some engaged in the slave trade, but many opted for a more comfortable life as plantation owners, drawing on enslaved labor to clear the land, plant clove and coconut trees, and harvest the produce. To finance their production, they would borrow from Zanzibar's Indian merchant class, putting up their plantations as security and, often, handing over the annual harvest of cloves and coconut.[36] It was in this way, and through their close relationships with Zanzibar's Busaʿidi rulers, that Arabs came to form something of a landholding aristocracy on the island, displacing the indigenous elite and establishing their place in a racialized hierarchy.

It is not insignificant that Al-Khariji described the clash's casualties by nationality. By the twentieth century, a general climate of racial animus had taken hold in Zanzibar. The socioeconomic stratification that had characterized much of coastal East Africa during the nineteenth century—the ways in which politics, land ownership, access to capital, and labor broadly segmented along ethnic lines—made for a tumultuous public sphere in the twentieth century. Racial tensions were only exacerbated by stagnating exports in the early decades of twentieth century and the onset of the global depression in the 1930s, which dealt an enormous blow to the wealth of much of the landed (and principally Arab) class. To avoid complete foreclosure on their plantations, many landowners had to sell off parcels of their estate, often to upwardly mobile members of other ethnic groups. The material foundations for many of the sociopolitical distinctions that had animated Zanzibar in the nineteenth century eroded under the economic pressures of the twentieth century, creating what one historian called a "racial state."[37]

The changing ethnopolitical calculus of Zanzibar was only made more volatile with the arrival of new immigrants from the Arabian Peninsula, many of whom came by dhow. When Villiers sailed on a dhow to East Africa, he described the different passengers they picked up from Oman and Yemen to take to Mombasa and Zanzibar; at least a few had no means to sustain themselves when they arrived and resorted to begging on the streets.[38] But begging was the least concerning activity. Immigrants from Arabia—the Wamanga (Omani Arabs) and Washihiri (Hadhramis)—were associated with all sorts of unsavory activities: theft, thuggery, and kidnapping. That they came from dhows and frequently dealt with mariners, many of whom had long been associated with the slave trade, only added to the air of suspicion surrounding them.

It was amid these tensions that members of the Zanzibari intelligentsia took to the press. They had long consumed newspapers published elsewhere in the Arab world: Zanzibar's sultans even offered financial support to journals like *Al-Hilāl*, which was published out of Cairo; Zanzibaris even contributed articles to it. In 1911, they began publishing their own newsletter, the thrice-monthly *Al-Najāḥ*, which formed the mouthpiece of the Al-Iṣlāḥ party, the political hub to a Muslim reformist movement in Zanzibar that sought to align the island more closely with pan-Arabist and pan-Islamist currents around the Arabian Peninsula and broader Middle East. By the late 1920s, Zanzibar could claim a healthy public sphere, boasting at least two regularly published newspapers—a number that would only grow over the next two decades, as Zanzibaris became even more politically active and sought to express themselves in writing.[39] Among the more active of these was the Arabic-language *Al-Falaq*, established in 1928, which the Arab Association published once a week. The newspaper included sections in both English and Arabic, publishing reportage on events both local and around the world, opinion pieces, advertisements, obituaries, literary pieces, and the occasional poem. During its twenty-five-year run, it was, for the Arab community of Zanzibar, the newspaper of record.[40]

Al-Falaq paid keen attention to political developments taking place around the Arab world, from Morocco to the Gulf. Virtually every issue included political news from Egypt, North Africa, Iraq, and beyond. Its editors seemed particularly concerned with ongoing developments in Palestine—the arrival of Jewish immigrants, their violent attacks on Palestinian villagers, and the implications of a partition plan more generally.[41] And at times, they reprinted articles published in newspapers in Cairo and elsewhere, merging the pages of *Al-Falaq* into a broader circuit of political news in the Arab world.[42] The newspaper also tracked events in Muscat and Oman especially closely. Together with news from Europe and the broader Middle East, editors of *Al-Falaq* included reports from the Arabian Peninsula that would have been of interest only to the island's Arab community. The newspaper reported on mutual visits among notables between Oman and Zanzibar, of marriages in Oman, and of the travels of the sultan of Muscat within his dominions. Earlier newspapers in Zanzibar like *Al-Najāḥ* did the same.[43]

Alongside news articles, *Al-Falaq*'s editors printed opinion pieces that expressed islanders' concerns about the decline of Arab identity in Zanzibar.

Column after column lamented the Arab elite's loss of their linguistic connection to their ancestral homelands, and voiced a desire to send Zanzibari students abroad to places like Iraq, Egypt, and Syria, where they might better acquaint themselves with the Arabic language and Arabic culture.[44] They often called for solidarity with their fellow Arabs in Palestine, Morocco, and Egypt, and at times explicitly invoked their shared language, blood, and culture—a shared "national consciousness" that they claimed was present everywhere in the Arab world.[45] Much of this was further couched in the language of Muslim ethics, though the editors' Arab nationalist sentiments seemed to absorb their pan-Islamic sympathies.[46] Through the pages of the newspaper, *Al-Falaq*'s editors situated the Arab community of Zanzibar within a broader arena of reenergized pan-Arab and pan-Islamic sentiments.

Al-ʿAsʿousi saw many of these themes echoed in the issues of *Al-Falaq* published during his time in East Africa. The newspaper's pages brimmed with news of developments in the European theater of the Second World War, which readers followed with enormous interest, together with war-related events in East Africa. At the same time, the newspaper conveyed news from the Arabian Peninsula, discussing attempts by the sultan of Muscat to modernize his army, and covering agreements between Ibn Saud and the Kingdom of Iraq. Al-ʿAsʿousi would have learned, too, of new agreements between the Saudis and the ruler of Kuwait regarding the freedom of trade between the two emirates and of skirmishes between different groups in Dubai.[47]

Al-ʿAsʿousi would have also seen echoes of discussions he heard in the years after the First World War, when Muslims around the Indian Ocean grappled with the loss of the caliphate and attempted to conceive of new forms of connection with one another. On and off the pages of *Al-Falaq*, Arabs and Muslims wondered aloud what the implications of the Second World War would be for them. On the front pages of one issue of *Al-Falaq*, dated May 11—when Al-ʿAsʿousi's dhow was still docked in Zanzibar, after his travels to Rufiji—the editors discussed a statement made by the Lebanese mufti Muhammad Tawfiq in the Indian press, in which he advocated for Arab and Muslim cooperation with the Allies, who in his view had articulated the principle of support for weaker states (and he saw Arabs as falling within that category). By contrast, the German pursuit of *Lebensraum* threatened the strong and weak alike. "What do we gain from an Allied victory as a nation?" asked the editors of *Al-Falaq*. For all their differences, all nations wanted the same thing—self-determination. The goal was not contingent on their support for the Allies: "The Indian leader [Mahatma] Ghandi did not flatter his government, respected Mufti," they wrote, "and he did not wait until after the war to press his demands."[48] One wonders how the nakhoda might have interpreted that message as he scanned the paper for news from the Gulf and Arabian Peninsula.

But *Al-Falaq*'s heightened sense of Arabness has to be read within the very local context of the island's politics. The Arab Association that published the newspaper emerged out of a desire to protect the interests of the island's Arab landowning class, whose increasing indebtedness to Indian financiers over the course of the nineteenth century placed them in a vulnerable position during the economic shocks of the 1930s. Adding to the tensions were legislative enactments by British officials in Zanzibar that made distinctions between actors based on ethnicity, and which targeted specific ethnic groups.[49] Much of *Al-Falaq*'s commentary was on these local dynamics—on the plight of the Arab planter vis-à-vis his Indian financier, and on the need for Arabs to diversify their economic activities so as to better withstand these shocks. *Al-Falaq*'s main work, the paper makes clear, was to advocate on behalf of the island's Arab community, within a colonial regulatory framework that placed them in competition with other communities for legislative concessions. British authorities in Zanzibar recognized the Arab Association, which published *Al-Falaq*, as the only legitimate representative of Arab landowners—one that existed alongside other similar organizations like the Indian National Association and the African Association, each of which also published their own newspapers.[50] The bulk of the editorials published in *Al-Falaq* and other periodicals coming out of Zanzibar spoke to these local concerns, within a climate of increasingly racialized regulations on economic and political life.

The notions of Arabism that newspapers like *Al-Falaq* helped platform were hardly new, of course. There had long been a latent discourse of Arab superiority throughout the Indian Ocean world, sometimes mediated through the language of Islam and at other times through more exclusivist idioms.[51] These were activated in moments of tension, like the ones that gripped Zanzibar in the early twentieth century. The racialized state that British regulations on the island helped bring into being accentuated Zanzibari Arabs' notions of difference, and Zanzibaris' access to news and editorials from around the Arab world amplified their sense that they were, above all, Arabs. But rather than see this as an inherent feature of the community, one that was at odds with trends in other parts of the Indian Ocean, we might read it as a time- and place-specific expression—a cosmopolitan political identity that was nonetheless rooted in very local anxieties that emerged out of a historically specific moment in East Africa's colonial history.

The forms of political consciousness that publications like *Al-Falaq* gave rise to were thus simultaneously global and local. The newspaper's contributors couched their local concerns—legislation, regulation, competition with other ethnic groups—within coverage of and engagement with a much broader Arab world. They drew a transregional Arab public into their understanding of local concerns and framed their local struggles within a much broader context of late colonialism. They were neither Arab cosmopolites engaged in a global anticolonial struggle nor resolutely local figures who could not see beyond their own immediate struggles. Through their engagement with the broader Arab world on the pages of their newspapers, they developed a political consciousness that collapsed the distinction between home and homeland, between the local and the global.

Al-Falaq formed just one node in an increasingly transregional public sphere that nakhodas like Al-ʿAsʿousi, Al-Failakawi, and Al-Khariji traveled through. Issues of *Al-Falaq* reached readers in Muscat, who consumed with interest its coverage of local and regional politics, and who fed it with news items. There, it intersected with publications from around the Indian Ocean: *Al-Kuwait*, which began publication in 1928; *Al-Bahrain*, established in 1939; *Al-ʿArab*, published out of Bombay in the 1940s; and many others, all of which made their way to Gulf readers. One of *Al-Falaq*'s editors, Hashil Al-Maskari, was highly peripatetic, traveling back and forth between Oman and Zanzibar. In Bombay, Al-Sudairawi subscribed to several different newspapers and journals in Beirut and Cairo, which he also forwarded to readers in the Gulf. His letters include correspondence with the editors and general managers of *Al-Iqbāl*, a weekly newspaper out of Beirut; *Al-Hilal* in Cairo; the Egyptian newspaper *Al-Muqaṭṭam*, to which he also bought a subscription for a friend in Kuwait; and *Al-Muʾayyad*, a newspaper that he subscribed to on behalf of several different Kuwaiti readers.[52] In the age of steam and sail, there was no keeping things local; news writing and editorials spread as quickly as people could carry them.

In Arabia, East Africa, and South and Southeast Asia, the proliferation of print media gave voice to an emboldened community of thinkers, writers, and political actors, all of whom addressed one another and syndicated one another's writings, speaking in a multitude of voices about issues that were of common interest to them all—namely, what shape their world might take amid the imperial convulsions that characterized the wake of the First and Second World Wars.[53] When Al-Khariji wrote, then, that he witnessed the clash between Omanis and the British "with his own eyes," his were among many thousands of eyes that saw the incident, telescoped through the writings of Omani Arabs in *Al-Falaq* and other papers, read out aloud on the decks of dhows and in coffeehouses as far away as Muscat, Kuwait, and Basra, and framed within a much broader moment in the formation of a transregional political consciousness.

• • •

If the political consciousness of nakhodas like Al-Failakawi partly took form in small spaces like coffeehouses, it was only possible because of the wide and far-flung cast of characters that helped shape it. Around the Indian Ocean, scholars, journalists, poets, and schoolteachers all conjured up a transoceanic public sphere that brimmed with books, pamphlets, newspapers, and other publications, all of which made their way around the region by steam and sail. Nakhodas contributed to it as well, seizing upon print technologies to publish manuals and almanacs that found circulation among fellow captains and mariners. Though they did not share a common agenda, they all began to imagine themselves as addressing a broader audience, within a moment in which it seemed that new futures—political, social, economic, and literary—were possible. European officials took notice of these currents because they knew that the colonial project was always a precarious one. And though they thought about censoring, banning, and otherwise containing the spread of critical voices into places under their control, there was no stymying the flow of print into the port cities of the Gulf.[54]

After nine days in Muscat, it was time to continue onward. With a printed copy of a navigational manual in hand, Al-Failakawi and the crew of the *Crooked* sailed out from Muscat on May 13, crossing over to the Persian coast, where they caught sight of Elijah's Mountain (*Al-Yāsh*). For the next ten days, they very slowly maneuvered their way through the Straits of Hormuz, anchoring overnight at different small ports and islands just off the coast of Persia. They pushed on until the end of the month, making slow progress: the wind was light, but the westward-running currents propelled the dhow along. Dhows bound for Kuwait kept sailing along the Persian coast, ringing the top of the Gulf before making it home for the summer. The end of the season's voyage was within grasp.

But not for the *Crooked*—at least not this year. A day after sailing past Qeshm Island, Al-Failakawi ordered his helmsman to turn the dhow seaward. "In pursuit of our livelihoods, we crossed to Bahrain," wrote the nakhoda. "The currents are running from the southeast [toward the northwest], and the winds are blowing from the south. Oh Allah, ease things upon us, and do not make them difficult, oh gracious one, oh Prince of Messengers; Amen."[55]

INSCRIPTION

Poems

If you desire to cross the ocean by ship // And how to reach the lands [*al-diyār*]
And knowledge of the stars and what follows // and going and returning is the decision [*al-qarār*]

Then return to the *jāh*, that *qutb* calls // and the two *farqads* reply to the *silbār*
And the *na'sh* responds, leave the *suhayl* // opposite him, as you calculate the course [*al-majār*][1]

So begin the opening lines to the first poem—an ode to the compass rose—that Mansur Al-Khariji inscribed into his notebook, onto its very first page. In a notebook full of navigational principles, eyewitness accounts, contractual templates, calculations, and marketplace arithmetic, he committed the first six pages to poetry. Indeed, there is no overstating the popularity of poetry among the nakhodas of the Gulf. Together, Al-Failakawi and Al-Khariji's notebooks both include more than a dozen pages crammed with verse. Though it was not their principal form of expression—they all wrote mainly in prose—the amount of space they devoted to poetry makes it clear that it enjoyed enormous aesthetic appeal among them. As nakhodas tapped into circuits of news and books that made their way around the Indian Ocean world, so, too, did they consume a world of expressive verse.

Scholars know all too well that poetry was an established medium of expression for nakhodas and navigators. Ahmad Ibn Majid and his student Sulaiman Al-Mahri both made extensive use of verse in their writings to map out the cosmos, charting the movement of stars and planets in the sky, the phases of the moon, and the transitions between months and seasons. They used verse to map out their routes around the Persian Gulf, Red Sea, and Indian Ocean too. These, many have pointed out, served a distinct pedagogical purpose: the verse form, with its meter and rhyme, furnished a convenient mnemonic device for conveying knowledge—much like Al-Khariji's poem above, on the points on a compass rose.[2]

But the vast majority of the poetry that Al-Failakawi and Al-Khariji inscribed into their notebooks was neither navigational nor didactic. It belonged to several different genres and expressed many different motifs, defying any attempt to impose a single category onto them. Whether devotional or didactic, read or recited, the nakhodas' collected poems point to a world of circulating verse—of aesthetic expression and performance on the move.

• • •

The poetry that Al-Khariji and Al-Failakawi wrote in their notebooks was, in a word, eclectic. Though much of it spoke to religious devotion and piety, and love and longing, that is a broad tent, as far as poetry is concerned. It's hard to tell why the nakhodas chose to write these poems down, but they clearly saw in them some aesthetic value: they appreciated the beauty of the verse itself, and the themes the poems spoke to seemed to resonate with them. Much of their interest appeared to be in collecting and anthologizing what appealed to them from what they heard or read over the course of their voyages. In inscribing the poetry, they captured a literary form in circulation.

Al-Khariji made a point of identifying the sources of the verses he chose to copy into his notebook, sometimes with very specific details. At least a handful of the poems Al-Khariji included came from someone who traveled on his dhow: "We heard these from Jassim bin ʿIsa bin Nasrallah, who sailed with us from Kuwait to Karachi on the *boom Fateḥ Al-Malik* [belonging to] Shaheen Al-Ghanim," he wrote; on the second and third poems, he noted that the voyage took place in 1360 AH—between 1940 and 1941. Nasrallah was a peripatetic figure: born in Bahrain, he worked up and down East Arabia during the 1920s and 1930s before setting sail with Al-Khariji. He stayed in India for more than seven years, where he learned Urdu and published in Arabic-language journals that came out of Bombay; he later became a celebrated literary figure in Kuwait.[3]

One of the Nasrallah poems that Al-Khariji copied down appears to have been composed during the voyage itself. Over thirty verses, Nasrallah gave thanks to Allah and the Prophet, then went on to describe the voyage and praise its captain, Al-Khariji. He described his arrival in Karachi, the inspection of his travel papers, and his feeling of trepidation at arriving in a foreign land—followed by his relief at Al-Khariji's reassurances that he would help the inexperienced traveler. The poem drew from a genre that conveyed the experience of travel, usually through mapping out routes and voyages.[4] But Nasrallah also tapped into the conventions of *madīḥ* poetry, a praise genre of poetry in which the writer heaps praises upon his patron—or, as is more often the case, upon the Prophet Muhammad, the best of men. Here, though, the object of his praise was the captain himself.[5]

The other three poems attributed to Jassim Nasrallah, however, were retellings of much older poems chosen from the literary canon of the medieval Islamic world—some from the early centuries of Islam and others from the thirteenth century. Indeed, *most* of the poems that Al-Khariji chose to record were sampled from across space and time—the "greatest hits" of Arab Muslim poetry. Al-Failakawi chose to inscribe poems that were well-known among literary connoisseurs (though he was far more elliptical about his sources). He included one stanza of *madīḥ* poetry, another work by an Andalusian poet, and excerpts from a variety of well-known devotional poems from across the Islamic world.[6] In some cases, the poem's circulation among its readers suggests itself. Among the dozen or so poems he wrote down, Al-Khariji included a stanza of *madīḥ* poetry that he attributed to his contemporary, the Yemeni poet ʿAbdulrahim Al-Barʿi, but which was originally composed by the twelfth-century Andalusian Sufi and poet Abu Madyan Al-Telemsani.[7] This was very much a cosmopolitan world of Arabic poetry, of multiple circulations and tellings.

Not all the nakhodas' verses were from an Arabic literary canon. Al-Khariji also included a selection of poems in Persian, from the ʿAjami tradition. One was a collection of couplets on the fleeting nature of life written by the seventeenth-century Persian Sufi poet Muḥayya, who spent much of his life traveling the Gulf, Balochistan, and India. The other expressed a longing for home by one who had been away for too long.[8] Unlike his Arabic poetry, Al-Khariji did not draw these from a Persian literary canon: these were lesser-known poems written in the coastal Persian vernacular—though very much a part of the oceanic world that Al-Khariji inhabited. Because of their coastal origins, their lifetimes were more ephemeral; they were pushed to the margins of a terrestrially framed literary tradition.

Indeed, in this multilingual world of poetry in motion, some had more staying power than others. Among the more recognizable pieces that Al-Failakawi copied into his logbook were selections from *Nafḥat Al-Yemen*, itself an anthology of poems, short stories, and maxims from Yemen, compiled by the Yemeni scholar Ahmad Al-Ansari Al-Shirwani sometime in the late eighteenth century. *Nafḥat Al-Yemen* earned the distinction of being among the first collections of poetry to be committed to print—in 1811, by the Hindoostani Press at Fort William, Calcutta, which at the time had been publishing a range of different Arabic and Persian texts for the educational benefit of East India Company officials. Al-Shirwani taught Arabic at Fort William, and the *Nafḥat* remained as the teaching text for the Arabic proficiency exams administered by the Company, and later by the Government of India, long after his death in 1840. Its widespread availability as a printed text helped ensure *Nafḥat Al-Yemen*'s place in an oceanic literary canon; it is no surprise that excerpts from it ended up in Al-Failakawi's notebook.[9]

Ultimately, what the nakhodas stitched together was neither an anthology of the works of a single poet, nor even a collection of poems on a single theme. It was an anthology of the poetic world of the dhow, in all its diverse forms and languages. By collecting and inscribing into their notebooks, Al-Khariji and Al-Failakawi threaded together the literary communities of the past and the present, bringing them all together both on the page and on the deck of the dhow.

• • •

But poems were not only meant to be read with the eyes; they were also meant to be read out loud and heard. Many of the poems in the nakhodas' notebooks would have been recited orally, during performances on board the dhow. They called not for pen and paper, but for a warming and tightening of the drum skin, a readying of the clapping hand, a tuning of the ʿ*oud* strings. The poem was embedded in a sonic world of rhythms and melodies; they could be appreciated in their written form but were rendered meaningful in their performance, against the backdrop of drums, claps, ʿ*ouds*, and wind instruments. "Reading," as one scholar reminds us, "is what you do with poems."[10] The aesthetics of the performance were integral to the poems themselves.

The rhythmic world of the performance was baked into the poem itself, which was written and recited along a meter, with verses that ended on the same vowel-consonant combination—a rhyme. The poems in the nakhodas' possession featured verses ending in "*ān*," "*īn*," "*kum*," "*lā*," and "*hā*," to take just some examples. The metered rhyme reproduced itself verse after verse, making for easy memorization, but also a more rhythmic recitation—one that lent itself very well to a musical accompaniment.

The work of stitching together verse, rhythm, and melody fell to the onboard musician, the *nahhām*. Though he was hired like a regular seaman, the *nahhām* was chosen for his musical prowess. A good *nahhām* had a reputation for agile finger work on his *ʿoud* neck, a powerful voice, and a deep catalog of poems to draw from as he serenaded the crew. Nakhodas worked hard to hire a good *nahhām*, who was seen as an integral part of the crew. On at least one of his voyages, Al-Failakawi hired the *nahhām* ʿAwadh Al-Doukhi, who would later go on to have a successful musical career.[11] The nakhoda ʿAli Al-Najdi, with whom Alan Villiers sailed, hired one "Ismael," whom he referred to as a "famous singer from Kuwait," who played the *ʿoud* and sang as the sailors beat drums and tambourines and danced.[12]

The performative dimension may be why very few of the poems that the nakhodas copied down aligned exactly with versions we have available to us today. They swapped out a word here and there, skipped over entire verses, and scissored others in. Performance, whether recitational or against a musical backdrop, requires improvisation—the ability to swap out one verse for another to speak to the pulse in the room. And the genre itself allowed it: the poetry they chose to copy down had deep historical roots and equally long genealogies, shapeshifting from one performance to the next.[13] What the nakhodas recorded, then, were not always faithful copies of poems in their purported original form, insofar as they could pin one down. Much of the time, the nakhodas inscribed tellings of those poems: versions they heard recited in particular contexts, most likely during musical performances on board their dhows. Al-Failakawi and Al-Khariji chose to anthologize those unique tellings; they chose to inscribe—and thus capture—a literary form in circulation.

By the 1920s (and in some cases, even earlier), it became possible to capture the performative soundscape in musical studios that emerged in Baghdad, Aden, and Bombay. The Kuwaiti musician ʿAbdullatif Al-Kuwaiti, who performed some of the poetry that Al-Failakawi inscribed into his notebook, recorded a performance of a poem from *Nafḥat Al-Yemen* at the HMV Studios in Bombay, where he was accompanied by the Kuwaiti poet and musician ʿAbdullah Al-Faraj.[14] As a result, some poetic performances could enjoy the kind of circulation and posterity that written poems did in the age of print; they could be consumed among friends in coffeehouses or from the comfort of one's home. Most never made it that far, though: they stayed on the decks of the dhows that breathed life into them to begin with.

• • •

Al-Khariji and Al-Failakawi's notes point to a transoceanic public sphere infused with poetry. Whether in manuscript or print, recitation or record, poetic forms circulated around the Indian Ocean world, on and off the decks of the dhow. In Al-Failakawi and Al-Khariji's poetic inscriptions, we see attempts to capture the uncapturable—to take a snapshot of an aesthetic world in circulation that we can only experience second or even thirdhand. With few exceptions, by the time the poems make it to us, the lived contexts that once animated them are completely gone.[15]

And yet, their writings remind us that these literary forms did enjoy a social life. Their inscription, recitation, and performance signaled relationships that are not at all apparent to the modern reader: relations between poets past and present, between the nakhoda and his passengers, and of course between the nakhoda and the onboard musicians. They touched on religious communities, but also on linguistic ones—for how else might we understand Al-Khariji's insistence on blending Arabic and Persian together? Though we can appreciate the poems primarily for their aesthetic value, contemporaries on board the dhow would have appreciated so much more: there is a whole world that operated behind the text that we simply cannot see.

This ultimately may be why Al-Khariji chose to include headings that signaled to its reader the who, what, and when of the poem, and why he chose to include it among notes on navigation, contracting, and other phenomena linked to the dhow trade. Poetry was not separate from the more mundane aspects of the dhow, it was wrapped up in it. A nakhoda was not only tasked with the management of the functional aspects of the voyage; he needed to know how to manage its aesthetic dimensions as well. Poetry had clear relations to specific modes of circulation and forms of discourse and address, both on and off the deck of the dhow. And though the twenty-first-century reader might not know how to read these relations, those immersed in that world could, and did.[16]

TEN

Bahrain

THE CONDITIONS WERE IDEAL for the *Crooked*'s Gulf crossing. The winds blew southward, and the currents ran in a westerly direction. The dhow's time on the open sea was short; within a day and a half, Al-Failakawi sighted Ras Rakn, the northernmost point of the Qatari Peninsula, and soon made landfall. He couldn't sail directly to Bahrain, though. The channel between Qatar and Bahrain was littered with reefs and shoals, and approaching the island from the east would have almost guaranteed damage to the ship. In fact, the entire island was ringed with reefs—all but one area, a channel between the main island and a smaller island to its west, which was, rather appropriately, called "Khor Al-Bāb," or "The Door."[1] Al-Failakawi and his crew sailed past the island and made landfall on the mainland strip opposite its western coast. On April 13, in the company of another *boom*, they crossed into the channel, making their way into the harbor at Manama a day later.

"We arose in good health and spirit in the port of Bahrain," he wrote in his logbook, repeating the phrase for roughly a page and half.[2] Bahrain was a good place to sell off some of the cargo he had picked up while in Malabar. With the Saudi blockade on Kuwait showing no signs of relenting, merchants would have eagerly searched for other outlets for their goods. Bahrain made sense: it was the area's biggest market, and served the Saudi mainland, too. Its marketplaces overflowed with wares that testified to its central place in the Indian Ocean economy: rice, textiles, spices, coffee, sugar, tea, and shipbuilding materials from India; grains, rugs, rosewater, nuts, and cattle from Persia; fruit and halwa from Oman; dates from Iraq and Hasa; and mangrove poles, cloves, and coconuts from East Africa.[3] Despite its small size, Bahrain had long tentacles.

Many other nakhodas would have also been selling their goods in Bahrain. That year, goods from India amounted roughly 80 percent of the total imports to the island. The harbor was thronged with mariners and stevedores unloading gunny sacks of rice, sugar, tea, grains, and pulses, whose earthy scent mingled with the briny, humid air that hung about the waterfront. Sailors shouted to one another as they carried out shipbuilding timber and coir rope, both in high demand among the island's shipbuilders. All this happened to the hollow rattle of mangrove poles being stacked on the waterfront, where they awaited purchase by home builders.

Merchants from around the Gulf walked along the harbor, eyeing the goods. They visited the island in anticipation of dhows returning from India and East Africa. They'd buy goods—cereals, sugar, coffee, and textiles—and carry them on their dhows to smaller ports.[4] These short-distance voyages, which nakhodas called *qiṭāʿa*, served the needs of smaller ports that seldom made it onto the nakhodas' itineraries. Many nakhodas engaged in this short-distance trading from Kuwait itself, but those with time to spare in ports like Manama could easily find cargoes to move to Qatar, the Trucial States, or the smaller ports of Persia.[5]

The three weeks that Al-Failakawi spent in Bahrain were busy ones for the inhabitants of the island. The summer pearl dive began in mid-May; some dhows would have already sailed out to the pearl banks for the early pearl dive, the *khanjiah*, while others were preparing to leave for the big dive. As Al-Failakawi walked along the waterfront and through the souk, he would have seen mariners coming and going, carrying with them sacks of rice, dates, and flour—the very goods that he himself carried—and loading them onto the pearling dhows that lined the harbor. In Bahrain, the pearl dive marked out the year's seasons and gave it its rhythm. Divers from all over the Gulf descended upon the island in the late spring, coming from Arab port cities, from Persia, from the interior of Arabia, and from as far away as the Gulf of Aden. Alongside them arrived large numbers of enslaved divers, sent to Bahrain to dive by their masters in Kuwait, Qatar, Oman, and elsewhere in the Gulf.[6] As these seasonal mariners poured into the island, the machinery of the marketplace cranked to life, as shopkeepers readied their bags of provisions and marked out new pages in their ledger books.

From the harbor, Al-Failakawi walked out into a covered portion of the marketplace, the *Sūq Al-Ṭawāwīsh*—the pearl merchants' market. There, he would have seen another scene with which he was intimately familiar. The offices were crowded with fellow nakhodas carefully going over the season's

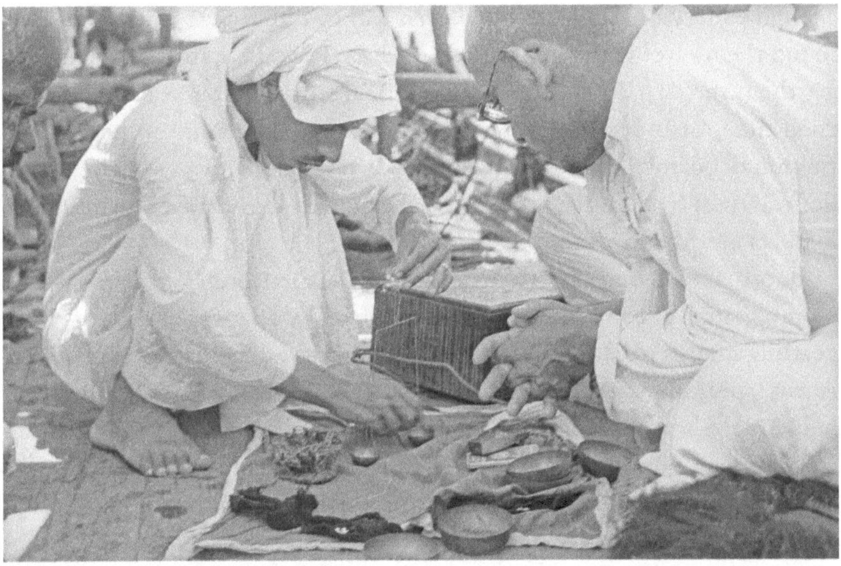

FIGURE 18. Pearl merchants inspecting and weighing a catch aboard a pearling dhow, 1939. Photo from National Maritime Museum, Greenwich, UK.

finances with different pearl merchants. It was peak credit season: nakhodas were taking on loans from merchants, negotiating over how to price the bags of Karachi rice and flour, Basra dates, Indian tea, Malabar coffee, and Javanese sugar that they needed to provision their dhows. These goods made their way to the nakhodas' mariners on credit. For most pearling mariners, the year was punctuated by three moments of borrowing. The one Al-Failakawi saw nakhodas working out was the *salaf*, a loan they would take just before the diving season began so they could provide for their families and purchase provisions for the dive itself—rope, protective gear, and extra food. During the dive itself, many asked for the *kharjiah*, a smaller loan they used to buy supplies from ships passing by. And then after the dive, there was the biggest loan of all—the *tisqām*, which they took in the early fall to tide them over until the next season.

All the divers hoped to earn enough after the sale of pearls to cover their debts. And they could certainly expect to earn something: after the nakhoda deducted the cost of provisions and handed over one-fifth of the net profits to the dhow's owner, he would apportion the rest between himself and the divers and haulers; a diver was always entitled to three shares of the earnings—as much as the nakhoda himself.[7] However, few of them ever did so well as to buy themselves out of debt. Most of the time they only earned enough to service the debts—and even if they did earn so much that they found themselves with cash in their pockets, it would hardly be enough to get them through an entire off-season. And so, year after year, whether in good times or in bad, everyone on the island could count on the seasonal rhythm of borrowing and diving, one rooted in regular waves of rice, sugar, flour, and tea from India, produced and exported in enormous quantities to markets just like Bahrain.

But this year, things were a little different in the *Sūq Al-Ṭawāwīsh*, a little more agitated. The rhythm had been thrown off by a battery of reforms to the pearling industry that had been recently introduced by the British Political Agent, speaking through the island's ruler. Everything from the customs house to the account ledger was being reconfigured in an attempt to centralize governance on the island in the hands of a nascent government. In this administrative revolution, the maritime economy was ground zero.

• • •

For millennia, Bahrain was the center of pearl fishing in the Persian Gulf, its waters studded with pearl banks. It also had access to one of the few constant sources of fresh water in the area, submarine springs that discharged fresh water into the saline waters of the Gulf—thus its name, Bahrain, or "two seas." Boats sailing near the island could easily replenish their drinking supplies; all a sailor would have to do is dive beneath the surface and fill sacks or other containers with fresh water. The same groundwater discharged onto the northern part of the island, forming several springs that nourished agriculture in fruits and vegetables, along with some date palms.

The pearl trade was a pillar of the Bahraini economy; really, its sole pillar. Though there was enough agriculture to feed many of the island's villagers, no activity rivaled the pearl dive, in value or scale. Year after year, the island's inhabitants sent out several hundred pearling boats, carrying many thousands of divers, haulers, and other mariners. The pearl trade attracted merchants from all over the coasts of Arabia and Persia but also scores of Indian traders. In the early fall, when the dive was over, the island would be thronged with merchants and sojourners from around the Western Indian Ocean. Its pearls mostly went to Bombay, though by the turn of the twentieth century, European buyers also visited Bahrain. Pearling was the island's lifeblood.

Given Bahrain's wealth in fresh water and pearls, and its defensible geography—one could only attack it by sea, and even that was not a straightforward proposition—there is little wonder that it became the object of so many political ambitions. At different times between the 1500s and 1700s, the island was occupied by the Portuguese, the Safavids, the Omanis, and tribes from the Eastern Arabian Peninsula; even the Ottomans tried to take it in the mid-1500s, though without success. With help from their ʿUtubi allies, the Al-Khalifa family conquered Bahrain from the vassals of the last shahs of Safavid Persia in the 1780s, but that only exacerbated, rather than resolved, the contests for the island. For the next eighty years or so, they faced constant threats by sea from neighboring tribes and political aspirants—and also repeated attacks at the hands of the rulers of Muscat, who wanted to integrate the island and its pearl fisheries into their expanding empire.[8]

Challenges came from within the family as much as they did from the outside; in fact, the two often aligned with one another. Shaikhs ʿAbdullah and Salman, the two sons of Ahmad bin Muhammad, Bahrain's first Al-Khalifa ruler, governed the island together and managed to maintain a united front against repeated Omani attacks. But once the Omani threat subsided in the late 1820s, the family members turned on one another: a protracted dispute emerged between ʿAbdullah and his grand-nephew Muhammad, each of whom sought out different allies among their relatives and the island's neighbors. The island descended into civil war for much of the 1840s, as the warring sides jostled with one another for forts, ports, and supporters.[9] Some of ʿAbdullah's supporters even sailed as far as Zanzibar, where they tried to entice the sultan to capture the island from Muhammad.[10]

ʿAbdullah's death in 1849 settled the war in favor of Muhammad, but his sons continued to challenge Muhammad's rule for the next two decades. The fighting between the ruling family members and their neighbors spilled out onto the pearl banks and sea lanes of the Gulf, much like they often did in this corner of the Indian Ocean world.[11] In earlier times, much of this would have passed with little comment; maybe a line or two in a local chronicle. In the 1860s, though, it came up against another emerging order, the Pax Britannica. In 1868, the Government of India sent three gunships from Bombay, with instructions to fine Muhammad and have him surrender his armed vessels. Upon learning of the news, Muhammad fled; his brother ʿAli settled the affairs with the British commander instead and was immediately recognized as the ruler of the island. His rule only lasted a year, though. Muhammad quickly returned, bringing with him supporters from the mainland. ʿAli died on the battlefield, and Muhammad's followers proceeded to plunder the towns of Manama and Muharraq.

The Gulf Resident, Lewis Pelly, moved decisively. He sailed out to Bahrain with two large ships and two gunboats and proceeded to bombard the island. Outmaneuvered and forts shelled, Muhammad gave himself up. Another one of ʿAli's sons, ʿIsa, sailed over to the island from Qatar and was installed as ruler. Meanwhile, Muhammad's chief supporters were rounded up and their properties were forfeited to the new ruler. They were summarily exiled to different British Indian possessions: first to Asirgarh, then Chunar—and then finally to Aden. Pelly would have been justified in feeling a sense of déjà vu: it was only seven years earlier that he oversaw another power transition between feuding members of a ruling family in the Indian Ocean—in Muscat and Zanzibar, between the sons of the recently deceased Saʿid bin Sultan.[12]

The island that ʿIsa bin ʿAli took over had been ravaged by decades of war. Many of its inhabitants had found themselves caught in the crossfire: the island's merchants had been extorted and robbed, and many had decided that it was better to take their business elsewhere, to port cities like Kuwait and Lingeh. The Baharinah—Indigenous inhabitants of the island who lived in the villages and worked primarily as cultivators—enjoyed no such luxury. Their property, boats, and animals were frequently seized under the flimsiest of pretenses, and even in better times, they found their crops subject to raid by roving tribesmen.[13]

And then there were the political wounds that had been left wide open in the wake of decades of strife. ʿIsa could plausibly win over merchants and cultivators. To consolidate rule within his branch of the family, though, the nascent shaikh married different cousins of his, all granddaughters and great-granddaughters of Shaikh Salman. And to build a base of political support within his extended family, he handed out privileges, like the right to tax villages and to run their own courts in them. Meanwhile, powerful tribes on the island, like the Dawasir, whose political weight could very easily tilt the scales, were rewarded with effective autonomy, including tax exemptions and marketplace credit.[14] Through a combination of gifts, titles, juridical concessions, and genealogical threading, ʿIsa bin ʿAli slowly patched the island back together into a durable, if somewhat polychromatic, quilt.

The new ruler had another tool in his kit: British protection. Having expended effort toward installing ʿIsa bin ʿAli as the ruler of Bahrain, the Gulf Resident was in no hurry to see him dislodged. The British position in the Gulf, too, was threatened by a growing Ottoman presence, particularly in Hasa and Qatar, and the possibility that the Qatari ruling family might lend material support to ʿIsa's antagonists. In December 1880, the Gulf Resident Edward Charles Ross concluded an agreement with ʿIsa that placed the island's defense and foreign affairs in British hands; it effectively rendered the island a British protectorate—a status formalized 1892.[15]

But though ʿIsa bin ʿAli received confirmation as ruler from outside powers, internally he headed an island of multiple, sometimes competing jurisdictions. More than a century of civil strife, coupled with waves of migration of tribes, merchants, nakhodas, mariners, and scholars, sedimented a legal regime in Bahrain that, depending on the context, could privilege virtually any marker of legal belonging under the sun. Adding to the legal confusion was the British assumption of jurisdiction over Indian merchants on the island, and, beginning in 1904, all Persians on the island as well.[16] Political stability came at the cost of juridical coherence. Which rules a person was subject to differed in the villages and in the marketplace, varied from Manama to Muharraq, necessitated a change in forum if they were Sunni or Shiʿa, and could change altogether if they could make a plausible claim to British subjecthood.

And then, in the pearl trade, things were completely different.

• • •

In the *Sūq Al-Ṭawāwīsh*, Al-Failakawi saw a robust marketplace that had benefited from a half century of political stability. Though its fortunes could vary from one season to the next, on the whole, the pearling industry witnessed an enormous efflorescence during the second half of the nineteenth century: a report in 1878 suggested that the value of pearls had doubled since the middle of the century, and by the early 1900s, Bahrain's pearl exports had soared by nearly 800 percent.[17] As the dust that clouded Bahrain's political atmosphere settled in the last quarter of the nineteenth century, economic life on the island followed a more regular, predictable beat.

There were well-established practices in the pearling industry too. When pearl merchants financed nakhodas, they were mostly after the catch itself. They extended credit to nakhodas and their crews with an expectation that they would be repaid in pearls; they effectively secured the season's haul by having paid for it in advance. Sometimes they negotiated with the nakhodas themselves, extending them lines of credit in the stores they owned. Other times, they sent agents to the waterfront to find nakhodas looking for credit—a loan of cash or goods for the season. This shore merchant—the *musaqqam*—would be waiting for them when they returned, his weighing apparatus and *chau* book in hand.

Pearl merchants were also after the manpower, perhaps just as much as the pearls themselves. To build a firm that could effectively produce wealth required human capital, and though it was in enormous supply during the pearling season, it varied in terms of quality. A good nakhoda—one who knew the pearl banks, but who also could attract a skilled crew of mariners—was hard to find. The merchants that Al-Failakawi saw bargaining with nakhodas in their offices knew that keeping a nakhoda meant granting access to credit and offering other enticements. Doing business in pearls was as much about building wealth in people as it was about accumulating the gems themselves.

But some operated on a wholly different level of trade. Of the offices that Al-Failakawi walked past, Yousef bin Ahmad Kanoo's would have been particularly busy; it always was. Kanoo was a well-networked man with broad-ranging business interests. He was, like everyone else, involved in the pearl trade, but he never got involved in directly financing pearling dhows. He occasionally dabbled in the sale of mother of pearl, which European companies expressed interest in, but even that was just a passing interest of his. Kanoo was, principally, a wholesaler. He supplied the Bahraini marketplace with dry goods that made their way into the shops.[18] He wrote frequently to business associates in India, arranging for them to send him sacks of flour, rice, tea, coffee, and sugar—as many as a thousand gunny sacks at a time, enough to feed an entire fleet of pearling dhows.

Wholesale merchants like Kanoo were critical to the workings of the Bahraini marketplace. Without a steady flow of dry goods, the entire pearling industry would grind to a halt. The sacks of rice and flour that Kanoo imported from India fed the divers when they were on board the dhows and provisioned their families while they were away. They also underpinned the financial arrangements that bound merchants in the bazaar to dhow captains and mariners; without them, there would be little basis for credit. The entire firm structure of the marketplace depended on there being access to provisions; merchants like Kanoo were the ones who kept the machinery going.

In addition to being a wholesaler, Kanoo was a banker. He kept ongoing accounts with Al-Sudairawi in Bombay, along with a host of other merchants up and down the Western Indian Ocean littoral. Those who traveled to India to sell the season's catch would deposit the proceeds into the accounts that Kanoo kept there with merchants like Al-Sudairawi. They would then draw from those accounts in Bahrain in the late spring when it came time to outfit pearling dhows for the summer dive. And those who simply wanted to travel to India to buy goods without the trouble of moving money often deposited cash in Kanoo's office; he kept a strong room next to his office, secured with thick steel panels, a heavy door, and locks, and even stationed a guard outside.[19] Whoever kept money with him could draw on Kanoo's accounts in any number of cities in India. And when the Eastern Bank Ltd. opened up a branch in Bahrain in 1920, linking the island with a company network that included branches across the Gulf and India, Kanoo adapted. Like Al-Matrook, he made use of Eastern Bank to transfer money between accounts and effect payments for transactions that took place around the Western Indian Ocean littoral.

Kanoo was hardly the only merchant able to move large quantities of goods and money. Other merchants on the island enjoyed business associates, accounts, and even sources of revenue in other parts of the Indian Ocean world.[20] These merchants found ready clients among members of the extended Bahraini ruling family. Shaikh ʿIsa often turned to people like Kanoo for access to goods on credit, which he doled out among different groups on the island—rival factions within the family, but also tribes, whose support for him was contingent on his largesse, either in the form of goods or in tax exemptions. Over time, his children, cousins, nephews, and other dependents all did the same: to shore up their own bases, they looked to merchants, who would open up their stores and ledger books to them. The revenues the shaikh and other family members received from taxing the pearl trade or from farming out the customs house were simply not enough to pay off the sprawling number of rival claimants within the family, and certainly not enough to reward different factions on the island for their loyalty, particularly when so few of them paid regular taxes. Squeezed on all sides by the webs of political obligations they found themselves enmeshed in, rulers sought lifelines from the merchants.

In return for loans of money and goods from merchants, Shaikh ʿIsa sectioned off the pearl dive as an arena of self-regulation—partly out of a recognition that those who engaged in the pearl business knew its rules best, but also in part as a concession to the very merchants whose contributions and loans fueled the state. He reinforced the authority of institutions like the *Sālifat Al-Ghawṣ*, a tribunal of merchants and nakhodas who meted out justice when it came to pearling disputes, and the *Majlis Al-ʿUrfi*, a tribunal of merchants who oversaw business disputes. As early as 1875, the Gulf

Resident noted that "the part taken by the local Government or regular courts in such [Salifah] cases is merely to enforce obedience to such decisions when appealed to."[21]

Law, as it was practiced in the *Sālifah* tribunal, was a highly localized regime that was deeply rooted in notions of custom, usages, and local knowledge; its content depended on its context. The *Sālifah* tribunal's ambit included, among other matters, the financial arrangements that underpinned pearling; the intricacies of the mutual rights and obligations of merchants, nakhodas, shipowners, and mariners, and the practices of profit sharing in pearling. For the most part, these arrangements rested on a direct knowledge of those involved, in a sector of economic life in which people knew one another well and encountered one another frequently. The institution's success lay in its ability to respond to the particularities of each circumstance; "custom" guided actors toward the resolution of conflicts in ways that restored the social relationships in which finance, production, labor, and exchange were embedded.[22]

As they oversaw the localized legal regime in which pearl finance and the pearl dive itself took place, merchants and nakhodas worked to keep other authorities out of it altogether. In part because of this, pearling emerged as a site of regular moral anxiety. People regularly voiced concerns over whether the rules that governed pearling ought to take precedence over others—the divers' religious obligations, for example, or Quranic injunctions that forbade usury.[23] Though merchants and nakhodas could often deflect these concerns with a reference to the intricate nature of the arrangements that allowed pearling to function, that was not always possible. And as the British presence in Bahrain grew, so, too, did their scrutiny of the privileges that pearl merchants and nakhodas guarded from others.

• • •

By the time Al-Failakawi arrived in Bahrain, new plans were afoot. Two years earlier, the Political Agent, Clive Kirkpatrick Daly, had sketched out an ambitious plan to reform the practices that underpinned pearling—specifically the systems of loans and accounting that animated the credit market. And that was just the beginning: Daly had his sights set on the political economy of the island more broadly. Taxation, customs, and justice—all were on the line, and all hinged on the reform of the pearl dive. Much of the island's pearling revenue, he suggested, leaked out from state coffers and into the pockets of the different merchants and firms that bled into the administrative apparatus. With its public finances on a firmer footing, the island could better reap the bounties that Daly firmly believed lay ahead. Unlike Muscat, which was "a dead city," he wrote, "Manama is a live and flourishing town, and has every prospect of a brilliant future."[24] In this imagined trajectory of Bahrain, the pearl dive, revenues, and the development of the state all moved in lockstep.

Other imaginaries animated the reform project too. As the pearl trade took off, Daly voiced concerns surrounding the situation of the divers themselves. In a letter to the Gulf Resident, he argued that reforms were necessary because of "the very serious abuses which have crept into the diving industry, rendering the divers, ninety percent of whom are foreigners and under British protection, slaves in all but the name."[25] The trope of practical enslavement went far beyond Daly. Visitors to the island would comment on it, and there were occasional press reports in Europe about the slave-like conditions of the divers—a development that caused no small amount of embarrassment. The Agent found himself confronted with an antislavery chorus that only seemed to get louder; his situation was rendered only more embarrassing by the fact that many of the divers were, in fact, enslaved.

But the divers were not the only group that Daly expressed concerns about. The Political Agent had serious qualms when it came to the treatment of Bahrain's Shiʿa population, whom he saw as being the victims of systematic oppression at the hands of the island's Sunni Arab elite. The Shiʿa villagers of Bahrain submitted one petition after another to him, complaining of the abuses they faced at the hands of the ruling family. Taxes from Shaikh ʿIsa himself were one thing, they argued; those were acceptable, and they could reasonably count on his support in return. However, in recent years the number of ruling family members who made demands on them had grown beyond his control. One petition, signed by dozens of village notables, claimed that the shaikh's sons, nephews, and cousins all made different claims on the villagers: they would "act despotically, according [their] own opinion," and engage in "oppression and cruelty." Men were murdered and property plundered, all on a whim.[26]

Together, the plight of the Shiʿa villagers and the perceived wretchedness of Bahrain's divers held the potential to cause enormous embarrassment for British officials. Daly was sensitive to the reality he was confronted with. Bahrainis were traveling in greater numbers to India, Iraq, and Egypt, and had seen people there press their leaders for political concessions. Daly noted, too, that the island's inhabitants were consuming newspapers from those places as well, which had "given rise to ideas about democracy completely opposed to the antiquated and autocratic rule of the Shaiks." He wasn't wrong: Bahrainis were avid readers of (and regular contributors to) periodicals coming out of Egypt, the Levant, and Bombay, weaving themselves into the channels of news and liberal thought that crisscrossed the Eastern Mediterranean and Western Indian Ocean. At the same time, Daly downplayed the degree to which the Bahraini intelligentsia was just as likely to be critical of the British presence on the island as they were the shaikhs. Faced with a potentially restive population, Daly saw reforms as the only way the British administration on the island might save face—and potentially stave off calls for real political reform.[27]

The first plank in Daly's reforms took on what he understood to be the most important issue: public finances. He began with the customs house, which he saw as a principal source of state revenues, but one that had become too leaky. The arrangement that the ruling family had with merchants, whereby the right to collect customs would be farmed out to the highest bidder, opened too many doors into what Daly marked off as the property of the state. He was quick to point out that the most recent customs director, Gangaram Tikamdas, did not even reside on the island, but was based in Karachi; he would only occasionally travel up to take care of business in Bahrain. More serious, however, was Daly's accusation that Tikamdas had been embezzling money and had blurred the distinction between state finances and those of his firm. In his telling, the island's leading merchants—including Kanoo, whom Daly had a particular disdain for—would pay their customs obligations irregularly, and "their payments were never credited to the customs books, but were credited in the books of the Director's private firm."

Daly was convinced that he could solve the problem of customs receipts by hiring a professional to manage the place. For the time being, the managers at Eastern Bank, which had just opened a branch on the island, agreed to take on the job (they would eventually hand it off to a professional customs director, a Belgian). But Daly also pointed out that the real issue was that there was no sense of regularity in state finances. Immovable property—homes and farmland—was not registered and could thus not form the fiscal basis of the state.[28] Moreover, the prevailing system of pearl dhow taxation was irregular; ships would pay varying amounts, and some—at times, entire groups—were exempted from paying anything.[29] The system allowed for political stability, but, in his view, could not nourish state finances. A registration tax on pearling dhows, he suggested, would bring regular income into state coffers and do away with the privileges and preferences that had long characterized tax collection.[30]

Daly envisioned a broad reform package that involved judicial reordering, land surveying, and building up administrative capacity on the island. By putting state finances on a firmer footing, he would be able to reform Bahrain's administration and have it pay for itself. At the same time, he would extricate the state from the marketplace. The revenues generated from the pearl tax and customs would fill state coffers and would reduce—and hopefully eliminate—the degree to which the ruling family had become dependent on merchants for fiscal stability. This came through clearly in Daly's attempts to establish allowances for different members of the ruling family, to which they would have to limit their spending. In his view, the prior arrangement by which members of the ruling family and their retinue would simply run up tabs in the bazaar, borrowing against future income, allowed too many merchants into the counting house and distorted state finances. An allowance, fed by dhow taxes and supplemented with customs revenues overseen by a professional, would effectively put an end to an arrangement in which rulers effectively mortgaged state revenues to the island's merchants and shopkeepers.[31] The marketplace would be a source of revenues, not advances.

By reducing the rulers' dependence on merchants, Daly was well-positioned to take on his main goal: the reform of the pearling industry itself. He hoped to see the proper maintenance of accounts between nakhodas and their divers and wanted nakhodas to produce those accounts in court in case of a dispute. Over time, the divers would be able to work off their debts, ultimately arriving at an "ideal system . . . in which the divers worked for a fixed wage and the pearls would belong to the Nakhuda." This would place the divers in a strong position to bargain: "The work is strenuous," wrote Daly, "and the divers would demand high pay."[32] If it succeeded, not only would it do away with the slave-like conditions of the divers and reduce the public embarrassment that the British government potentially faced, but it would also establish Bahrain as an attractive destination for maritime labor.

Though it seemed like a straightforward goal, reforming accounting was an enormous task. Nakhodas simply knew more about the accounts than their mariners did. But also, and like in the dhow trade, accounts in the pearling industry coursed through the marketplace, yoking together mariners, nakhodas, merchants, shopkeepers, and rulers. And Daly could see at least parts of this market-wide fabric, particularly when it came to dispute resolution. As it was practiced in the *Sālifah*, "law" reinscribed the ties that bound together different actors in the maritime economy and marketplace and was rooted in local knowledge and circumstance. But from Daly's perspective, this was a corrupt system that was too entrenched in local power relations to administer the kind of impartial justice he valorized; there was no chance that divers would find redress in a forum headed by the very class of people they sought justice against. Reforming accounts necessitated a reform of the legal system that underpinned the pearling industry.

To replace the *Sālifah* tribunal, Daly proposed a diving committee (whose members he would appoint along with the ruler) that would oversee the issuing and repayment of loans in public, checking accounts against a central registry. Alongside the diving committee, he envisioned the expansion of Bahrain's Joint Court, which he and the ruler both presided over, so that all cases involving Shiʿa would go directly to it, along with any appeals from the diving committee. He suggested yoking the court to British oversight and training Bahrainis within it, on the principles of Indian law.[33] The expense of staffing the court with professionals would come from receipts from the reformed customs house, which in turn would be fed by a humming pearl diving industry that was bolstered by British oversight, which would ultimately add to the island's revenues.

The diving-revenue-justice loop that Daly proposed was meant to be self-sustaining. He envisioned a system that took the job of ordering the marketplace out of the hands of economic actors and placed it in the hands of trained, vetted bureaucrats who would adjudicate according to externally derived legal principles. In a sense, what he was proposing was no less than the disembedding of the marketplace from society itself.

• • •

In its journey from the mind of the Political Agent to its implementation, the reform proposal hit several speed bumps. The biggest obstacle was the ruler himself. Shaikh ʿIsa, who had long cooperated with the British and derived benefit from their support, was in no mood to throw his weight behind Daly's reforms. He built the entire state edifice around his need to balance his political obligations: removing merchants from the customs house and chipping away at the authority that nakhodas had over their divers threatened to erode the very foundations of his government. Daly read the ruler's position as a product of his personality. Shaikh ʿIsa, he wrote, was "quite incapable of taking any decision, of understanding the situation in his own dominions, or of performing any of the functions of a ruler."[34]

Daly's solution was subtle, though awkward. In late May 1923, the Gulf Resident met with Shaikh ʿIsa and intimated to him that it was time to go into voluntary retirement. The shaikh initially demurred but then relented; "the whole of Bahrain was not worth a cigarette," he sighed, as he assented to the Resident's plan. Ten days later, a public meeting was called, and the reins of the government were transferred into the hands of the more pliable heir apparent, Shaikh Hamad, ʿIsa's son. Daly was present at the meeting, and took the opportunity to lecture those present on the need for governmental reform and "the absolute necessity for strong rule of one shaikh of the Al Khalifa family and not the irresponsible rule of 30 or more persons appointed by and responsible to themselves alone."[35] Together with the Gulf Resident, Daly was able to reconfigure the palace to his liking. Against many odds, he had managed to realize a more streamlined and centralized Bahraini government.

Out in the marketplace, though, the situation was altogether different.

. . .

When Al-Failakawi arrived in Bahrain in April 1925, he witnessed a marketplace humming with activity but also teeming with discontentment. Daly's reforms to the pearling industry were in the process of being implemented, and Al-Failakawi's fellow nakhodas grumbled over the new rules. As he gathered with them in the tea shops and on the benches outside of the merchants' offices, he would have heard them all talking about one of two things: the new account books that the government had recently issued to all of them, at 6 annas per book, and the proclamation recently made by Shaikh Hamad announcing a battery of new regulations surrounding the recording of loans they gave out and the process by which they could sell the season's catch. Both had implications for the work they were doing in the lead-up to the summer dive, but also for the governance of the pearl trade more generally.

The first half of Shaikh Hamad's proclamation was straightforward enough: nakhodas were to record accounts in the new books, with line-by-line entries showing very clearly what a mariner was being debited for—rice, dates, tobacco, and other provisions. Alongside these, the nakhoda was to enter in the amounts the pearls fetched and what the mariner's share amounted to; near the bottom of the page, he was to deduct one from the other and arrive at the balance due to either side. This was standard practice in the marketplace, though it may have been manipulated by some unscrupulous nakhodas in the past. Of the nine articles in the proclamation, six addressed rules surrounding accounting; the young ruler was keen on ensuring that he covered the topic well.

However, accounting was not the main concern. The bulk of the proclamation's text was taken up by the last three articles, all of which involved new rules surrounding the financing and sale of pearls. A nakhoda who took on an interest-free loan from his merchant financier was obliged to sell his season's catch to him at a discount of no more than 20 percent below market value. If he could not agree on the sale price, he could take it to court for arbitration—and if he did not agree to the court's decision, he could choose to sell the pearls to whomever he pleased, though he had to repay his debt to the merchant in full. By contrast, a nakhoda who took on the standard advances from a merchant (the *tisqām*, which the proclamation set at no more than 20 percent interest per annum, and the *salaf*, which was set at 10 percent per annum) had no obligation to sell to him, but could do so if he pleased, though at no discount. In both cases, nakhodas needed two-thirds of the pearling crew to consent to the sale of the pearls.[36]

It was this last point that set the nakhodas off. When the pearling season came to a close in September 1925, a few dozen prominent pearling nakhodas got together and drafted a lengthy petition to the ruler. Due to the reforms, they wrote, there were "no more merchants giving out money and no nakhodas readying dhows." Nakhodas groused that the rule requiring the divers' consent to the sale of pearls interfered with an already-established arrangement in which they acted as the divers' representatives at the bargaining table. Indeed, they perceived the entire reform package as intruding on their authority: the petition brimmed with reassertions of the nakhoda's right to discipline his divers at sea and to take charge of their financial affairs on land. The government, the petitioners made clear, had no business regulating the affairs of the pearl dive. The reform effort would send reverberations throughout the marketplace.

Not all nakhodas stuck around to hear the shaikh's response. As their peers gathered to draft their petition, nine nakhodas decided they would permanently relocate to Qatar in protest. One British official wrote that these were "men who before the introduction of the reforms indulged in the worst forms of oppression"—though it did not stop him from asking after them, no doubt in part out of an anxiety that their move might encourage others to do the same. The ruler of Qatar, who for his part stood to gain dependents and revenue with the nakhodas' secession, feigned ignorance: "Truthfully, I know nothing of the matter," he wrote, perhaps as a sly smile crept across his lips.

The nakhodas' reactions threatened to diminish the standing of Shaikh Hamad, who was already quite anxious about the damage the entire reform process might do to his legitimacy. The petition was a blow to his authority, and he looked to the Political Agent to back him up. The Agent called the nakhodas into the Joint Court and, in Hamad's presence, reprimanded them for not having discussed things with their ruler before committing their complaints to writing. The meeting produced the desired effect: "They all agreed that they had done wrong, and Hamad's prestige was enhanced a little by this action," he noted.[37]

After the meeting, the nakhodas presented a watered-down version of their petition. They withdrew their insistence on the return to the old ways of doing things and were careful to couch their authority as being contingent on the support of the government. Their ability to punish recalcitrant divers and pursue runaways was all, they acknowledged, subject to the approval of the government. However, they held on to at least some of their requests. First, they argued that the two-thirds diver representation rule was practically impossible; four or five representatives from among the divers ought to suffice. Second—and perhaps just as important—was that there needed to be some sort of committee of individuals steeped in the customary affairs of the pearl dive who would help arbitrate claims between divers and nakhodas—something like the old *Sālifah* tribunal. They suggested that this committee would only hear claims that the government referred to it but were adamant that at least some part of the juridical process needed to remain in their hands. Governance over the pearl dive, the nakhodas made clear, ought to remain in the hands of those who knew it best. At stake was nothing less than the marketplace itself: "We will neither be able to give nor receive in pearling until we find a just solution."[38]

• • •

A year later, the waterfront spoke again. On December 31, 1926, a group of two hundred divers marched into the marketplace at Manama and began looting some of the shops there; they went on to the house of a pearl merchant, where they "helped themselves to his rice, and destroyed his records."[39] One Bahraini merchant described the riot in a letter to his trading partners in Kuwait: "The pearl divers have attacked the marketplaces of Muharraq and Manama, and have taken foodstuffs, rice, and dates," he wrote. "In Muharraq, they attacked the shops of the *sayyids* [the members of the ruling family] and took from them rice, and also tampered with their safes but were unable to open them. They then tore up whatever they could find from the safes and ledgers, and other [papers]."[40]

As the divers made their way to the bazaar in Muharraq, they encountered a small guard of Indian soldiers employed by the Agency, and quickly scattered. "The divers seem to have behaved like naughty schoolboys rather than men determined to be nasty," wrote the Political Agent. Their riot, he suggested, was most likely in reaction to a government decision to place limits on the advances mariners could take from their nakhodas. The divers, the Agent wrote, "have never had any connection with the pearl trade, and know nothing of the value of their catches." They relied wholly on the loans they took from their nakhodas.[41]

When the divers ransacked the bazaar that day, they engaged in symbolic action that spoke to processes that had been unfolding for nearly three years. The limits the ruler and Political Agent placed on the loans, premised on the notion that the mariners might one day no longer be working off debts, translated most immediately into a lack of access to credit—to the sacks of rice and dates the divers helped themselves to at the merchant's house, the very lubricant of the credit economy. By taking rice and tearing up ledger books and papers belonging to members of the ruling family, they underscored the core of the maritime enterprise: bonds of credit and obligation that ran from the ruler down to the diver, pulsing through so many gunny sacks of food. Though the reforms that Shaikh Hamad assented to sought to wrench the marketplace out from the social world in which it was embedded, the divers' actions reminded him that society was still present in the marketplace.

• • •

Al-Failakawi left Bahrain on May 7, nearly eighteen months before the divers looted the bazaar, but he would have heard about it in Kuwait. Based on what he saw in Bahrain, he could have predicted the outcome. He knew all too well that printing new account books and putting divers in the merchant's office when it came time to sell the catch would not pull the marketplace out from the social and political world in which it was embedded. He also knew how deeply the bonds of debt and credit ran through the bazaar, across the deck of the dhow, and into the ruler's palace. One couldn't pick apart the bonds between the state and the marketplace without taking virtually everything else with it. He also would have immediately been able to read the symbolism behind the looting, the tearing of the records, and the taking of rice and dates.

As he sailed out from the harbor in Manama, alongside the pearling dhows that were beginning the summer season and amid the tensions surrounding the reforms, he likely felt some unease. Kuwait was no Bahrain, and his dhow was not a pearling dhow, but he may have gotten the sense that his turn was coming—that in other ports, on other routes, he and his fellow seagoing nakhodas were going to find themselves in somebody's crosshairs.

Other changes were afoot too. While Al-Failakawi was in Bahrain, a geologist from New Zealand, Frank Holmes, was moving about the island, digging water wells. But Holmes was not interested in water; it was only a means to an end. On December 3, 1925, just seven months after Al-Failakawi left Bahrain, Holmes signed an agreement with Shaikh Hamad for Bahrain's first-ever oil concession.

INSCRIPTION

Accounts

To show a half of a ninth, i.e., a half of a third of a third, so the share is as follows:	Among 18 10 180	Divided into half of a third of a third	For example 180 010
Multiply the share by the number of heads, and the result corresponds with the original.			

There are about thirty of these bewildering formulas in Al-Khariji's manuscript, ranging from the simple (dividing in halves, quarters, eighths, etc.) to the more complex (thirds, sixths, sevenths, and fourteenths). But he gives us no reason for why he included them; the notes are about as opaque as they come. Why have such a long discussion of division?

There is a clue, though: his mention of a share (*sahm*), which was the standard unit in maritime accounting. Whether in pearling, fishing, or trading, the dhow enterprise was at its core a partnership between the dhow's owner and its crew, including the nakhoda. Voyages of the sort that Al-Failakawi and Al-Khariji would have captained were partnerships in freightage. The nakhodas took the net profits of the voyage—that is, the total freightage earned, along with proceeds from the sale of mangrove poles, minus the cost of provisioning and ship repair—and split the earnings into two halves, one for the dhow owner (who in most cases was a merchant but could sometimes be the nakhoda himself) and the other for the crew. Within each were designated shares: each sailor received one share, but officers received more. The nakhoda was normally allotted 4 shares, the first mate (*mjaddimi*) 2, his assistant 1.5, the carpenter 1.5, the helmsmen and singers 1.25, and so on—all customary proportions, though they were open to negotiation.[1] Regular sailors could expect one share each. In addition, 4 shares were set aside for the dhow's owner from the crew's half; shares were set aside for the nakhoda and the dhow's officers from the dhow owner's half, in addition to other bonuses. Nakhodas divided up the net profits from a season's voyage by the total number of shares to get the value of a single share. Thus, they distributed the normal profits and losses of the season equally among stakeholders in the venture: the value of a share rose or fell alongside the season's earnings, and everyone shared in the risks.

Division thus mattered, and knowing how to divide into sometimes awkward proportions was a necessary skill. Al-Khariji's note was in effect a series of mathematical shortcuts for a nakhodas looking to divide up the profits of a voyage among its different participants. It was a system that would have been familiar to anyone on the dhow, sailors included—and not just because they had participated in it before. They would have seen it in the realm of inheritance, where wealth, however paltry, was similarly divided into shares, with each heir receiving a designated proportion of the estate.[2] Like joint heirs in a house or farm, the mariners and nakhodas were all stakeholders in the enterprise of the voyage.

It ran deeper than shares, though. Rooted beneath the associations and profit sharing was a whole substratum of accounts. Behind every share was a back-and-forth of different monetary figures, all arranged into neat rows and columns in account ledgers. In them, we see a commercial society bound together in flows of goods and money—sometimes between partners, sometimes mutual agents, and sometimes people who seemed to have nothing to do with commerce at all. From their ledger books emerges a picture of a world in which money, goods, and debts circulated between individuals, often in far-flung locations. And in those circulations, we can see the production of a Gulf commercial society.

• • •

It is telling that Werner Sombart, who was partly responsible for elevating capital from a noun to a phenomenon, associated it so closely with bookkeeping. The advent of double-entry bookkeeping, Sombart wrote, transformed assets into abstract values and thus made clear the path to "the rationalistic pursuit of unlimited profits." The relationship between bookkeeping and capitalism was "that of form to content," he asserted; "capitalism and double-entry bookkeeping are absolutely indissociable."[3]

Sombart's declarations touched off generations of scholarship in history that produced (and reproduced) a triumphalist narrative. The story goes something like this: With the Commercial Revolution that swept Europe in the wake of the Crusades, merchants faced mounting challenges in managing the flows of goods and money. There existed different forms of accounting in circulation, including double-entry bookkeeping, which merchants in Venice had used. However, it was Luca Pacioli's *Summa de Arithmetica*, published in 1494, that both popularized the practice and gave it its theoretical and mathematical moorings. Translations of and reflections on Pacioli's writings on double-entry bookkeeping appeared in various European languages; among the more notable is John Mellis's *A Briefe Instruction and Maner how to Keepe Bookes of Accompts*, published in 1588. With the spread of double-entry bookkeeping around Europe came the triumph of rationalization in commerce, of an ability to measure reality in new ways, and to plan business activities based on observable data.[4]

Even those who were wary of the Whiggish narratives surrounding accounting could not help but acknowledge the growing popularity of quantitative methods in European and American society from at least the sixteenth century onward. Whether by merchants hoping to declare their honesty to a distrustful public or political actors looking to discredit the fiscal schemes their opponents proposed, there emerged a growing trust in numbers as a way of measuring and representing social, economic, and political phenomena. Here, accounting became not just a tool for bookkeeping, but the reigning paradigm for political economy, and the chessboard on which the game of politics played out.[5]

Others have taken a different tack, focusing on how bookkeeping translated into managerial power. Historians of slavery have been particularly enthusiastic about this approach: with hundreds of ledger books from the slave trade and plantations at their disposal, they pointed out how quantification brought with it control of human bodies and their productive capacities.[6] All these accounts, whether they unfold in the counting houses of Europe or the plantations of the American South, reflect Sombart's assertion that bookkeeping and the rise of (Euro-American) capitalism are inextricably bound up in one another.[7]

Much of this literature presumes that over time, humans came to embrace an undersocialized view of the world, one in which numbers guided human action and, more importantly, in which social relationships came to be understood as quantifiable. It is, in short, a world in which we all somehow unwittingly became economists. Without dismissing that view altogether, I want to suggest that we might see things differently. What if we instead imagined accounting to serve human relationships? Rather than think of accounting as a paradigm that holds powerful sway over human beings and reconfigures social relationships, I want to suggest that, at least in the sphere of Indian Ocean commerce, it remained auxiliary to the very relationships it served. In a sense, I want to reembed accounting in commercial society and suggest that none of the abstractions that accounting deals in exist separately from the commercial associations and obligations that sustain them. As obligations ran through Indian Ocean commercial society, so, too, did accounts.

• • •

The account ledgers that survive in Muhammad bin ʿAbdullah Al-Matrook's papers span an unusually long period, at least as far as private collections in the Indian Ocean go. The first surviving sheet—the early material comes in loose pages rather than bound volumes—dates from near the beginning of January 1903, and the last dated page of the last volume is dated December 15, 1944. Not all forty-one years of records have survived: there are gaps of two to three years here and there, and one yawning gap of seventeen years, between 1921 and 1938, and another between 1945 and 1956, when Al-Matrook passed away.

There was also more variety than I had initially anticipated. What I'm calling "account ledgers" groups together at least two, perhaps three, different subgenres of accounting. The first is what we might call a daybook, in which Al-Matrook recorded transactions he entered into, day by day. The entries are indiscriminate: he entered goods being given to and coming in from different individuals, and purchases and sales made, all in a given day.

The second subgenre groups—or perhaps regroups?—the transactions by the associate. Typically, Al-Matrook would declare at the top of a page that the entries were "A Description of Accounts with X, Beginning in Month Y." Going down the center of the page are transactions of different sorts—money and goods coming in and going out, with a description of the transaction, the amount it involved, and the month and day in which it took place (the ledger books were already organized by year). In a column on the right, Al-Matrook kept a running sum of the transactions. He listed his transactions in chronological order, breaking the pattern only when he needed to keep another running sum for the same associate, but using a different currency. Krans, riyals, and rupees all appeared in his ledgers—though by the late 1930s, he lists all transactions in (almost certainly Iraqi) dinars and fils.

Alongside the pages listing transactions by associate were others that Al-Matrook referred to as *taqāṭī*ʿ, a term which we might approximate as "fragments"—apt, as it refers to transactions involving individuals with whom Al-Matrook seemed to have no deep commercial ties. These may have involved more than one-off transactions, but they were not people who Al-Matrook kept running accounts with; hence, "fragments."

None of these involved double-entry bookkeeping. There were credits and debits, but they were not listed on separate pages; they were, for all intents, single-entry accounts. There were double-entry accounts, though. One ledger book from 1913 includes a table of contents with two pages for each agent—one for money and goods coming in, and the other for the same going out, all arranged chronologically. The entries did not necessarily follow the double-entry conventions; not every credit mirrored a debit on the opposite page, and transaction dates did not neatly correspond to one another. What the book suggests is that these were ongoing transactions—money and goods went in and out at different times, and though they sometimes ended up balancing out at the end of the season, they more often did not. Still, in many of the pages in the ledger book, Al-Matrook scrawled out the word *khāliṣ*, or settled, over the accounts in blue or red pencil, indicating that they had somehow been squared away.

It's tough to tell whether the daybooks, per-agent accounts, *taqāṭī*ʿ, and double-entry ledgers all worked together or in parallel with one another, in part because we don't have a full set for any single year. However, there are hints that these acted as nested genres, that one fed into the other. Al-Matrook regularly crossed out transactions in the per-agent accounts and *taqāṭī*ʿ, either because the amount had been repaid, or more likely because the sums had been copied into a double-entry ledger book; by striking them, he ensured that he didn't end up counting them twice.[8] We might imagine the *taqāṭī*ʿ strikethroughs as indicating repayment, fragmentary as they are, whereas the longer-running transactions would have been copied into a separate ledger book, where they would have been reconciled and settled. Only the daybooks seem to stand apart, neither referenced nor crossed out.

Though there is considerable variation in the different accounting genres, they all share a common feature: they revolve around people. Despite the

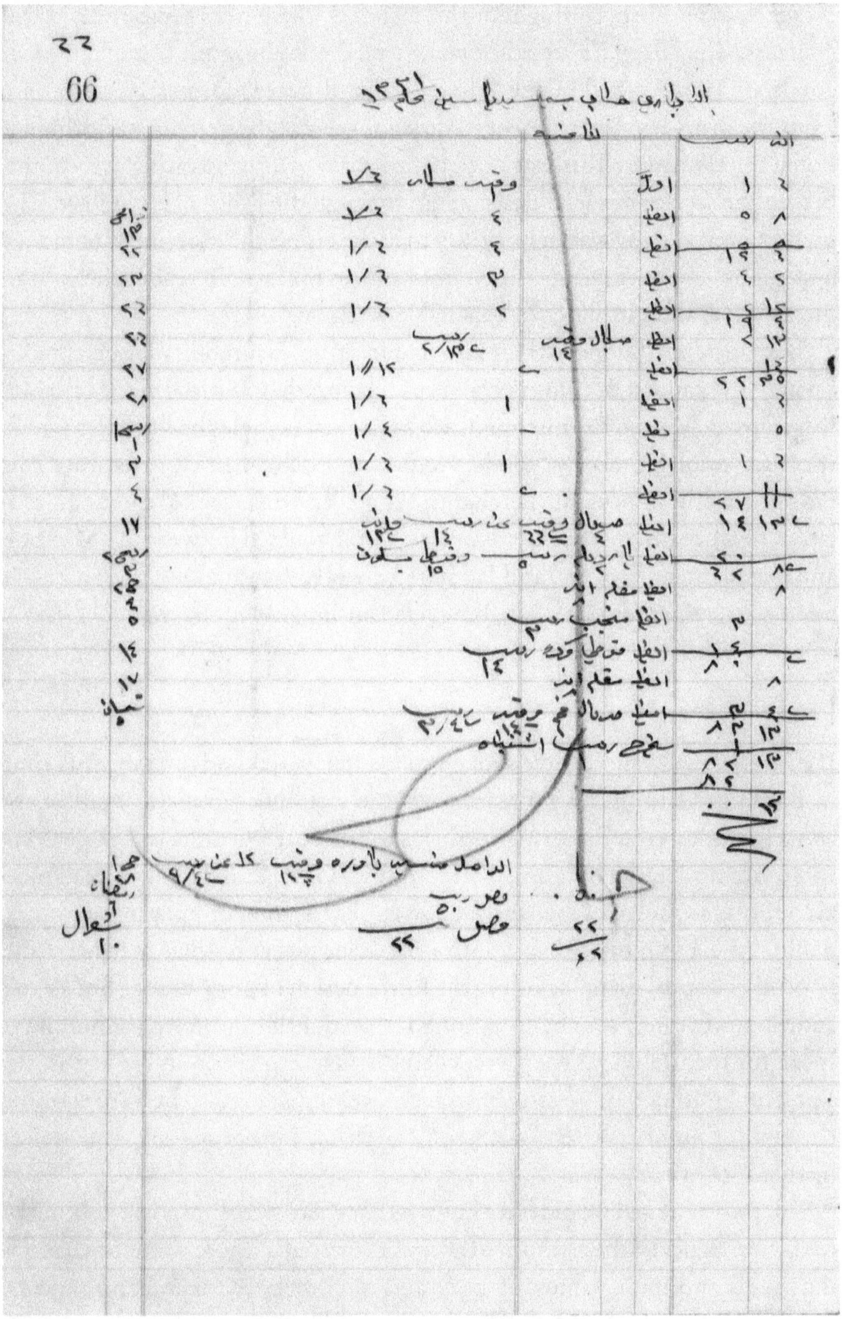

FIGURE 19. A page of struck-out accounts for the Hijri year 1331 (1912–1913 CE) by Muhammad bin ʿAbdullah Al-Matrook. Photo from Aslan Al-Matrook.

presence of numbers throughout the accounts, these were profoundly social artifacts. They depict a world of commercial relationships, of exchanges of goods and money within the framework of different forms of association, from the arms-length fragments to the longer-standing, more multidimensional relationships. This was not quantification run amok; they were numbers in the service of social ties that ran through the world of commerce. The power of accounting was in its ability to make visible the movement of money and goods, and to transmute them into a form that would render them commensurable, and ultimately reconcilable, with one another.

But accounts did more than just enable commensurability and reconciliation in the context of commercial relationships; they reinforced the relationships themselves. To the historian, accounts make visible the degree to which the circulation of money, goods, credits, and debts helped constitute this far-flung world of oceanic merchants. And even when capital did not circulate across oceans and borders—even when it coursed between Al-Matrook and the planters, brokers, and laborers he worked with along the Shatt—it circulated, nonetheless, reinscribing relationships along the way.

• • •

If Al-Failakawi were to have spoken to his fellow nakhodas during his time in Bahrain—and there's no reason why he wouldn't have—he would have come to understand the British reforms to the pearling industry as speaking to a particular understanding of accounting. Officials like Daly understood accounting as a form of managerial power: that is, by manipulating accounts, nakhodas could more closely control the labor power of their mariners. The reforms they proposed, including the introduction of new account books and auditing processes, sought to rationalize accounting: to have the numbers speak for themselves, as they wrenched the marketplace from society. It was a project grounded in what they would have understood to be a story of the triumph of numbers over society—a story that Europeans told themselves many times over.

But Al-Failakawi would have known that there were limits to this kind of narrative. Beneath the stratum of accounts existed a whole world of circulating goods and money, moving back and forth between merchants, agents, brokers, planters, nakhodas, and mariners, binding them all together in different enterprises in which everyone had a stake—or, following his teacher Mansur Al-Khariji, a share. As Al-Khariji's long note on short division reminds us, numbers did not speak for themselves: they worked to apportion rights and to divide up profits of enterprises involving many people. Accounting was a technology in service of relationships, not an ontology in and of itself. In the great ledger book of economic life, there were no atomized, rational individuals—only joint enterprises. To divide, then, was not to conquer, but to collaborate.

ELEVEN

Returns

THE VOYAGE BACK TO KUWAIT was brief, and one can tell that Al-Failakawi was getting eager to return. He crammed all his entries on this last stretch of the voyage into the top third of a page in his logbook, so much so that it is tough to tell where one ends and the other begins. Leaving Bahrain, he sailed out into the Gulf, along with the currents that ran in a northwesterly direction. A day and a half later, he could spot the port of Jubail, baking under the afternoon sun. All that kept him from home was a long, unbroken stretch of yellow-white sandy beach, punctuated by the occasional headland. The currents pushed the dhow westward, turning slightly north with the coast. It was smooth sailing—or perhaps the nakhoda just didn't have the patience to note any minor inconveniences.

By May 11, the *Crooked* began approaching Kuwait, one coastal marker at a time. First came the headlands at Al-Khafji, just south of Kuwait, and by the evening they passed Ras Al-Ardh, another headland across the bay from the town; they would have been able to see its lights in the distance. As they caught glimpses of Kuwait, the sailors warmed up their hand drums and, "in great excitement, rushed about getting out their best gowns and their most heavily blued headcloths, trimming their moustaches, cleaning their teeth," all in anticipation of seeing their loved ones.[1] Night fell, and the *Crooked* drew closer to the town; what was once a dim dot on the horizon expanded, and the fuzzy lights gradually took the shape of buildings, illuminated streets, and wharves. The sailors would have been too excited to sleep.

As the sunrise filled the night sky with the pale blue of dawn, the *Crooked* pulled into the harbor, where buildings crept right up to the edge of the water. After having spent eight months at sea and traveling many thousands of miles, Al-Failakawi could finally write, "We have safely reached the port of Kuwait." He may have had to be brief this time: on the dhow, the drums were going and the deck was packed with men clapping rhythmically and singing their praises to Allah. They sang praises as they lowered the longboat, and sang praises as they rowed it ashore.

The shore was crowded with the sailors' wives, mothers, children, and extended family. Family and friends could spot the *Crooked*'s sail from miles away and knew it by its misshapen hull; they knew who the dhow belonged to, and who sailed on it. They eagerly anticipated the dhow's return so that they might spend time with their loved ones. For months, many of them had clustered into matrifocal households, with the mariners' wives and children staying with their mothers and sisters, all drawing from the same stores of food. For a brief moment, their households would fill out again, gender roles would be reperformed and reinforced, the unmarried could get married, and life could inch forward once more.

But first, there was work to do. The day after arriving, the *Crooked*'s sailors were back on its deck, taking down its mainmast. With a signal from Al-Failakawi, the sailors hoisted the jib sail and raised the anchors just above the waterline. Slowly, to the beat of drums and clapping, and against a rising chorus of chants, the dhow entered the harbor, into its owner's basin (*nigʿah*). Sailors leaped from the bow into the sea and swam to the *nigʿah*, fastening a rope. Others lowered the sail, and together they swung the *Crooked* around to its place in the *nigʿah*. The drumming and clapping stopped as the crew worked to untie the sails, bring down the halyards, and pack up their belongings. To a rhythmic chant—"*hay-la, how-la*"—they pulled on different ropes, let down anchors, and steadied the dhow in its place. As children crowded around the dhow, paddling around its stern in dugout canoes, the sailors slowly lowered the masts and laid them and the yards down on the deck, along the length of the dhow. They then rolled them up in reed mats and carried them down to the shore, along with the chains and anchors. At low tide, they leaned the *Crooked* up against its props and then scraped and oiled its hull. The dhow would spend the rest of the summer in the basin, surrounded with coir matting strips to protect its hull from the relentless sun.[2]

Emotional as they were, these returns were always brief. The sailors, many of whom were itinerant, would be back at sea soon enough—some almost right away, as the summer pearl dive beckoned. Every return, whether in the winter or the summer, felt a little like the one before it: greet, disassemble, unload, visit, get ready to head out again—a return to the sea. People and things came in; people and things went out. Rinse, repeat. And yet, slowly, often imperceptibly—and at times quite palpably—things did change. With every return there was a new wrinkle in the fabric, a new change, a new tension. Al-Failakawi's voyage that spanned the fall of 1924 and spring of 1925 felt like the one just before it, but also like the last of its kind. Every time they sailed back out, things felt just a little more different than they did before. One could not sail the same sea twice.

• • •

The *Crooked* was one of many dhows that arrived in Kuwait that day. As the late spring turned to summer, dhows made their way back from around the Western Indian Ocean—from Karachi, Porbandar, Bombay, and Calicut; from Aden, Shihr, Mukalla, and Muscat; from Lamu, Mombasa, and Zanzibar; and from around the Gulf. The *Crooked* would have been among the first to return from India that season; waves of arrivals sailed into the harbor day after day, often until late June.

After months of relative quiet, the dockyards hummed with activity. Dhows were still being floated in and laid in their summer berths next to the scores of others that had already arrived. Up and down the waterfront, mariners unloaded goods from their dhows, alongside hired Persian porters. Piled up around the dockyards were stacks of lumber from Calicut and mangrove poles from East Africa, bales of Mangalore coffee and Malabar tamarinds, gunny sacks of rice and sugar from Karachi and Bombay, tins of kerosene from Persia, and wheat and barley from Basra—and more than fourteen thousand coconuts. India traders like Al-Failakawi brought in the lion's share of goods that year—more than 70 percent of the port's total imports; the ships coming in from Persia and Iraq trailed far off in the distance.[3]

As the mariners unloaded the dhows, merchants walked around piles of goods, inspecting the cargoes their nakhodas had brought in. Were the tamarinds of the right grade? Were there as many *korjas* of poles and *khandis* of teak as they had expected? Did the shipments of coffee and rice bear the right marks of the sender? They needed to sort out the piles and know what belonged to whom so they could properly adjust the accounts they kept with their business associates. If everything looked okay, business went on as normal. But at times, they didn't, and they would write to their partners and let them know of the mistakes they found, either in the quantity of goods sent, the quality—had they been damaged along the way?—the listed price, or in the money that was due to them.

Much of what had come in was already earmarked for someone else. The lumber that mariners carried out of Al-Failakawi's dhow most often went straight to the shipyards, where groups of shipbuilders anxiously awaited fresh supplies of building materials. Using simple tools—pencils, adzes, saw drills, and hammers—a master shipwright and his team could fashion a large plank of teak into a keel or rib and rework a tree trunk into a mast or yard. With enough effort, a pile of Calicut lumber and coir could be turned into a large oceangoing dhow, capable of carrying a few thousand packages of dates in roughly two months.[4] With any luck, they could have a dhow ready in time for the next sailing season.

Alongside the large oceangoing dhows that unloaded their cargoes were those that were waiting to load them. Smaller vessels, *baggarahs* and *jalbuts*, from small ports and islands along the Persian and Arabian coasts all crowded into the harbor. They knew that the oceangoing dhows were arriving and that they would be carrying goods that were in demand in their home markets. They also knew that a dhow returning from India was most likely not going call at ports like theirs. Here and there, nakhodas from around the Gulf bargained with merchants in the hopes that they would be able to buy at a wholesale price, and hopefully on credit. Among them were Kuwaiti nakhodas who owned their dhows and ran the shorter routes within the Gulf, often carrying sacks of rice, tea, sugar, and pulses, along with bales of Indian textiles. These voyages would last a month or two, and a nakhoda could do several a year.[5] Through bazaars like Kuwait, the wares of India, Yemen, and East Africa funneled into marketplaces around the Gulf, shaping home life around the region.

Outside the marketplace, in the open-air square of Al-Safat, groups of Bedouins milled about with their herds of camels, goats, and sheep. They, too, anticipated the arrival of the Indian Ocean fleet. In better years, before Ibn Saud's blockade on Kuwait, there would have been buyers coming from the desert as well. In those days, Kuwait was an entrepôt for the Najdi interior: it supplied caravans coming in from the desert with the rice that formed the foundation of their meals and the coffee on which so many social rituals in the interior rested. Through Kuwait, the foodways of India made inroads into the Arabian Peninsula. Bedouin merchants would strike deals with their Kuwaiti counterparts, taking goods on credit and returning months later to service their running accounts with goods from the interior—fuel, animals, dates, and cloaks—and money they had made selling coffee, rice, and pulses in marketplaces further inland. At least some of these were professional caravanners from among the ʿAqil traders, whose expansive trade network connected the Gulf and Iraq to markets in the Eastern Mediterranean.[6] Even during Ibn Saud's blockade, desert caravanners would creep through into the marketplace, but in 1925. trade with the Najdi interior had fallen to a quarter of what it had been in the past.

Others were waiting too. The pearl diving season would begin in just a few short weeks; some dhows would have already gone out for the *kharjiah*, the early dive that took place in April. The market would have been running low on provisions, too low, at least, to meet the demands of the country's pearling fleet. Tea, coffee, sugar, rice, dates, and flour, all staples of the pearling vessel, were all in high demand in the marketplace. The sacks of dry goods that came off the large oceangoing dhows made their way into the storehouses kept by the town's merchants, only to go back out again into the hold of another dhow.

For the mariners who served on the *Crooked*, these early weeks in Kuwait were a time for family, if any were around, but also a time for accounting. It was then, after they unloaded the cargoes off the dhow and leaned it against its props for the summer, that they could figure out how much they might have earned. Al-Failakawi and other nakhodas tallied up the amounts they were owed for carrying goods in different ports, along with what they might have earned on the side in mangrove poles and other cargoes, and divided it between the ship owner and his crew, and then again between the different crew members. He moved a share here and another there, all while subtracting whatever advances the crew member received over the course of the voyage. Thus, the profits generated from eight months of work traveled from the dhow's hold to the marketplace, and to the ledger book. They then circulated back out again, as the nakhoda apportioned out loans of cash and credit, both for the summer and in anticipation of the next sailing season, which would begin in just a few short months.

For the mariners' wives, mothers, and sisters—for the women of the town more generally—this was a time of reprovisioning. They had spent months administering their households in the men's absence, and many had to anticipate a few more months on their own as their fathers, husbands, brothers, and sons readied themselves for a summer out on the pearl banks. With the men's brief return and the revving back up of the marketplace, the women could replenish their homes' stores with the essentials and settle their accounts with the shopkeepers with whom they ran up a tab over the last several months. Everyone, of high and low standing alike, took advantage of

FIGURE 20. Women buyers crowding around goods near the harbor, 1939. Photo from National Maritime Museum, Greenwich, UK.

the return of their husbands, fathers, sons, and brothers to cross out a page in the ledger book and open a new one.

For other women, the arrival of oceangoing dhows infused their own enterprises. These women ran businesses in the marketplace, hawking different wares along a walkway known as *Souq Wājif*—the "standing market," also called *Souq Al-Ḥarīm*, or the Women's Market. Shortly after the dhows arrived, they would visit merchants here and there to buy the bolts of cloth, shoes, perfumes, and other consumer goods they sold over the course of the year to supplement the household income. Some may have even placed orders with sailors coming back from cities like Bombay and Aden. Others, who ran businesses as seamstresses, tailors, and embroiderers, also made their rounds to merchants, captains, and sailors bringing back fabrics and threads.[7] Through them, the textiles of India—the printed cottons, colored silks, scarves, and taffetas—made their way into wardrobes around town. It was through their purchases, too, that merchants knew what was fashionable and what might sell. That sort of information was valuable to their trading partners in Bombay, who could let textile mills know what consumers in the Gulf wanted to wear.[8]

The arrival of the dhows thus called into motion an entire ecosystem of porters, traders, seamstresses, shopkeepers, and shipwrights, all of whom had some sort of stake in the cargoes that came out of the dhow's hold. As things came in, they went out, pulsing through the arteries of the marketplace and along the veins of society. It was a scene that Al-Failakawi was deeply familiar with: he had experienced it many years before, and it set the stage for the season that lay before him. Except, of course, that things changed, much like they always did.

. . .

Al-Failakawi sailed for another twenty years, up and down the coasts of the Western Indian Ocean. He changed ships twice: in 1935, he turned in the *Crooked* for the more aspirationally named *Tayseer* (*Good Fortune*), and in 1945, on his last voyage, he captained the dhow *Al-Ḥarbi* (the *Warship*)—appropriate for a dhow sailing the waters of the Indian Ocean during the Second World War.[9] On these dhows, he would retrace many of the routes he had taken before, sailing from the Gulf down to India, to Yemen, to East Africa, and back. And yet, for all the repetition in Al-Failakawi's motions, in the twenty-two years between Al-Failakawi's first voyage as nakhoda and his last, the Gulf and Indian Ocean witnessed enormous change, from the global scale down to the very local.

The political map of the Indian Ocean littoral had already changed, even before his first voyage. On the Persian side of the Gulf, the quasi-autonomous principalities that had long dotted the coast were all being swallowed up by an expanding Iranian state under Reza Shah Pahlavi. Beginning at the turn of the century, port rulers in Lingeh, Balochistan, and Bushehr all saw their autonomy gradually erode at the hands of a slowly territorializing Persian central government, until their authority washed away altogether. For Al-Failakawi and Al-Khariji, both of whom had ties to Kharj Island, the changes heralded a much broader transformation.[10]

But there was a much closer domino to fall. In 1925, Al-Khariji wrote that he heard from a colleague in ʿAbadan "of Reza Shah's seizure of Mohammerah and all that remained of the dominions and wealth of Shaikh Khazʿal, son of Shaikh Jaber."[11] In his final days, Shaikh Khazʿal, a longtime friend of the Al-Sabah family and ruler of Mohammerah and broader ʿArabistan, pleaded with British consuls, oil company representatives, and international diplomats to defend his claims to the emirate. However, he could not convince them to stand up to the shah, who amassed troops at the gates of Mohammerah. Faced with military pressure, he agreed to meet the shah, who had him kidnapped and sent to Tehran in 1925, just as Al-Failakawi returned from India. The erstwhile ruler spent the next decade under house arrest, and his emirate was quickly abolished; Khaz'al was assassinated in 1936.[12]

Other changes were afoot too. Basra saw a rapid succession of changing overlords: the British Mandate of Iraq ended in 1932, but the newly independent Kingdom of Iraq enjoyed no respite. For the next decade, the kingdom would witness one military coup after another, as pro- and anti-British jockeyed with one another around the monarchy. Further down the coast, things were more stable but no less dynamic. The expansion of the Saudi Arabian state reached its zenith over the preceding two decades, beginning with Ibn Saud's capture of the Hejaz in 1925. The ruler then took on an aggressive policy of internal consolidation: he reined in his Wahhabi commanders, ultimately executing them in 1929, and united the Kingdoms of Hejaz and Najd, which he had ruled separately.[13] India, too, would undergo a similar process in which the expanding nation-state swallowed up hundreds of smaller Princely States and colonial enclaves, many of which dotted its western coast.[14]

Kuwait's merchant elite found themselves galvanized by the spirit of their times. They traveled around the region and observed political events in Iraq, India, and Aden. Moreover, they were voracious readers of Arabic newspapers and literature published in Iraq, India, Egypt, and the Levant. They contributed the profits they made from their trading activities to the development of libraries, clubs, and schools in both Kuwait and India, and they stayed plugged into cultural, political, and literary circles in other parts of the Arab world.[15] And it was amid this world of political transformation—amid a world of many possibilities—that they began to organize around a different vision of the political arena. In Kuwait, the date trader Hamad Al-Sager headed a merchant-led advisory council in 1921 that sought to intervene in matters of succession among members of the ruling family and to advise on matters of administration. The council ultimately faltered for different reasons, but the stake merchants planted in Kuwait's civic landscape remained.[16] They took on positions on the boards of schools and municipalities, and they continued to finance communal institutions like mosques, schools, libraries, and gathering halls.

The competition for influence in the political arena reached its peak in 1938. That summer, as Al-Failakawi readied himself to sail to India, a group of merchants combined to form a legislative assembly, backed by dissident members of the ruling family. Chief among these was ʿAbdullah Al-Sager, Hamad's son and cousin to Yousef Al-Sager in Calicut. ʿAbdulmuhsin bin Nasser Al-Kharafi, Al-Matrook's business partner, participated in the movement as well—though Al-Matrook himself, like many of Kuwait's Shiʿa merchants, largely stayed out of it. When elections were held for the first legislative council in Kuwait's history, half of the fourteen seats were won by scions of the transregional mercantile elite.[17]

Not everyone in the political movement that ultimately produced the council ended up participating in it. As the ruling family cracked down on political opposition, some decided to flee the country—many to Basra, where they owned property and could plug into political currents more closely aligned with their own. Many of these were the children of the merchants and nakhodas who had shaped Al-Failakawi's world. Hamad Al-Sager's son ʿAbdullah, who had studied in Bombay and spent time in Aden managing the family business, participated in Kuwait's Majlis movement but ultimately left for Basra, where he engaged closely with Arab nationalist thought.[18] The nakhoda ʿIsa Al-Qitami's son ʿAbdulwahhab, who voiced demands for political participation in Kuwait, also left Kuwait for Basra, taking his family with him in the still of the night. He and the others only returned when the ruler issued a general amnesty for the political opposition in 1944.[19] Their continued activity in the arena of politics highlighted the degree to which the transregional world of the Indian Ocean reverberated into local politics, across generations.

Despite the crackdown, the Majlis scored some important victories, pushing through a battery of legislation within the first two years of its existence, interrupted though they may have been by its dissolution and reconstitution. In its Basic Law, the Majlis declared that it would establish several other laws: the Law of the Budget, which regulated state incomes and expenditures; the Law of Justice, which addressed the administration of justice in the country; the Law of Public Security, which took on matters of policing and defense, both within and outside the town; and similar laws for education, public health, urban improvement, and states of emergency. It further asserted that "whatever law it is found necessary to establish in the interests of the public, it is the right of the Majlis to establish it."[20]

At the same time, the Majlis took control over revenues from the customs house, introduced a series of reforms aimed at abolishing monopolies that the ruler gave out, targeted privileges that Al-Sabah family members had been exercising in the marketplace, reduced rents and taxes, and canceled export duties and other taxes. It also expanded Kuwait's educational system, and went about replacing individual office holders—judges, customs officers, and self-appointed officials—with new, bureaucratized departments.[21] The prospect of increased oil revenues, animated by recent concessions agreements, only heightened their development-oriented concerns; at stake was the economic future of the nation itself, and the merchants wanted a say in it. An arena of legislative politics was set in motion.

And so, as Al-Failakawi plied the same routes he had his whole life, his small port city—indeed, the entire region—was on the precipice of major political and economic transformation. He was likely interested in what was going on, and even though he was away in India and elsewhere, he undoubtedly heard all the details from his fellow nakhodas as they gathered in the harbors, cafés, and merchant offices abroad. And then there were matters that the Majlis took up that were of direct interest to them—matters in which life at sea and life on land intersected in ways that directly impacted their livelihoods.

· · ·

Among the piles of books and pamphlets that made their way into Al-Failakawi's teak chest was a small, yellowed booklet entitled *Qānūn Al-Safar fi Al-Kuwayt*—the Kuwaiti law for *al-safar*, the oceangoing dhow trade—promulgated by the Legislative Council in 1940. A note by Al-Failakawi on the first page states that it entered into his possession that same year. In the preamble to the law, just under Al-Failakawi's note, the text declares that the Emir of Kuwait assented to the law, which had been presented to him by the head of the Majlis, out of a desire to "reform the country and its people." Behind the boilerplate legislative language lies a reality that was a little more contentious, but perhaps also much more mundane.

Sixty-one articles long, the *Qānūn Al-Safar* took on a host of different issues related to the dhow trade, and in exacting detail. The bulk of the articles outlined the prerogatives that the nakhoda enjoyed as commander of the vessel: the right to appoint mariners; to promote or demote them; to award them with bonuses or subtract from their shares; and to alter routes or jettison cargo as he saw fit to preserve the vessel and the lives of his crew. A handful of articles detailed the responsibilities a nakhoda had to his merchant financier and the owner of the vessel. The law also included articles that outlined a mariner's duties to his nakhoda and fellow crewmembers, and which traced out the web of financial and legal obligations that crisscrossed the dhow trade and pearl dive. It was easily the most detailed and comprehensive compendium of the rules of the maritime economy that had been laid out in living memory.

Rules had existed before, of course, but they mostly circulated within the dhow trading community itself. If a nakhoda, merchant, or sailor had any grievances with another, they would take them before an expert in the customs of the trade, the *Rāʿi al-Sālifah*, who would adjudicate based on circumstance, local knowledge, and a long experience with dhow-related matters. The dhow trade and pearl dive each had their own *Sālifah* men, each grounded in the customs of their respective sectors. In Al-Failakawi's time, the *Rāʿi al-Sālifah* for oceangoing dhow crews was his own employer, the dhow trader Shaheen Al-Ghanim; Al-Ghanim's son Muhammad, who worked alongside his father, took over after him. Rules were enshrined through consensus and practice; they were not written and never codified.[22]

It's not clear what the precipitating event behind the dhow trade law might have been. For all the ink contemporaries spilled on the merchants' movement of 1938 and the subsequent standoff between the ruling family and the merchant class, the context for the 1940 dhow trade law—and its immediate predecessor, the pearl trade law of 1939—remains obscure. The laws have been printed and reprinted in various books on the history of Kuwait's dhow trade, but invariably without comment or clarification. They stand as a portal to the pre-oil past—a luminous compendium of a world of customary practice that had long eluded the written word and legislative act. And yet, there is much to suggest that the laws were the product of a particular moment in Kuwait's political and economic history—that they captured a moment of struggle over the right to govern, rather than codifying all the customs that had animated the maritime economy before their promulgation.

There was some context to it, of course. The decade leading up to the pearling and dhow trade laws was devastating for the Gulf's maritime economy. This came in waves: the first blow was the arrival of the Japanese cultured pearl onto the scene in the 1920s. As early as 1921, one Gulf merchant based in Bombay wrote to a partner in Bahrain that people were unloading the real pearls they had and were "buying the fake pearls that come from Japan, and in London the fake pearls fetch twenty thousand pounds, which is almost 3 lakhs of rupees."[23] Most pearl merchants were not enormously concerned about the Japanese pearl, though; it was a different product, and the discerning consumer could distinguish between the two in the thickness of their nacre, which gave pearls their luster and brightness. Gulf pearl merchants could still count on their buyers to find niche markets for their goods; the power of conspicuous consumption among the global elite was still strong enough to fuel the pearl trade.

Strong enough, that is, until the consumers could no longer afford luxury items. The global economic depression that swept the 1930s sent markets around the world on a downward spiral. The pearl market was no exception. In 1930, Bahrain sent out more than nineteen thousand divers on over five hundred ships; six years later that number had dropped by more than half, and the value of pearl exports in 1936 was less than a third of what it had been in 1930.[24] The outbreak of the Second World War decimated what little demand there was left in Europe and North America. If anyone was buying pearls at all, they were certainly not going to buy natural ones.

Gulf merchants who had invested in the pearling business were left devastated. The pearls they held in stock, into which they had sunk so much of their capital, were suddenly worth less than what they had paid for them—and that was if they could sell them at all. One pearl merchant wrote from Bombay that the market was so weak that Bahraini merchants he knew there were left with more than two thousand *chau* without a buyer in sight.[25] Those who had mortgaged property to finance their pearl business faced lawsuits and foreclosures at the hands of their creditors. They lost ships, farms, shops in the marketplace, and even homes.[26]

Some were able to ride out the worst troughs of the 1930s. Merchants who owned agrarian property in Basra enjoyed a diversified portfolio; they did not have nearly as much capital tied up in pearls and could more flexibly serve marketplaces around the Western Indian Ocean. Of course, Basra dates were not unaffected by the global slump: the price of dates fell by nearly half, and many Basra landowners found themselves deeply indebted; many agrarian laborers could no longer justify working in the date gardens. To combat the growing social unrest, the Iraqi government set in motion a battery of regulations aimed at preserving the position of landowners and tenants and stabilizing the price of dates as much as possible.[27]

Still, Gulf merchants who continued to pour their money into the dhow trade did reasonably well. Al-Matrook's accounts show that his business continued to thrive despite the worldwide depression, much like other date merchants who engaged in long-distance trade. As late as 1939, Villiers observed that while the pearl trade was depressed, the Indian Ocean trade was still booming; shipwrights were turning out two to three new oceangoing dhows every month, and used ones were being sold up and down the Gulf to other buyers.[28] Even if there was no more appetite for pearls, people still needed to eat—and merchants needed ships to bring food to the marketplace.

What Villiers witnessed was the beginning of a brief efflorescence in the dhow trade. As Europe and the Pacific descended into another world war, the British government requisitioned privately operated steamships for the war effort. Shipping moguls like the British India Steam Navigation Company were left with a minimal fleet, and the weekly steamer service between India and the Gulf had all but stopped, leaving a chasm in regional transport that merchants and nakhodas were eager to fill. For dhows, the war opened the doors to a shipping boom: crews stood to make a lot of money freighting goods that would normally be sent on steamships.

With the opportunity to make money also came the pressure to make it. The nakhoda ʿIsa Bishara recalled that in the early 1940s he would do two return voyages to India in the time it took his father and grandfather to do just one: "If the *farman* breaks, Mohammad Al-Thunayyan [Al-Ghanim] will give you another!"[29] Faced with the promise of good freightage, nakhodas pushed their ships and crews to the brink—sometimes with disastrous consequences. In 1943, the nakhoda Bilal Al-Sager set out from Basra to Bombay too late in the season. Eager to return on time to pick up another cargo, he left Bombay without paying enough attention to the change in seasons; his dhow capsized and sank, and the nakhoda and many of his sailors drowned. And in 1945, as Al-Failakawi was finishing out his final voyage, another over-eager nakhoda, ʿAbdulkarim bin Ghaith, sailed back from Karachi too late in the season and also sank; only one sailor made it onto the longboat, and he died of thirst before the boat could wash up onto the shore. There were many others who suffered similar fates.[30] Al-Failakawi himself was hardly averse to risk-taking. He knew of the dangers, but also of the opportunities to make good money on freightage.[31]

There were also other, less licit opportunities to make money as well. As the British government in India placed controls on the import and export of foodstuffs and gold to ensure a supply of provisions and money for the war effort, there emerged lucrative opportunities for smuggling between the Gulf and South Asia. Tea, flour, rice, sugar, and specie were all whisked away under the cover of night as dhow captains maneuvered their way through the different harbors and inlets that concealed them from official view. Secondary ports like Dubai emerged as principal entrepôts for smuggled goods—

particularly gold.[32] Under normal circumstances, much of what was smuggled would have constituted the cargoes that nakhodas carried as a matter of course. Under a new regulatory regime that limited the amounts that merchants could ship out, though, the difference between 100 sacks of flour and 150 amounted to more than just space in the dhow's hold, as nakhodas navigated the line between handsome earnings in freightage and a heavy fine.

Amid the opportunities carved out by the Second World War were new wrinkles in Gulf economic life. The wealth of the Gulf nations was slowly shifting toward the land—or, more accurately, beneath it. Different individuals and companies had already begun prospecting for oil in the region from the early twentieth century and were beginning to find what they were looking for. In Bahrain, a subsidiary of Standard Oil struck oil in 1932, and on the eastern coast of Saudi Arabia, the company discovered commercially viable deposits of oil in 1938. In Kuwait, the ruler, Shaikh Ahmad Al-Jaber, signed a concession at the end of 1934, giving prospecting rights to the Kuwait Oil Company, a joint venture of the Anglo-Persian Oil Company and the Gulf Oil Company—companies that would later be known as British Petroleum and Chevron. The company struck oil in 1938 and began exporting in commercial quantities almost immediately after the Second World War ended.[33]

The shape of the ocean itself was changing, too, as the oil companies tasked with prospecting for oil on land began to express interest in the sea. In Kuwait, the process began with the bay adjoining the town, which British officials and oil company personnel were keen on establishing as belonging within the ruler's territorial jurisdiction. It soon expanded beyond that: around the world, governments scrambled to define the extent of their jurisdiction over the shores adjacent to them, gradually extending beyond the three-mile limit they had previously defined as territorial waters and carving out jurisdictions further out into the sea. Over the late 1940s and 1950s, Gulf rulers, prodded by American and British companies, proclaimed their jurisdiction over seabeds and continental shelves, and litigated over rights further offshore. If nakhodas always had to be mindful of the jurisdictions they passed through as they sailed along the shorelines of the Indian Ocean, they now had to contend with much more abstract claims to resources they could hardly even see, let alone disturb.[34]

With a burgeoning oil industry attracting working-age men from around the Gulf, Al-Failakawi and other nakhodas had to cast further out for mariners to man their vessels. A partial crew list from 1940 includes sailors from Basra, Bahrain, and Yemen, and longer lists from 1944 and 1945 identify sailors from Qatar, Oman, Yemen, and Kharj Island. Other nakhodas sailing in the 1940s and early 1950s similarly list crew members from Oman and different port cities in Yemen.[35] The opportunities, though, were still there for those who wanted to go out to sea—and if anything, they were increasing. The rise of oil work didn't displace the maritime economy; it existed in tension with it.[36]

Similar issues emerged in Bahrain, where pearling nakhodas had to contend with the pull of oil work among their divers beginning the mid-1930s. There, the negotiations over the transition to an oil economy took place not in the legislative halls, but in the courtroom, where merchants and nakhodas tried to work out what they owed, and what was owed to them, as former mariners left for work on the oil rigs. By 1940, the Bahraini government realized that the pearl dive sat in tension with the work obligations that the oil industry entailed. Former mariners who were still indebted to their nakhodas were caught between two masters. To address the matter, the government, in its annual proclamation fixing the sums that nakhodas were allowed to loan mariners, added clauses stating that mariners who did not want to sail with their nakhodas could pay a *faṣl*—a severance—that would release them from their obligations for the season. It further stated that mariners "who are in the employ of the Bahrain Petroleum Co. Ltd. . . . shall pay Rs. 5 monthly to their Nakhudas [*sic*]," adding that "the Nakhuda has no right to ask for the dismissal of his [mariner] from employment for the purpose of taking them for diving." The proclamation only granted this exception to employees of the oil company and the government itself; employees of any other institution could not expect to be released from their diving obligations.[37]

Court proceedings in Bahrain between nakhodas and their divers often followed a regular script. Nakhodas would file a complaint with the court, asking that their former divers appear for their summer diving obligations or pay the *faṣl*—"as per the government decree," they were sure to include in their petitions to the court. The rest was often uncontentious: mariners rarely denied the nakhoda's claim, and most seem to have been glad to have the court mediate between them and their former employer. The court would rule in favor of the nakhoda for the amount of the *faṣl* plus court fees, and the mariner would agree to pay over a certain period. Another kind of *faṣl* released a mariner from his obligations for the whole year for Rs 60. For oil workers, this was a more predictable arrangement: the court would simply garnish Rs 5 a month from their salaries, gradually paying off whatever they owed their nakhodas. Cases like these rarely filled more than a couple of pages—petition, proceedings, judgment, and all.[38]

And yet, ongoing litigation surrounding divers made it clear that the pearl dive had not yet ceased. Over the 1930s and 1940s it faltered, sputtered, and all but collapsed—but still constituted a regular feature of economic life in the Gulf, even amid a growing oil industry. Nakhodas had not yet given up on pearling: in 1939, which was the worst season on record for that decade, nearly two hundred boats went out for the pearl dive. And although the mariners no longer saw diving as a regular occupation, they did take to it "as a temporary employment when all else fails." Later seasons proved more optimistic, as buyers in India preferred to convert some of their cash into pearls as a safeguard against wartime inflation. Divers could stand to make enough money to last them several months.[39] The number of boats that went out fluctuated, but they continued to go out in scores, season after season, whether under a nakhoda or via an independently organized voyage.[40] The workforce that pearling and oil converged on was not finite, nor was it singular in its capacity; there were always people willing to go out and dive.

Nakhodas were willing to keep sailing, too, and maintained their strong attachments to the sea. Throughout the 1940s, they continued to ply the waters of the Gulf and Indian Ocean, even well into their old age. We catch glimpses of an aging Mansur Al-Khariji through the writings of the English traveler Wilfred Thesiger, who traveled on Al-Khariji's dhow from Abu Dhabi to Bahrain in 1945. By that point, Al-Khariji was "an old man, nearly blind, who spent most of his time asleep on the poop." His first mate would relay what he saw from the deck of the dhow, and the nakhoda would give him orders. As the dhow approached Bahrain, Thesiger marveled as the old man with failing eyes took his ship into the crowded roadstead at full sail, smashing through the choppy waves and bringing it within twenty yards of another dhow.[41] For people like Al-Khariji, the age of sail was still alive and well, even as the first oil tankers began making their way out of the Kuwaiti harbor.

And so, this may be how we might read the dhow trade law that Kuwait's Legislative Council passed in 1940. Although it is tempting to see the law as something of a swan song for an industry that was on the cusp of extinction, doing so would commit the error of reading the outcome into the process. At the time it was drafted, there was nothing to suggest that the dhow trade was going to die out within the next few years. Deposits of oil had been found around the country, but they had not yet generated visions of an oil-based economy. The dhow trade was still alive and well—and in fact, was amid a wartime boom. When oil companies needed to move containers of oil, they hired dhows.[42] Even the pearling trade, which had suffered an enormous blow over the preceding decade, still constituted a regular feature of Kuwaiti economic life. The council's desire to regulate emerged from a sense of optimism about the town's maritime future, articulated by those who were the most involved in its operations. For those involved in the oceanic trade, business amid oil was going to be business as usual.

• • •

Al-Failakawi had little time to spare to think about the distant future. There was hardly any time to celebrate anything now. He had a voyage to prepare for: he would leave again barely four months later, in mid-September, for a trip that would take him down to the Malabar coast and back in just over three months—a quick turnaround that would leave him enough time to make another India-bound voyage in the winter months. There was much to do in the meantime: profits to pay out, freightage agreements to work out, sailors to hire for another season, and provisions to secure. And then there was family to visit too.

For Al-Failakawi, that was it: there was no climax to his arrival, no dramatic ending. It was just another year. His logbook didn't skip a beat. He left no room for such matters. His last entry from the voyage he had just completed and the first entry from his new one merged seamlessly into one another, so much so that it was hard to tell that any time had lapsed at all. Barely three months after getting back home, it was time to go back out again—back across the Indian Ocean.

Epilogue

TRIUMPH AND LOSS

WHEN THE *FATEḤ AL-KHAIR* (the *Triumph of Fortune*)[1] sailed into the harbor at Kuwait in the summer of 1952, its nakhoda, ʿIsa Bishara, and his crew were returning from Zanzibar, carrying a load of mangrove poles. It had been a particularly busy season: the nakhoda and his crew had initially set out in mid-September, just after the beginning of the sailing season, and after spending three weeks loading the dhow with dates near Basra, they sailed out to Bombay, which they reached by the first week of November. Instead of coasting around India and eventually making their way back to Kuwait like they often did, the crew of the *Fateḥ* decided to push their luck: they sailed back to Basra to load dates one more time, and by the beginning of the new year set out once again—this time to the Swahili coast, stopping in Yemen along the way. By the time they reached Zanzibar, it was mid-February; they had barely two months there to load their cargo before having to turn around and head to Kuwait. The return trip itself took nearly two months, during which the dhow crawled up the Somali coast, sailed across the northwestern Indian Ocean to Oman, and then ambled along the Gulf until it reached Kuwait. The *Fateḥ*'s arrival in Kuwait that June would be its last; just a couple of months later, its owner sold it off to an Iranian nakhoda.[2]

As it turned out, the 226-ton *boom* had at least one more return left in it. It was the middle of June when the dhow pulled into the harbor at Al-Doha, on the far western end of Kuwait City. The hot air was thick with dust, much as it had been in summers past; every year, the *shamāl* winds would blow from the northwest, kicking up sand particles from Jordan, Syria, and Iraq, and blowing them across Kuwait and the rest of the Gulf. That year, however, the *Fateḥ Al-Khair* was going to be the only dhow to pull into the Kuwaiti harbor. It was 1994, and more than forty years had passed since the dhow was

FIGURE 21. The *Fateḥ Al-Khair*, moored in the Kuwait Scientific Center. Photo by author.

last there. That February, the preeminent maritime historian of Kuwait Yacoub Al-Hijji learned that the dhow was still shuttling goods between Iran and Dubai. He went to see it with his own eyes, and was astonished to hear its captain recount the long list of previous owners and skippers. After several rounds of pictures, round trips, verifications, and funding proposals, the *Fateḥ* began its journey back to Kuwait. It would take almost another two years before the *Fateḥ* would be even recognizable as a dhow: it was in bad shape, and needed new masts, yards, and rigging—all of which had been removed years before, when the dhow was motorized. The work took place under the watchful eye of the master shipwrights Hassan and ʿAli ʿAbdullah ʿAbdulrasul, among the last remaining shipwrights in the country.

On April 9, 1996, the dhow was unveiled to the public, and to much fanfare: the Emir of Kuwait, Jaber Al-Sabah, visited the dhow, accompanied by its former captain, the nakhoda ʿIsa Bishara. For the next several weeks, newspapers teemed with stories of the dhow, its rediscovery, and its return home. After five years of poking and prodding by curious tourists and schoolchildren on field trips, the *Fateḥ* was moved to a water basin outside the Kuwait Scientific Center, where it was joined by a handful of smaller vessels of more recent construction. The celebrations began anew: on February 10, 2000, a group of sailors assembled at the Scientific Center, singing and dancing as they welcomed the dhow to its new home. Together, they worked to raise the masts and yards, and to fasten its rigging.[3]

As it turned out, the exuberance was fleeting. Though the new location was more accessible than the old shipyard, which was more than an hour's drive from the city center, it was the beginning of a decidedly undignified end to the dhow—even for one that was no longer in use. At the Scientific Center, access to the dhow was restricted: people could look at the dhow from a distance but were not allowed to climb aboard the vessel. Over time, it sat there looking sullen, its hull and deck dried out, and the plaque recalling its name and past owners fading. There are no images to accompany it, no narratives of its past, no live recordings of the last remaining mariners to have sailed on the ship. All the pasts, the lives, the histories, the ideas, the concepts, the writings—all the different dimensions of the world of the dhow that made it come to life—all of them were missing. The *Fateḥ* was a fish out of water—a dead artifact, a limb severed from its body and put on display as an object that might speak for itself.

• • •

Four men shuffled into a plain room, taking their seats in wooden armchairs around a low coffee table. Three of the men were older, and dressed elegantly in *dishdashas* and cloaks, dark sunglasses over their eyes. The fourth was much younger and wore a brown blazer over his white *dishdasha*. It was 1974, and the young man, Redha Al-Fayli, had been working for several years on a television program, *Ṣafḥāt Min Tārīkh Al-Kuwayt*, or *Pages from Kuwait's History*, a series of interviews with Kuwait's merchants, nakhodas, statesmen, and other famous personalities—a dive into Kuwait's past. The three men who sat across from him had spent much their lives trading around the Gulf and Indian Ocean: the eighty-four-year-old Khaled Al-Hamad, who had spent most of his adulthood trading out of Aden; the eighty-nine-year-old ʿAbdulrahman Al-Bahar, who had imported goods from around the Indian Ocean into Kuwait; and the eighty-two-year-old Mishʿan Al-Khudair, who once did business out of Bahrain and whose extended family, the Al-Khaled, owned farmland in Basra and also engaged in the date trade. They were all prominent businessmen and had also actively participated in the Kuwaiti political arena from the 1930s onward.[4] They were well positioned to speak to Kuwait's commercial past.

Except that two of the three could only speak about Kuwait's maritime commerce in very vague terms. Al-Khudair began the interview by speaking generally about Kuwait's trade with India and Africa, emphasizing its flourishing under Mubarak; he spoke with only a little more specificity about the pearl dive. Al-Bahar added a little more, mostly on Kuwait's trade with its immediate neighbors—Arabia and Mohammerah. As he and Al-Khudair tried to recall different aspects of Kuwait's commercial past, Al-Hamad would frequently interject, jogging their memory and feeding them different names, facts, and talking points. Of the three, only he held forth with detail on Kuwait's maritime commerce and the degree to which it was embedded in a broader Indian Ocean world—so much so, Al-Khudair chimed in, that a change in the price of rice in Karachi would impact the marketplace in Kuwait.[5]

As an oral history project, *Ṣafḥāt Min Tārīkh Al-Kuwayt* documented a Kuwaiti past that was very quickly dying.[6] The interview with the three merchants was one of dozens: Al-Fayli interviewed merchants, nakhodas, fishermen, political actors, and more (all men) about the work they did, their memories of particular moments in Kuwaiti history, and their opinions on its transformation. And though *Ṣafḥāt* aimed to educate its viewership, it may have also served to educate the interviewers themselves. In episode after episode, Al-Fayli displayed a curious disconnection from Kuwait's maritime history: the geographies seemed foreign to him and the practices antiquarian. Even the language appeared to be almost completely incomprehensible to him: he spoke in a measured, standard Arabic to interview subjects who often responded in the vernacular.[7] In Al-Fayli's interviews, we see a shifting cultural terrain—a world of words and signs that were losing their meaning, and a moment in which Kuwaitis were already becoming tourists to themselves. And in *Ṣafḥāt* more generally, we see an attempt to capture a past at a moment in which it was already slipping away.

The *Ṣafḥāt* project was not Al-Fayli's alone—and by some accounts, not even principally his. The series is more closely associated with the historian Saif Marzouq Al-Shamlan, who conducted at least as many interviews as Al-Fayli did, and who in 1958 published *Min Tārīkh Al-Kuwayt*, a survey of Kuwaiti history that quickly found its place in the canon. Al-Shamlan's narrative of Kuwait's past in *Min Tārīkh* is effectively a terrestrial one: a group of settlers from Central Arabia established Kuwait, building the town, and as the settlement grew, it fended off attacks from the interior. The history proceeds one ruler and after another, one raid followed by the next, all on the dry land of the Arabian Peninsula. But for one discussion of merchants—the much-cited migration of pearl merchants to Bahrain in protest of Mubarak Al-Sabah's conscription policies—and scattered references to battles in Mohammerah, there is nary a ship nor a port in sight. Kuwait's history, the book makes clear, played out in what modern readers would understand to be Kuwait proper.[8] His interviews traced the same narrative: he mostly asked his guests what they remembered about this battle or that personality.

Al-Shamlan's position is a little surprising, given that he would later publish a two-volume history of pearl diving in Kuwait, one that drew heavily on his own family members' experiences as pearl merchants; if anyone ought to see the place of the sea in Kuwait's past, it would be him. And yet, Al-Shamlan's writings and interviews relegate the sea to a backdrop. The sea was an incubator for the nation: Kuwaitis sailed out with a national consciousness that was already fully formed, and their fortitude and bravery were tested in unruly waters. Even in Al-Shamlan's conversations with nakhodas, he asked about the routes they sailed, but was far more interested in discussing shipboard life, dangers at sea, and stories of famous shipwrecks.[9] Through a combination of tragedy and triumph, Al-Shamlan drew on the sea to tell the story of a nation of heroes.

More broadly, in Al-Shamlan and Al-Fayli's work—and really, in much of the historiography, with some exceptions—we see a tension between *maritime* history and *oceanic* history. The differences between the two are subtle, and my focus on a dhow and its voyages most likely blurred the distinctions between the two—but they are nonetheless different. In maritime history, at least in its older guises, the sturdy container of society sails out into the sea (often by way of empire) and encounters other similar containers that it bounces off from. The sea is effectively an arena for staging the nation, and when the container returns home, it returns whole and uncontaminated. By contrast, an oceanic history (and an Indian Ocean history in particular) proceeds from a different epistemological standpoint: it assumes no such container. Oceanic societies are unbounded, spread out over distances, and entangled with and connected to others. Inasmuch as there is a model, it is a circulatory one: goods, people, institutions, ideas, plants, animals, diseases, and more all circulate between geographies and across and between empires and other political formations. This does not negate the idea of a nation or some other communal group, of course, but it allows it to remain more open-ended and less bounded by territorial delimitations.[10]

When viewed from the standpoint of oceanic history, the narrative that *Ṣafḥāt* put forward reads paper-thin. Kuwaiti seafarers and their Gulf neighbors were not at all insulated from the circulations and entanglements that constituted the Indian Ocean arena, especially empire; they were deeply bound up in them, and indeed constituted through them. However, the methodological nationalist paradigm that Al-Fayli and Al-Shamlan operated within, together with other historians (virtually all of whom studied in an educational landscape shaped by Arab nationalists), almost necessitated that these processes be pushed to the margins of the page, if not omitted altogether. And though Al-Shamlan and Al-Fayli can't fairly be put on trial for a literature they knew nothing about, it may explain at least some of their puzzlement—their inability to grasp what had only very recently passed. Read from the standpoint of a community that had already embraced a distinctly national history, the oceanic past was illegible.

• • •

"And then, with the help of Allah, and the good fortune of Kuwaitis and their good leaders, when doors closed—when pearling, as you know, left, and voyaging left, and the captains' ships also all left—Allah opened up the door to oil. He does not close a door without opening another hundred," declared Khaled Al-Hamad to Rida Al-Fayli.[11] It was a pregnant statement. What Al-Hamad remembered in 1974 was a narrative of linear progress: first, there were pearls and dhows, and then they went away, and then came oil, with all the benefits it brought to the realm of commerce.

He was partly right, of course. The dhows did leave, though not very far away. We can take the *Fateḥ Al-Khair* as an example: under its new owner, the ship enjoyed a brief second wind as an oceangoing dhow, and was later outfitted with a motor and used to shuttle goods around the Persian Gulf.[12] The *Crooked*, too, followed a similar path: it was sold to a trader in Dubai and plied the routes between the United Arab Emirates and the Iranian coast. Al-Failakawi's son ʿAbdulrahman recalls seeing it in 1984, beached near a creek in Ras Al-Khaimah, which he visited after learning someone there had bought it. He took pictures of it and thought about buying it from its owner but ultimately decided not to.[13] He doesn't know what happened to the *Crooked* next, but if it followed the same pattern as other dhows, it most likely ended up being sold again and outfitted with a motor. As Iranian ports like Bandar Kung established themselves as regional shipping centers in the age of oil, Kuwaiti ship owners sold their vessels off to merchants there, who would then outfit them with motors and have them ply the trade between Iran and the Arab coast.[14]

FIGURE 22. The *Crooked*, beached in Ras Al-Khaimah, the United Arab Emirates, in the 1980s. Photo from ʿAbdulrahman Al-Failakawi.

The coming of oil, too, was a boon to many of these merchant families. Al-Hamad, along with others, formed the National Bank of Kuwait, and remained on its board of directors until his death at the age of 111; his family members are involved in real estate, government administration, law, and international economic development. The Al-Sagers transitioned into manufacturing, oil services, and real estate; members of their family also remained active in politics and journalism. The Al-Kharafi family went on to establish a business empire, including one of the largest construction companies in the Gulf. ʿAbdulmuhsin's grandson Nasser was worth $10 billion at the time of his death in 2011; his brother spent many years as the speaker of Kuwait's National Assembly.[15] Muhammad Thunayyan Al-Ghanim's son ʿAli went on to establish a business conglomerate engaged in construction, contracting, real estate, and automotive sales; he married ʿAbdulmuhsin Al-Kharafi's granddaughter Faiza, and their son, Marzouq Al-Ghanim, also served as the speaker of Kuwait's National Assembly, as did his maternal uncle. And these are just a few of the business families that reaped the windfall of Kuwait's transition to oil; countless others around the Gulf and Arabian Peninsula went on to establish big businesses.[16]

Nakhodas crossed the watershed into oil as well, much like everyone else did. Members of the Al-ʿAsʿousi family either went into business—the family had accumulated enough capital to—or worked in a burgeoning landscape of government institutions. Bishara started work in a government position, and then, with credit from his contacts in the merchant community, went into business on his own. And two years after his last voyage, in 1947, Al-Failakawi moved his family from Failaka to the mainland, where he took up work as a pilot of an oil ship. He worked for the oil company until he died, never having fully left the sea.[17]

But this Whiggish ships-to-oil narrative overlooks the other stories: the lost narratives and lives of the world of the dhow that found no easy place in the newly told story of the nation. Muhammad bin ʿAbdullah Al-Matrook, whose work, web, and writings were so central to the Indian Ocean story I've told here, has all but been written out of it. Indeed, his story came to a rather abrupt end: in 1956, Al-Matrook was found murdered on his Basra property. His business passed down to his son ʿAbdullah, but at the wrong time in history: two years later, a group of Iraqi military officers overthrew the government of King Faisal and oversaw the execution of the members of the royal family. Over the next few years, Brigadier ʿAbdulkarim Qasim, who assumed the role of prime minister, directed much of his antipathies toward Iraq's southern borders: he disputed Iran's claims to the Khuzestan region (formerly known as ʿArabistan), and in 1961 he asserted that Iraq had a territorial claim to Kuwait itself. By that period, then, there was no longer room for people to move between Basra and Kuwait as there had been in the past; the borders were far too hotly disputed. Though ʿAbdullah managed to insert himself into the growing business landscape of Kuwait through investments in real estate, the family found itself pushed to the margins of the official narrative of Kuwait's maritime past, one dominated by the country's Sunni elite.[18] Al-Matrook met an unusually graphic end, but his fate within the national narrative is one shared by many others whose transition from the age of sail to the age of oil was not nearly so linear.

People throughout the Gulf have been left to grapple with the detritus of these lost pasts. Descendants of these Indian Ocean merchants still point to family properties and former offices in Basra, Bushehr, Bombay, and Calicut—buildings that bear the traces of their ancestors, physical traces of a past they can identify, but can no longer identify with. Those who dig a little deeper would find graves, endowed mosques, and even long-lost relatives. On one of my early trips to Mumbai, I was told by the former Kuwaiti consul Hussain bin ʿIsa that a member of my family still lived in the city. A few days later, I found myself sitting in a room with an old woman, Fatima Bishara, who spoke in the old dialect of Kuwaiti Arabic but struggled to remember her own father's name as she propped herself up in her bed.

"Bishari," she said they called him. She wasn't too far off: when I spoke to my grandfather later that evening, he told me that she was most likely the daughter of his uncle Mishari, who migrated to Bombay with his wife before my grandfather was even born in 1919. In that room, I was confronted with the phantoms of a past that I did not even know existed. And yet how many other Fatimas were there out there, how many severed branches of a society that once spread out across an oceanic world?

Even those with a tangible, knowable past—Letters! Ledgers! Logbooks!—sometimes find themselves grasping their way through materials that face in directions other than that of the terrestrial nation. As many hold up property deeds that situate their family within the walls of old Kuwait—a practice that has become commonplace in a historiographical arena plagued with discourses of autochthony—others find themselves at a loss. When everyone is trying to ground themselves within the boundaries of Kuwait, it's unclear what value a cache of letters from Basra or a stack of cheques from Bombay might do, much less a logbook that traced lines across an ocean. It's of course possible to use these materials to stake a claim of rootedness in the terrestrial nation-state, but doing so requires admission by the gatekeepers of the nation's past; they are just as easily dismissed as trifling materials of no real consequence to our understanding of that history.

Take Al-Khariji's notebook, which I've drawn on throughout Al-Failakawi's journey for its sharp reflections and theorizations on the world of the dhow. In 2007, the Center for Research and Studies on Kuwait published a transcription of the book, edited by the Yemeni maritime historian Hassan Shihab. It was clear, though, that Shihab did not know what to make of the text in front of him. He notes that Al-Khariji bought it from another Failaka nakhoda, Hussain bin ʿAli Jinaʿ and added to it. Shihab further observed that the manuscript "is not free of grammatical errors, and while it is clear and reads easily, its language is poor." He adds, too, that "it is not methodical in its treatment of the principles of navigation, nor in its organization" and is full of "matters that are unconnected to navigation," which Shihab derisively called "filler."[19] What Shihab looked for in Al-Khariji was another link in an imagined Arab genealogy of navigational writing that descended from Ahmad ibn Majid down to ʿIsa Al-Qitami and beyond. The manuscript that lay before him defied his expectations and understandings of the past, and he was left unable to do much with it.

Dhows face the opposite problem. They are already enshrined in the national iconography, but few of them survived: those that were not sold off faced less fortunate fates. As he rattled off the names of several dhows that were sold in the summer of 1947, the nakhoda Mufleh Al-Falah noted, in passing, that "the *boom* of Al-Qudhaibi, *Al-Neem*, burned."[20] Many suffered similar fates. Indeed, the only pre-oil dhow that did remain in Kuwait, the nakhoda Hussain bin ʿAbdulrahman Al-ʿAsʿousi's *Al-Mohallab*, was placed outside of the Kuwait National Museum; it was burned down during the Iraqi invasion of Kuwait in 1990. For a country that had inscribed the dhow as a central artifact of its culture, placing it on its currency and its state emblem, this was a problem.

Which brings us back to the *Fateḥ*. The discovery of the dhow offered up the possibility of filling the ship-shaped hole in their national consciousness, giving government officials and patriots alike a tangible symbol of pre-oil Kuwait to hold onto. But the problem, of course, is that the *Fateḥ* can't stand in for its own past; it is part of a fabric too broad for a national narrative to grasp. Though the *Fateḥ* has been recovered, refurbished, and put on display, it sits as a severed artifact, among the detritus of Kuwait's oceanic history. Much as the Kuwaiti sun dries out the *Fateḥ*'s planks, so, too, does turning it into a national symbol require a process of desiccation.

What I've tried to do in this book is submerge the dhow back into the salty water in which it belongs—not the sectioned-off basin at the Scientific Center, but the wide, watery expanse of the Western Indian Ocean. To bring it back into a world of circulation. To recognize that the very planks of wood it was made of were brought over from Malabar, measured in units that circulated between Calicut and Kuwait, and passed through communities of cutters, brokers, merchants, and conservation officials, none of which have a place in the stories that are told about it. To situate it within an oceanic world in which people, food, textiles, and objects all circulated by sea, on its deck, and in its hold. In sum, to embed Gulf history back into the Indian Ocean world.

As this fuller, richer, more deeply textured *Fateḥ* returns to the ocean, it opens new horizons for those writing from other shores. As Al-Failakawi's travels remind us, circulation and movement don't happen on their own; they require active thinking. Through the travels of the dhow, we see a world of texts and ideas—a world of transoceanic thinking, reading, and writing on navigation, on commodification, on contracting, on accounting, on politics and empire, on aesthetics, and so much more. These ran alongside the dhow, propelling it across the water, and weaving it through so many societies, markets, and pasts. To connect the dots around the Indian Ocean, to stitch these worlds together, we need both needle and thread. On the deck of the dhow, in the nakhoda's chest, in the margins, we find both.

. . .

I had planned on writing these final sentences in the shadow of the *Fateḥ*, if only to smell the dusty, leathery teak as I typed—or maybe even to break a few rules and hop onto its deck for a few minutes before the security guards threw me out. "But my grandfather captained this ship!" I'd shout out as they escorted me back to my car. It would have been a fitting ending. In some ways, my journey with this book began with the *Fateḥ*, which I spent time on all those years ago, and which remained in the inner recesses of my mind as the years went by. I wanted to understand its world—my grandfather's world. I wanted to bring it back to life, and to guide others through it.

But my journey couldn't start with the *Fateḥ*, which was simultaneously too close to home to think through and too far away as a historical object. It started with the *ruznamah*, which carved a pathway through the world I wanted to write about and opened vistas onto the intellectual labor that made it all knowable, and ultimately possible. I could hear my grandfather's whispers throughout the book, but the words were not his: they belonged to our nakhoda ʿAbdulmajeed, the son of the Mulla Ahmad Al-Failakawi. Read together with the words of Al-Khariji, Al-Matrook, Al-Sudairawi, the Al-ʿAsʿousis, the Al-Sagers—but also the rulers, jurists, British officials, sailors, the pearl divers, the enslaved, the mothers, the sisters, and so many others—the *ruznamah* came to life. And with it, so did the *Fateḥ*. That the *ruznamah* declared itself to have entered Al-Failakawi's possession in Calicut, alongside the dhow *Fateḥ Al-Khair*? Well, that was just a delicious coincidence.

What I've traced in this book is one narrative—one nakhoda and his world, one season, one itinerary, though told from several perspectives and toggling between different scales. Had I started in a different place and at a different time, had I chosen different perspectives, the shape of the story would have changed. There are dozens of others routes I could have followed, scores of nakhodas I could have grounded the voyage in, and hundreds of other lives and lifeworlds that I could have narrated. From Bengal to the Red Sea, the Indian Ocean teems with stories like these.

Let me finish with one more. In Muscat, sometime in the early twentieth century, the British consul approached the sultan, Faisal bin Turki

Al-Busaʿidi. For years, British cruisers had been chasing after dhows from Sur for suspected involvement in the slave trade. When they'd catch them, the Suris would beguile them by producing French flags and passes. The British consul at Muscat pressed his French counterpart for a list of names of dhows that had been suspected of slave-running, along with the names of their owners, which he then presented to the sultan. One of the dhows, owned by Jumʿa bin Saʿid and Khalfan bin Ahmad, was named *Fateḥ Al-Khair.* Nonplussed, the sultan returned the list to the British consul. "God knows," he exclaimed, "there are twenty people in Sur named Jumʿa bin Saʿid or Khalfan bin Ahmad, and twenty dhows named *Fateḥ Al-Khair*!"

NOTES

PROLOGUE: THE LOGBOOK

1. I draw these outlines from Muhammad, *Al-Nawkhithā ʿAbd Al-Majeed Al-Mulla*, 17–18.

2. Authors of these histories are admirably showcased in Potter, *The Persian Gulf in History*.

3. Prins, *Sailing from Lamu*, 57.

4. Chaudhuri, *Trade and Civilization in the Indian Ocean*; Pearson, "Littoral Society"; Das Gupta, *The World of the Indian Ocean Merchant, 1500–1800*.

5. Ho, "Empire through Diasporic Eyes."

6. Ho, "Inter-Asian Concepts for Mobile Societies."

7. Polanyi, "Ports of Trade in Early Societies"; Chaudhuri, *Trade and Civilization in the Indian Ocean*, 98–118.

8. Among the canonical works in microhistory are Ginzburg, *The Cheese and the Worms*, and Davis, *The Return of Martin Guerre*. For an excellent overview of microhistory on a larger scale, of the sort I envision here, see Ghobrial, "Seeing the World Like a Microhistorian."

9. Lepore, "Historians Who Love Too Much."

10. Muir and Ruggiero, *Microhistory and the Lost Peoples of Europe*.

11. In what is easily the most comprehensive biographical dictionary of Kuwaiti nakhodas, running nearly six hundred pages of biographical entries, some of which take up three or four pages, Al-Failakawi receives but two paragraphs. Al-Hijji, *Nawākhithat Al-Safar Al-Shirāʿī fī Al-Kuwayt*, 529–30.

12. Levi, "On Microhistory," 98.

13. See also Khalili, *Sinews of War and Trade*.

14. Huber, *Channeling Mobilities*.

15. For scholars, the bazaar is a physical site, but also an analytic idea: a marketplace of buyers and sellers characterized by a high velocity of exchanges, often very local in their dimensions, and characterized by clientelism and bargaining. The standard reference in writings on the bazaar is Geertz's "The Bazaar Economy."

Geertz's study gave rise to other studies in different settings. See also Keshavarzian, *Bazaar and State in Iran*. Historians have been inclined toward an understanding of the bazaar as a one of the frontiers of global capitalism. Rajat Kanta Ray positioned the bazaar as an intermediate site, one that bridged the capital markets in metropolitan colonial centers in India to lesser-known (and less serviced) markets in the broader Western Indian Ocean. See Ray, "Asian Capital in the Age of European Domination," 554. Other studies have illustrated how the world of Indian Ocean trade rested on a foundation of firms, instruments, and institutions that facilitated the circulation of goods and capital, and mediated between producers and consumers as they linked together far-flung markets. See also Barendse, *Arabian Seas*; Bose, *A Hundred Horizons*; Machado, *Ocean of Trade*; McDow, *Buying Time*; Mathew, *Margins of the Market*; Prestholdt, *Domesticating the World*; Amrith, *Crossing the Bay of Bengal*; Hopper, *Slaves of One Master*; Yahaya, *Fluid Jurisdictions*; Bishara, *A Sea of Debt*; Bishara and Wint, "Into the Bazaar"; Wint, "From *Desh* to *Desh*."

16. Lipartito, "Reassembling the Economic."

17. For some useful examples, see Parthasarathi, *Why Europe Grew Rich and Asia Did Not*; Pomeranz, *The Great Divergence*; Frank, *ReOrient*; Kuran, *The Long Divergence*.

18. Mathew, *Margins of the Market*, 44–51.

19. Chaudhuri, *Trade and Civilization in the Indian Ocean*, 221–22.

20. Martin and Martin, *Cargoes of the East*, 225–26.

21. Chakrabarty, *Provincializing Europe*, 47–71.

22. See, for example, Ali, *A Local History of Global Capital*; Prestholdt, *Domesticating the World*.

23. Thesiger, *Arabian Sands*, 277–79.

24. This one is for you, Laurie Benton.

25. I chose 1924 because it marks the first full year's voyage in the logbook. The two preceding logged voyages give less of a sense of a dhow's itinerary over a year. The first only records Al-Failakawi's return voyage from the Indian coast in 1920, while the second only involves a half year's travel, in what would be the winter season's voyage (*muṭrāsh*) that Gulf nakhodas embarked upon when they were able to complete two return trips in a calendar year.

26. Tanaka, "History without Chronology," 172–73.

CHAPTER ONE: KUWAIT

1. "Dhow" is a generic term used to describe a number of different lateen-rigged sailing vessels in the Indian Ocean. Nakhodas never used the term; they called their ships *booms*, *baghlahs*, *sanbuqs*, *bateels*, etc., depending on their size, shape, and function. When they used a generic term, they would use either *markab* ("ship") or *khashab*, the Arabic word for "wood."

2. A *mann* was a measure used to calculate weight; in turn-of-the-century Kuwait, it equaled roughly 76 kilograms. The cargo capacity for each 1,000 *mann* of dates equaled roughly 75 tons. Al-Hijji, *Kuwait and the Sea*, 56.

3. Al-Hijji, *Al-Nashaṭāt Al-Baḥriyya*, 109–11.

4. Al-Hijji, *Ṣināʿat Al-Sufun*, esp. 55–56.

5. Kuwait housed a number of different *sūqs*, the most prominent of which were the *sūq al-tujjār* (the merchants' *sūq*), where merchant-wholesalers maintained their offices, and the *sūq al-dākhilī* (the indoor *sūq*, named so for its palm frond and mangrove pole covering), which began as a general goods market but became more specialized as the marketplace itself expanded in the early twentieth century. The rest of the *sūq*, historian Farah Al-Nakib points out, was "segmented into smaller inner markets" that were "specialized according to commodity, trade, or both"; Al-Nakib, *Kuwait Transformed*, 46–47. More detailed information on the various *sūqs* can be found in Jamal, *Aswāq Al-Kuwait*, 137–312.

6. During the summer months, shopkeepers kept a running tab with merchants, which they would service on a weekly basis. Jamal, *Aswāq Al-Kuwait*, 36–38, 253–54.

7. Farah Al-Nakib beautifully describes the deeply textured social life in Kuwait's various pre-oil neighborhoods and the self-regulation that characterized them. See Al-Nakib, *Kuwait Transformed*, 58–62.

8. Lorimer, *Gazetteer*, 2:1051; Al-Nakib, *Kuwait Transformed*, 74. A British traveler passing through Kuwait in 1831 suggested that there were "no natives of any other country resident in the place." Stocqueler, *Fifteen Months Pilgrimage*, 1:20. It's unclear, though, whether he would have been able to distinguish natives from non-natives.

9. On the caravan trade between the Gulf and Arabian Peninsula, see also Pétriat, "Caravan Trade in the Late Ottoman Empire," 38–72.

10. Its rulers capitalized on this advantage: in exchange for a fee, they arranged with neighboring tribes to furnish protection to overland travelers. As early as 1758, the English traveler Edward Ives negotiated with the ruler of Kuwait to provide him with protection as he crossed from Basra to Aleppo (though he ultimately took the route to Baghdad instead). Ives, *A Voyage from England to India*, 222–23.

11. Al-Hijji, *Kuwait and the Sea*, 124–29.

12. "Koweit, 1908–1928," India Office Records, British Library, London (hereafter, IOR), IOR/L/PS/18/B395, 9–10.

13. IOR/L/PS/12/3734. See also Toth, "Tribes and Tribulations," 150–55.

14. Al-Nakib, *Kuwait Transformed*, 44–45, 57.

15. Jamal, *Aswāq Al-Kuwait*, 98–100.

16. The walls themselves had to be demolished and rebuilt several times as the town's population expanded during the late nineteenth and early twentieth centuries. Lorimer, *Gazetteer*, 2:1050; Al-Nakib, *Kuwait Transformed*, 22–25; Khalifoh, *Sūr Al-Kuwait*.

17. Al-Nakib, *Kuwait Transformed*, 55.

18. Failaka also formed an important settlement on the maritime trade routes that linked the ancient civilizations of Mesopotamia and Dilmun (situated in what is today Bahrain). See also Crawford, *Dilmun and Its Gulf Neighbors*; Potts, *The Arabian Gulf in Antiquity*, 2:154–96.

19. Lorimer (*Gazetteer*, 2:512) suggests that the population of Failaka numbered no more than five hundred, of which two hundred were men, whereas one Kuwaiti historian suggests that there were two hundred households and a total population of roughly one thousand. Muhammad, *Al-Nawkhithā ʿAbd Al-Majeed Al-Mulla*, 12.

20. Lorimer, *Gazetteer*, 2:513; Muhammad, *Al-Nawkhithā ʿAbd Al-Majeed Al-Mulla*, 13.

21. Lorimer, *Gazetteer*, 2:513.

22. Lorimer, 2:514–15. Al-Khiḍr was thought to be a contemporary of Moses—and, in Failakawi lore, a giant who could cross the entire bay in a single step. The *maqām* of Al-Khiḍr in Failaka was not his tomb; rather, it was believed to be a place he walked through, leaving an impression of his foot. It was the site of a range of different supplications and rituals, mostly related to fertility, and a pilgrimage destination for travelers from around the Persian Gulf and Arabian Peninsula. The *maqām* and the practices surrounding it became the subject of contention among clerics from around the region, and in the early 1980s the Kuwaiti government razed the shrine and built a military observation post in its place. For a useful discussion of Failaka's saints and Al-Khiḍr, see Ashkanani, *Middle-Aged Women in Kuwait*, 183–200. Shrines devoted to Al-Khiḍr exist in other parts of the Islamic world.

23. Most agree that the proximate cause of the ʿUtub's migration from Central Arabia to the coast was conflict between them and other tribes in the area. Al-Rushaid, *Tārīkh Al-Kuwayt*, 30–34; Abu Hakima, *The Modern History of Kuwait*, 1–5; Al-Shamlan, *Min Tārīkh Al-Kuwayt*, 106–11; Khazʿal, *Tārīkh Al-Kuwayt Al-Siyāsī*, 1:40–41.

24. Al-Rushaid, *Tārīkh Al-Kuwayt*, 61–72.

25. Al-Rushaid, 54.

26. Al-Rushaid, 56–5.

27. Bayly, *Imperial Meridian*, 35–52; Hodgson, *The Venture of Islam*, 3:134–61.

28. Mandaville, "The Ottoman Province of al-Hasā," 486–513; Anscombe, *The Ottoman Gulf*.

29. Abu Hakima, *History of Eastern Arabia*, 128–43.

30. Fattah, *The Politics of Regional Trade*.

31. See also Abdullah, *Merchants, Mamluks, and Murder*, 71–72; Abu Hakima, *History of Eastern Arabia*, 86–89.

32. Brucks, *Memoir Descriptive of the Navigation of the Gulf of Persia*, 566.

33. Stocqueler, *Fifteen Months Pilgrimage*, 2. By Al-Rushaid's time, *baghlahs* were rare; Kuwaiti shipbuilding had shifted more decisively to the construction of the *boom*, a double-ended ship, distinguished by its straight, sharp-pointed stemhead. Unlike the ornate *baghlah*, the *boom* was a more rough-and-ready sort of ship and took less time and money to build. It was also smaller, lighter, and faster. Agius, *Seafaring in the Arabian Gulf and Oman*, 15–16.

34. Al-Ghunaim, *Qirā'āt fī Wathā'iq Usrat Al-Nuṣuf*; Al-Ghunaim, *Ḥadīth Al-Wathā'iq*.

35. Muhammad, *Al-Nawkhithā 'Abd Al-Majeed Al-Mulla*, 17.

36. 'Abdulwahhab bin 'Abdulrahman Al-'As'ousi to 'Abdulrahman bin Hussain Al-'As'ousi (10 Rabi' Al-Akhar 1348 / 15 September 1929), Al-'As'ousi Collection, Kuwait (hereafter, ACK); on Rashid bin 'Ali, see Al-Hijji, *Nawākhithat Al-Safar Al-Shirā'ī fī Al-Kuwait*, 117–19.

37. Al-Hijji, *Nawākhithat Al-Safar Al-Shirā'ī fī Al-Kuwait*, 123–24. Al-Hijji notes that 'Isa Bishara worked as a navigator (*mu'allim*) under the nakhoda 'Abdullah Al-'Uthman; Al-Hijji, 202–3.

38. Al-Hijji, *Nawākhithat Al-Safar Al-Shirā'ī fī Al-Kuwait*, 113–19.

39. See, for example, Zanzibar National Archives, Zanzibar, Tanzania (hereafter, ZNA), AM 9/4, 92, 118, 164, 382; AM 8/2 Deed 2 of 1893; IOR R/15/2/2018, 26.

40. Villiers, *Sons of Sindbad*, 347.

41. Muhammad, *Al-Nawkhithā 'Abd Al-Majeed Al-Mulla*, 17–19.

42. Al-Hijji, *Nawākhithat Al-Safar Al-Shirā'ī fī Al-Kuwait*, 200.

43. Accounts from 1 Rajab and 9 Rajab 1359, Al-Failakawi Collection, Kuwait (hereafter, AFC).

44. Al-'Uthman, *Ruznāmat*, 67, 200, 219, 243, 270, 296, 331, 353, 383, 403, 426.

45. Bishara, *Ruznāmat*, 304, 408, 470, 498, 529, 598.

46. Al-Hijji, *Kuwait and the Sea*, 56.

47. Al-Nakib, *Kuwait Transformed*, 76; Fuccaro, *Histories of City and State*, 35, 155.

48. Rediker, *Between the Devil and the Deep Blue Sea*, 116–62.

49. Lienhardt, *Shaikhdoms of Eastern Arabia*, 140–41; Prins, *Sailing from Lamu*, 214–16; Villiers, *Sons of Sindbad*, 353, 401–2.

50. Muhammad Thunayyan Al-Ghanim interview on *Ṣafḥāt Min Tārīkh Al-Kuwayt*. The names are akin to the names of the enslaved that one might find in an Atlantic or American context—Thursday, Jupiter, and the like.

51. Hopper, *Slaves of One Master*, 81, 86–9.

52. On enslaved mariners in East Africa, see Glassman, "The Bondsman's New Clothes," 291–92; Gilbert, *Dhows and the Colonial Economy*, 49.

53. Al-Rifai, *Al-Nehmah wal-Nahham*, 121–202; Al-Habad, *Fann Al-Sangani*.

INSCRIPTION: DEBTS

1. Villiers, *Sons of Sindbad*, 346–47. I wrote about one form of this kind of acknowledgment, the *iqrar*, in my previous book, *A Sea of Debt*.

2. I draw all the discussion above from my book *A Sea of Debt*.

3. Barwa by 'Ali bin Saleh Al-Dabbus, 1307 AH, Ali Ra'is Collection, Kuwait (hereafter, ARC).

4. Barwa by Nahit bin 'Ali for Hussain Al-Balushi, 1313 AH, ARC.

5. The discussion here draws from *barwas* held at the Center for Research and Studies on Kuwait. I have not been able to ascertain whether the arrangements involving

partial repayment were a long-standing feature of the Gulf maritime economy or whether they were particular to the 1930s, from which all the partial *barwas* emerge.

6. On dhow firms, see Prins, *Sailing From Lamu*, 214–16.

7. Guyer, "Wealth in People, Wealth in Things"; Guyer and Belinga, "Wealth in People as Wealth in Knowledge."

CHAPTER TWO: THE SHATT AL-ʿARAB

1. Villiers, *Sons of Sindbad*, 389; Dowson, *Dates and Date Cultivation*, 3:40.

2. Lorimer, *Gazetteer*, 2:97; Dowson, *Dates and Date Cultivation*, 1:4–5, 41.

3. Villiers, *Sons of Sindbad*, 390.

4. Lorimer gives the value of Basra dates exported at that time as £330,000; I arrived at this number using his own figure of roughly £7 a ton as an estimate (for *sāyer* dates this would have been as low as £5; at other times he suggests it was as high as £9 per ton). Lorimer, *Gazetteer*, 1:2307. Dowson, *Dates and Date Cultivation*, 1:44, 3:18.

5. The bulk of this information comes from Dowson, *Dates and Date Cultivation*, pt. 3. See also Al-Hijji, *Kuwait and the Sea*, 86–87.

6. Dowson, *Dates and Date Cultivation*, 1:31–33, 3:33–35, 61, 64, 67, 80.

7. Al-Hijji, *Kuwait and the Sea*, 61–62.

8. Al-Matrook, *Min Tijārat Al-Māḍī*, 9–10.

9. On steam shipping in the Gulf, see Jones, *Two Centuries of Overseas Trading*, 79–110; Mathew, *Margins of the Market*, 21–51.

10. On "MK" dates, see also Thunayyan Al-Ghanim to Muhammad bin ʿAbdullah Al-Matrook (17 Jumada Al-Thani 1355 / 4 September 1936) and ʿAbdulrazzaq Al-ʿAdwani to Muhammad bin ʿAbdullah Al-Matrook (20 Muharram 1340 / 23 September 1921) Al-Matrook Collection, Kuwait (hereafter, MCK).

11. British officials had established a telegraph station in Al-Faw by 1865, connecting the upper Gulf to stations in western India via a network of submarine cables; from Al-Faw they ran overland to Baghdad. Kelly, *Britain and the Persian Gulf*, 555–63.

12. The bank also had branches in other major cities in India, and around Iraq. "German War: Banks at Basra &c.—the Eastern Bank," IOR L/PS/10/529.

13. See the letters from Thunayyan Al-Ghanim to Muhammad bin ʿAbdullah Al-Matrook, MCK.

14. ʿAbdullah bin Muhammad Al-Fozan to Muhammad bin ʿAbdullah Al-Matrook (2 September 1921), MCK.

15. Dowson, *Dates and Date Cultivation*, 1:6–7, 20, 27–29, 33–37, 46–47.

16. This was the case for at least three different seasons between 1888 and 1905. Lorimer, *Gazetteer*, 2:2299–302.

17. Though many laborers worked for a wage, *fellāḥs* took a share of the produce. In the 1920s, their share amounted to half of all vegetables grown, and anywhere between one-eighth to one-fourth of the produce on the palm trees. Dowson, *Dates and Date Cultivation*, 1:37.

18. Reza Sayegh to Muhammad Khalil Al-Sharif (28 July 1906), Ali Akbar Bushihri Collection, NYU Abu Dhabi (hereafter, ABC).

19. Dates receipts and accounts held by Al-Matrook in MCK.

20. Lorimer, *Gazetteer*, 2:544; Curzon, *Persia and the Persian Question*, 2:335–36.

21. On the Ottomans and Safavids, see also Dale, *The Muslim Empires of the Ottomans, Safavids, and Mughals*, and Hodgson, *The Venture of Islam*, 3:16–58, 99–133.

22. See also Floor, *The Persian Gulf*, chaps. 3, 5, and 8.

23. Abdullah, *Merchants, Mamluks, and Murder*, 28–34.

24. On the Shatt as a borderlands region, see Cole, *Empire on Edge*. See also Abdullah, *Merchants, Mamluks, and Murder*, 39–56; Fattah, *Politics of Regional Trade*, 185–206. On Kuwait and Mohammerah specifically, see Abu Hakima, *History of Eastern Arabia*, 64–65; Al-Najjar, *'Arabistān*, 159–60. I take on the question of trade and warfare in much greater detail in chapter 3.

25. Fattah, *Politics of Regional Trade*, 91–102, 113–18; Çetinsaya, *Ottoman Administration of Iraq, 1890–1908*, 66–71; Al-Hulw, *Al-Aḥwāz*, 3:7, 10–11, 19–28; Al-Najjar, *'Arabistān*, 88–89.

26. Kelly, *Britain and the Persian Gulf*; Peterson, "Britain and the Gulf," 277–93.

27. Kelly, *Britain and the Persian Gulf*, 483.

28. Al-Hulw, *Al-Aḥwāz*, 3: 92–95; Cole, *Empire on Edge*, 330–53.

29. Al-Najjar, *'Arabistān*, 161–62.

30. On Mubarak and his Shatt land, see Cole, *Empire on Edge*, 258–85; Lorimer, *Gazetteer*, 2:536–44. Some Kuwaiti rulers bought land along the Shatt privately, while others were awarded property in return for their political support: when Jaber Al-Sabah (1814–1859) gave refuge to the shaikh of the Muntafiq tribe during one of the latter's open revolts against the Ottoman government in Basra, the shaikh granted him three large plots in Al-Faw. Al-Reshaid, *Tārīkh Al-Kuwait*, 97. Al-Reshaid also recalled an episode in which Rashid Al-Sa'dun gave Jaber's son Mubarak a large plot of date palms in Al-Ma'amir as gift to mark the Eid celebrations; Jaber, however, forced his son to return the gift. Al-Reshaid, 101.

31. Lorimer, *Gazetteer*, 2:106. Mubarak's land purchases also became the subject of extensive litigation during the 1920s and 1930s. "The Date Gardens in Iraq of the Shaikhs of Koweit and Mohammerah," IOR L/PS/18/B468, 31–32.

32. Memorandum from the Political Agent, Kuwait (21 November 1931), IOR L/PS/18/B468, 18.

33. On Mubarak's father Jaber and voluntary contributions, see Al-Reshaid, *Tārīkh Al-Kuwayt*, 100–101; Al-Reshaid, however, notes in several instances that Jaber Al-Sabah and his son 'Abdullah received an annual salary of dates from the Ottoman government and the shaikhs of Mohammerah in return for political support that they had lent them, and that they drew income from their properties in Al-Faw and from the Al-Zuhair's date palms in Al-Sufiya. Al-Reshaid, 98, 101–3, 109. 'Abdullah had also received a stipend of dates from Jaber bin Mirdaw, Khaz'al's father, for helping the latter put down a rebellion in Al-Qasba. Al-Hulw, *Al-Aḥwāz*, 3:35.

34. Al-Hulw, *Al-Aḥwāz*, 3:138–39.

35. Memorandum from the Political Agent, Kuwait (21 November 1931), IOR L/PS/18/B468, 18.

36. Al-Khariji, *Al-Qawā'id wal-Mīl*, 84.

37. See also Metcalf, *Imperial Connections*, 89–101; Dodge, *Inventing Iraq*; Tripp, *A History of Iraq*, 30–74.

38. See also Carter, *Sea of Pearls*, 141–81; Hopper, *Slaves of One Master*, 80–104.

39. American imports nearly doubled between the 1890s and 1900s, and then more than doubled again from 32 million pounds. in 1920 to 79 million pounds. in 1925. Hopper, *Slaves of One Master*, 58; Hopper, "The Globalization of Dried Fruit," 167–69.

40. Lorimer, *Gazetteer*, 2:102, 104, 107–8. On the Lynch Brothers, see also Cole, "Precarious Empires," 74–101; Shahvanaz, *Britain and the Opening Up of South-West Persia, 1880–1914*.

41. Cole, *Empire on Edge*; The Bahrain-based merchant 'Abdulnabi Safar was one of these: at the time of his death in 1884, Safar owned property in Bahrain, Bushehr, Basra, and Bombay, which he left to his six children. Petition from Muhammad Sadiq Safar to the Ottoman Shahbandar of Basra (15 May 1911), ABC. Onley, "Transnational Merchants," 63–71. Decades later, Safar's descendants were still actively buying up property in Basra. Letter from A. Latif to Muhammad Khalil Al-Sharif (10 August 1930), ABC.

42. The bulk of this information is distilled from Al-Harbi, *Tārīkh Al-'Ilāqāt Al-Kuwaytiyya Al-Hindiyya*, 194–99; Al-Ibrahim, *Min Al-Shirā'*, 83.

43. The literature on the family firm is enormous. Examples that draw on the Indian Ocean world of trade include Markovits, *The Global World of Indian Merchants*, 156–84; Aslanian, *From the Indian Ocean to the Mediterranean*, 121–65; Wint, "Credible Relations"; Onley, "Transnational Merchant Families"; McDow, *Buying Time*, 117–44. On family firms in the Atlantic, see also Chandler, *The Visible Hand*, 15–19; Hancock, *Citizens of the World*, 40–84; Doerflinger, *A Vigorous Spirit of Enterprise*, 45–69. On family business and inheritance in the Islamic world, see also Kuran, *The Long Divergence*, 78–96; Hanna, *Making Big Money in 1600*, 39–42, 53–69. For an alternative interpretation, see Doumani, *Family Life in the Ottoman Mediterranean*.

44. Al-Harbi, *Tārīkh Al-'Ilāqāt*, 196.

45. Writing at the turn of the century, Lorimer described Al-Saraji as "a village, about 2 miles up a large creek" made up of two thousand inhabitants and roughly sixty thousand date palms. Lorimer, *Gazetteer*, 2:102.

46. I draw all this from a deed that collated property purchases by 'Abdulmuhsin bin Nasser Al-'Abdulkarim and Lulwah bint 'Isa Al-'Abdulkarim, along with *waqf* declarations by Lulwah bint 'Isa and her mother Tarqa bint Nasser, and *qadi* notes, all on a single, remarkable document held by the Al-Ibrahim family in Kuwait. The 1839 note refers to 'Abdullah as a minor, though he would have been twenty-four years old at the time.

47. Sale deed dated 16 Jumada Al-Awwal 1272 AH (24 January 1856); sale deed dated 21 Shawwal 1275 AH (24 May 1859); sale deed dated 13 Rabi' Al-Awwal 1273

AH (11 November 1856); sale deed dated 28 Jumada Al-Awwal 1277 AH (12 December 1860); sale deed dated 1 Muharram 1287 AH (3 April 1870). All from "Wathā'iq wa Asanīd," *Bayt Bin Ibrahim*.

48. Al-Harbi, *Tārīkh Al-'Ilāqāt*, 197.

49. Lorimer, *Gazetteer*, 2:1032.

50. See estate accounts dated 9 Rajab 1292 (11 August 1875), "Wathā'iq wa Asanīd," *Bayt Bin Ibrahim*. He also left behind a daughter, but I have no information on her fate.

51. Al-Ibrahim Family, *Al-Shaikh Jassim bin Muhammad Al-Ibrahim*, 10.

52. Lorimer, *Gazetteer*, 2:114; Al-Qasimi, *Bayān Al-Kuwayt*, 28.

53. When Mubarak assassinated his half brothers on the night of May 6, 1896, Yousef thrust himself into the political spotlight. In the wake of the assassination, Yousef worked to defend the interests of his nephews, the sons of the slain rulers—particularly to see to it that their shares in their fathers' properties along the Shatt would be preserved. Yousef and Mubarak both appealed to Ottoman officials in Basra and Istanbul, British agents in the Persian Gulf, and other Gulf rulers. See also Anscombe, *The Ottoman Gulf*, 92–109.

54. Al-Wazzan, *Tijārat Al-Naql Al-Baḥrī*, 37–49.

55. Al-Wazzan, 89–91.

56. Lorimer, *Gazetteer*, 2:7.

57. Rashid bin 'Ali Al-'As'ousi to 'Abdulrahman bin Hussain Al-'As'ousi (19 October 1924 and 7 November 1924), ACK.

58. 'Isa 'Abdullah Al-'Uthman interview on *Ṣafḥāt Min Tārīkh Al-Kuwayt*.

59. Rashid bin 'Ali Al-'As'ousi to 'Abdulrahman bin Hussain Al-'As'ousi (24 November 1921), ACK.

60. Debt acknowledgment from 'Abdulrahman bin Hussain Al-'As'ousi to 'Abdulmuhsin Al-Kharafi and Muhammad Al-Matrook (3 March 1931), ACK. The debt was for 143 *mann* of flour, to be deducted from the ship's freightage in six months.

61. Rashid bin 'Ali Al-'As'ousi to 'Abdulrahman bin Hussain Al-'As'ousi (27 November 1922), ACK; 'Abdulrahman bin Hussain Al-'As'ousi to Rashid bin 'Ali Al-'As'ousi (3 December 1922), ACK.

62. 'Abdulrahman bin Hussain Al-'As'ousi to Rashid bin 'Ali Al-'As'ousi (23 November 1921), ACK.

63. Al-Hijji, *Al-Nashāṭāt Al-Baḥriyya*, 139–40.

64. Al-Failakawi, Logbook 1, 122, AFC.

INSCRIPTION: FREIGHTAGE

1. The contract was reprinted in Al-Wazzan, *Tijārat Al-Naql Al-Baḥrī*, 57.

2. On a Kuwaiti nakhoda's responsibilities for losses of cargo at sea in the 1930s and 1940s, see also Al-Hijji, *Kuwait and the Sea*, 152–66; Villiers, *Sons of Sindbad*, 376–77.

3. Al-Wazzan, *Tijārat Al-Naql Al-Baḥrī*, 185.

4. Contract between Al-Matrook and Al-Kharafi, and Muhammad bin ʿUthman (10 November 1935), MCK. I thank Hollian Wint for translating both Gujarati texts.

5. Historians have identified a similar phenomenon in East African contracts that were in both Arabic and Gujarati; see Bishara and Wint, "Into the Bazaar."

CHAPTER THREE: THE GULF

1. I draw most of what follows from ʿIsa Bishara's interview with Rida Al-Fayli on Kuwait Television.

2. Lorimer, *Gazetteer*, 2:1473–74.

3. On Qays, see Margariti, "Mercantile Networks, Port Cities, and 'Pirate' States"; more generally, see Prange, "The Contested Sea."

4. "Piracies and Irregularities at Sea," IOR R/15/5/51, 58–113.

5. "Piracies and Irregularities at Sea," IOR R/15/5/51, 15–49.

6. "Piracies and Irregularities at Sea," IOR R/15/5/51, 148–212.

7. Abu Hakima, *History of Eastern Arabia*, 36–37; Fattah, *Politics of Regional Trade*, 28–41; Abdullah, *Merchants, Mamluks, and Murder*, 29–37; Matthee, "Between Arabs, Turks, and Iranians."

8. Axworthy, "Nader Shah and Persian Naval Expansion in the Persian Gulf, 1700–1747."

9. See also Perry, "The Zand Dynasty," 86–93. For a detailed account of this period in Gulf history, see Floor, *The Persian Gulf.*

10. Abu Hakima, *History of Eastern Arabia*, 64, 82–83; Floor, "The Rise and Fall of the Banū Kaʿb."

11. Historians have given a few, not necessarily conflicting, reasons for the move: first, to place some distance between them and the Bani Kaʿb, who had been harassing them for decades; second, to assert a greater degree of autonomy than they could under the Al-Sabah; and third, to situate themselves closer to the rich pearl banks of the southern Gulf. Abu Hakima, *History of Eastern Arabia*, 65–66; Al-Rushaid, *Tārīkh Al-Kuwayt*, 90; Al-Qinaʿi, *Ṣafḥāt min Tārīkh Al-Kuwayt*, 17.

12. Buckingham, *Travels in Assyria, Media, and Persia*, 299–300.

13. "Sketch of the Proceedings of Rahmah bin Jaubir, Chief of Khor Hassan," *Selections from the Records of the Bombay Government* (1856), IOR R/15/1/732, 528. The most complete treatment of Rahmah to date is Roberts, "A Tolerated Terror," in which he situates him at the peripheries of an expanding Busaʿidi Empire.

14. Fuccaro, *Histories of City and State*, 52–53.

15. I draw all this information from Robert Carter's excellent *Sea of Pearls*, 109–82.

16. Letter from Francis Warden, Political Department, Bombay Castle to Acting Political Agent at Qishm (1 August 1821), IOR R/15/1/24, 15–16.

17. Loch, *Diary in Persian Gulf and Sketch Books*, 88–95.

18. *Cambridge History of Iran, Vol. 7*, 86–87; Al-Qasimi, *Power Struggles and Trade in the Gulf*, 46–50.

19. For good surveys of Omani expansion into East Africa, see also Bhacker, *Trade and Empire in Muscat and Zanzibar*; Bishara, *A Sea of Debt*; McDow, *Buying Time*; Nicholls, *The Swahili Coast*; Sheriff, *Slaves, Spices, and Ivory*; Wilkinson, *The Arabs and the Scramble for Africa*.

20. Kelly, *Britain and the Persian Gulf*, 126; Wilkinson, *The Imamate Tradition*, 58.

21. Ibn Ruzayq, *Fatḥ Al-Mubīn*, 2:349–50.

22. Wilkinson, *The Imamate Tradition*, 48; Abu Hakima, *History of Eastern Arabia*, 34.

23. Ibn Bishr, *ʿUnwān Al-Majd*; Al-Rasheed, *A History of Saudi Arabia*, 14–22.

24. Fattah, *The Politics of Regional Trade*, 41–52.

25. Loch, *Diary in Persian Gulf and Sketch Books*, 125–26.

26. Buckingham, *Travels in Assyria, Media, and Persia*, 210–14.

27. Kelly, *Britain and the Persian Gulf*, 102–3, 110; Ibn Ruzayq, *Al-Fatḥ Al-Mubīn*, 353–54.

28. Fattah, *The Politics of Regional Trade*, 56.

29. On this, see Halevi, "Arabians for Guns." For the place of oceanic trade in the early years of the Saudi state, see Roberts, "Oceanic Wahhabism."

30. Buckingham, *Travels in Assyria, Media, and Persia*, 227.

31. Davies, *The Blood-Red Arab Flag*, 91–125.

32. Kelly, *Britain and the Persian Gulf*, 116–21.

33. Davies, *The Blood-Red Arab Flag*, 107–23, 300–319.

34. Kelly, *Britain and the Persian Gulf*, 139–54.

35. See also Elliott and Prange, "Beyond Piracy: Maritime Violence and Colonial Encounters in Indian History"; Layton, "Hydras and Leviathans in the Indian Ocean World"; Subramanian, *The Sovereign and the Pirate*.

36. See also Layton, "Discourses of Piracy in an Age of Revolutions."

37. Loch, *Diary in Persian Gulf and Sketch Books*, 305–8.

38. Loch, 354.

39. Loch, 350–51.

40. Dods, *The Works of Aurelius Augustine*, 1:139–40.

41. Kelly, *Britain and the Persian Gulf*, 154–59.

42. Aitchison, *A Collection of Treaties*, 7:249–51.

43. Kelly, *Britain and the Persian Gulf*, 156.

44. Aitchison, *A Collection of Treaties*, 7:249–54.

45. Johnson, *General T. Perronet Thompson*, 98. Drafts of the treaty were produced by Thompson's wife.

46. Johnson, *General T. Perronet Thompson*, 101–2.

47. Brucks, *Memoir Descriptive*, 541–42, 575–76.

48. Kelly, *Britain and the Persian Gulf*, 157, 193–259. For a clear case of piracy, tried in Bombay, see Maharashtra State Archives, Mumbai (hereafter, MSA), PD 1835–36 Vol. 31/687.

49. In 1841, however, the ruler of Kuwait did agree to abstain from maritime hostilities for a year. Kelly, *Britain and the Persian Gulf*, 369.

50. Aitchison, *A Collection of Treaties*, 7:258–59; Kelly, *Britain and the Persian Gulf*, 357–68.

51. Aitchison, *A Collection of Treaties*, 7:259–60.

52. This is similar to the argument made by Al-Naqeeb, *State and Society in the Gulf and Arab Peninsula*, but draws the timeline further back than Al-Naqeeb, who identified a political cycle that froze with the protectorate agreements that came later in the nineteenth century.

INSCRIPTION: PASSAGE

1. Al-Khariji, *Al-Qawāʿid wal-Mīl*, 82.

2. Foreign Office, 93/137, 27–30.

3. *Qawl* by Majid bin Sultan Al-Busaʿidi (13 Shawwal 1284), ZNA AA 7/2.

4. The writers refer to the documents as *buyurldu*, a term that denotes a decree by a member of the Ottoman bureaucracy (though not the sultan himself) that could include matters of safe passage but was not limited to them.

5. Al-Wazzan, *Tijārat Al-Naql Al-Baḥrī*, 25; *qawl* issued to Hussain bin Muhammad Al-ʿAsʿousi (2 December 1859), ACK.

6. "File 32/7 Certified copies of deeds and documents from 1 January 1927 to 1932," IOR R/15/5/89; *qawl* issued to ʿAbdulrahman bin Hussain Al-ʿAsʿousi (15 August 1917), ACK.

CHAPTER FOUR: THE SEA OF OMAN

1. Al-Failakawi, Logbook 1, 121, AFC.

2. On debates around fishing in the Sea of Oman, see Erich, "A Deeper History of the World's Largest Dead Zone."

3. Tibbetts, *Arab Navigation*, 169.

4. Tibbetts, *Arab Navigation*, 202, 217–24.

5. The Da Gama episode was first described by the Ottoman historian Qutb al-Din Al-Nahrawali, who Tibbetts suggests might have been motivated by personal or family-related differences with Ibn Majid. The myth was later picked up by early scholars working on the *Kitab Al-Fawāʿid* and circulated between then. Tibbetts, *Arab Navigation*, 9–11; Hourani, *Arab Seafaring*, 83–84; Hall, *Empires of the Monsoon*, 179–83.

6. Villiers, *Sons of Sindbad*, 186.

7. See also Tibbetts, *Arab Navigation*, 9–12; Subrahmanyam, *The Career and Legend of Vasco da Gama*, 121–28.

8. There are too many examples of this trope to list, but for a clear statement see Pannikar, *Asia and Western Dominance*, 29–30.

9. Tibbetts, *Arab Navigation*, 209; Alpers, *The Indian Ocean in World History*, 74; Ibn Majid, *Al-Nūniya Al-Kubrā*, 19–20.

10. Among many others on this, see also Subrahmanyam, *The Career and Legend of Vasco da Gama*; and Subrahmanyam, *The Portuguese Empire in Asia*.

11. Hall, *Empires of the Monsoon*, 214–26; Chaudhuri, *Trade and Civilization*, 68–69; Floor, *The Persian Gulf*, 91–97.

12. Maloni, "Control of the Seas." See also Benton, *A Search for Sovereignty*, 104–6; Alexandrowicz, *An Introduction to the History of the Law of Nations in the East Indies*, 64–65. For a succinct explanation of the *cartaz* system, see Subrahmanyam, *The Portuguese Empire in Asia, 1500–1700*, 82.

13. Thomaz, "Precedents and Parallels of the Portuguese Cartaz System."

14. De Barros, *Decadas I*, bk. 6, chap. 1.

15. Marcocci, "Saltwater Conversion," 239–40; Cattelan, "Iberian Expansion over the Oceans."

16. Casale, *The Ottoman Age of Exploration*; Özbaran, *Ottoman Expansion Towards the Indian Ocean*.

17. See also Vieira, "Mare Liberum vs. Mare Clausum"; Armitage, *The Ideological Origins of the British Empire*, 100–124.

18. Grotius, *The Free Sea*, 20–37.

19. Grotius, 15–17, 34, 38–39, 52.

20. Grotius, 14, 34–35, 54. Grotius engaged with the Spanish jurist Francisco de Vittoria's reflections on the Indigenous people of Americas, published in 1557 as *De Indis De Jure Belli*.

21. Knight, "Seraphim de Freitas"; Alexandrowicz, "Grotius versus Freitas."

22. Welwod, "Of the Community and Propriety of the Seas," in Grotius, *The Free Sea*, 63–74.

23. Selden, *Mare Clausum*.

24. Cunha, "Oman and Omanis in Portuguese Sources."

25. Al-Azkawi, *Kashf Al-Ghumma*, 101, 106.

26. For a full account of this history, see Bathurst, "The Yaʿrubi Dynasty," 111–48.

27. Nichols, *The Swahili Coast*, 21; Wilkinson, *The Arabs and the Scramble for Africa*, 24; Wilkinson, *The Imamate Tradition of Oman*, 48; Bathurst, "The Yaʿrubi Dynasty," 179–87.

28. Bathurst, "The Yaʿrubi Dynasty," 179–209.

29. Al-Shaqsi, *Manhaj Al-Ṭālibīn*, 2:608–9. One has to be careful with these materials, as it is often unclear how many of these discourses are time- and place-specific; many of Al-Shaqsi's writings transmit discussions from earlier periods, without attribution.

30. Al-Shaqsi, 9:138–39.

31. Al-Shaqsi, 9:141–45.

32. Al-Shaqsi, 13:77–86.

33. Al-Kindi, *Bayān Al-Sharʿ*, 39:3–16, 148–51. We know that Al-Shaqsi was deeply familiar with the discussions presented in *Bayān Al-Sharʿ* because he often

cited it, and sometimes copied entire passages from the text, which enjoyed widespread popularity in Omani juristic circles. Discussions surrounding *ḥarīm*, though, can be found in any number of texts of jurisprudence.

34. For a useful take on the family dynamics that informed the Canning Award, see Bhacker, "Family Strife and Foreign Intervention."

35. See Bishara, *A Sea of Debt*; McDow, *Buying Time*.

36. Bishara, *A Sea of Debt*, 107–68.

37. Ministère des Affaires étrangères, France (hereafter, MAE), Mascate Vol. 28, 19–21.

38. Erman, *Almost Citizens*, 85–87.

39. I discuss the case and its broader implications in Bishara, "No Country but the Ocean." For a detailed treatment of the nakhodas of Sur, see Al-Ghaylani, *Asyād Al-Biḥār*.

40. Petition by Muhammad bin Sulaym bin ʿAbboud (7 July 1900), MAE, Mascate Vol. 28, 21.

41. Gwadar remained an Omani enclave until 1958, when the sultan sold it to the Government of Pakistan for Rs 5.5 billion, financed in large part by the Aga Khan.

INSCRIPTION: GUIDES

1. *Al-Mukhtaṣar Al-Khāṣṣ lil-Musāfir wal-Ṭawwāsh wal-Ghawwāṣ* and *Al-Khāliṣ min Kul ʿayb li-Waḍʿ al-Jayb*. Of Al-Qitami's three known publications, only the *Dalil* has been republished a number of times.

2. Al-Qitami, *Dalil*, 15.

3. Alan Villiers recalled a visit to a small dhow belonging to a nakhoda from Dubai, whose only aids to navigation were a compass, a map, and "a well-thumbed copy of Isa Kitami's [*sic*] Arab directory of the eastern seas." Villiers, *Sons of Sindbad*, 272. Yacoub Al-Hijji writes that the *Dalil* "became a resource and guide for many of the nakhodas of the Gulf in their voyages to India and Africa and Yemen." Al-Hijji, *Nawakhithat Al-Safar Al-Shiraʿi fi Al-Kuwayt*, 89. This is in part confirmed by Dionysius Agius, who in his discussion of routes wrote that "practically all the Gulf and Omani nakhodas and navigators I interviewed kept referring in the course of the conversation to what Ibn Majid and contemporary authors (namely al-Qutami) said." Agius, *Seafaring in the Arabian Gulf and Oman*, 175.

4. See Varisco, *Seasonal Knowledge and the Almanac Tradition*.

5. Al-Khaduri, *Maʿdan Al-Asrar*. For just one tantalizing example, see Acevedo, "A New Arabic Nautical Manuscript in Lisbon."

6. Al-Qitami, *Dalil*, 15.

7. Al-Qitami, *Mukhtaṣar*, 8.

8. Al-Qitami, *Dalil*, 15–16. Al-Qitami also includes information on the Bay of Bengal, but the details he offers are much sketchier than those on the Western Indian Ocean.

9. Al-Qitami, 81.

10. Al-Qitami, 20.

11. Al-Qitami, 28.

12. Al-Qitami, 31. On headlands and routes, see Horden and Purcell, *The Forbidding Sea*, 125.

13. Casale, *The Ottoman Age of Exploration*, 96–98.

14. Al-Qitami, *Dalil*, 49. On the Treaty of Seeb, see Wilkinson, *The Imamate Tradition of Oman*, 260–73.

15. Wilkinson, *The Arabs and the Scramble for Africa*, 223–27.

16. Rosaldo, *Ilongot Headhunting*, 48, 55.

17. Wick, *The Red Sea*, 116.

18. Braudel, *The Mediterranean*, 1:278.

CHAPTER FIVE: KARACHI TO KATHIAWAR

1. Al-Failakawi, Logbook 1, 120, AFC.

2. I gleaned most of this information from US Navy Hydrographic Office, *Sailing Directions*, 403.

3. Burton, *Sind Revisited*, vii.

4. See also Jakes, *Egypt's Occupation*, 90–92, in which officials also compared Punjab and Egypt, though in less felicitous terms.

5. Dewey, *Steamboats on the Indus*.

6. Roy, *Business History of India*, 88–89; Banga, "Karachi and Its Hinterland," 342.

7. Burton, *Sind Revisited*, 1:33–38.

8. Banga, "Karachi and Its Hinterland," 351–52.

9. Lari, *The Dual City*, 9.

10. *Quarterly Journal of the Department of Agriculture, Bengal* 4 (1911): 137.

11. One nineteenth-century Kuwaiti ruler, Jaber Al-Sabah, was so closely associated with this act that he came to be known as "Rice Jaber" (*Jaber Al-ʿAysh*). Al-Qinaʿi, *Ṣafḥāt min Tārīkh Al-Kuwayt*, 19–20; Al-Rushaid, *Tārīkh Al-Kuwayt*, 94.

12. I draw this from a survey of Al-Marzouq's letters to Muhammad Salim Al-Sudairawi in Bombay, SCK.

13. I gathered this from letters to Al-Matrook from Al-Kharafi, Thunayyan Al-Ghanim, and Muhammad Al-Marzouq, MCK.

14. Jassim Muhammad Boodai to Hamad Al-Sager (6 September 1923), reprinted in Al-Wazzan, *Tijārat Al-Naql Al-Baḥrī*, 191.

15. Trivellato, *The Familiarity of Strangers*, 181.

16. Hamad Al-Sager entered into several of these with agents in Aden. Al-Wazzan, *Tijārat Al-Naql Al-Baḥrī*, 127–30.

17. It looked much like the *ṣuḥba* arrangements that medieval Jewish merchants in the Mediterranean and Indian Ocean entered into with one another. See also Goldberg, "Choosing and Enforcing Business Relationships."

18. Pearson, *The Law of Agency in British India*, 12–13.

19. For two useful surveys of work on organizations, hierarchies, and networks, see Lamoreaux, Raff, and Temin, "Beyond Markets and Hierarchies," and Lipartito, "Reassembling the Economic."

20. Nasser Al-Kharafi to Muhammad Salem Al-Sudairawi (9 December 1910), Al-Sudairawi Collection, Kuwait (hereafter, SCK).

21. Al-Kharafi to Al-Sudairawi (26 November 1912), SCK; Al-Kharafi to Al-Sudairawi (28 December 1915 and 22 January 1917), SCK.

22. Jassim and Muhammad Al-Marzouq to Al-Matrook (13 and 16 October 1921), MCK.

23. Tomlinson, *The Economy of Modern India*, 55.

24. Nasser Al-Kharafi to Al-Matrook (19 December 1917), MCK.

25. Nasser Al-Kharafi to Al-Matrook (27 February 1918), MCK.

26. *The Karachi Handbook and Directory for 1922–1923*, Karachi: Daily Gazette Press, 1922, C4.

27. Nasser Al-Kharafi to Al-Matrook (29 October 1921), MCK.

28. See also Bishara, *A Sea of Debt*; Bishara and Wint, "Into the Bazaar."

29. *The Karachi Handbook*, D93.

30. Lalji Lakhmidas to Muhammad Salim Al-Sudairawi (4 October 1915, 17 October 1915, and 29 September 1918), SCK.

31. Allen, "Ratansi Purshottam." By Al-Kharafi and Al-Failakawi's time, Ratansi's son Lalji had taken over the firm but continued to operate under his father's name. On a dispute between Ratansi and the Kuwaiti trader ʿAbdulkarim Behbehani (an associate of Al-Matrook's), see IOR R/15/5/87, 69–71.

32. Al-Wazzan, *Tijārat Al-Naql Al-Baḥrī*, 196–97. Hirji Tulsidas to Muhammad Al-Matrook (30 September 1921, and another undated), MCK.

33. "File 32/2 Cases before Foreign Judicial Courts," IOR R/15/5/87, 33–64.

34. Goswami, *Globalization Before Its Time*, 186–242.

35. See also Simpson, *Muslim Society and the Western Indian Ocean*.

36. Burton, *Sind Revisited*, 16–17.

37. *The Gazetteer of Bombay Presidency, Vol. 8: Kathiawar*, 304–14; Ramusack, *The Indian Princes and Their States*, 76–80.

38. Al-Khariji, *Al-Qawāʿid wal-Mīl*, 53, 93.

39. Watson, *Statistical Account of Porbandar*, 7–13; Chand, Mishra, and Bole, *Maritime Trade of Gujarati's Princely States*, 44–73.

40. The only letter ʿAbdulmuhsin sent from there was to Bombay, in 1912, and involved matters of banking. ʿAbdulmuhsin Nasser Al-Kharafi to Al-Sudairawi (13 November 1912), SCK.

41. Rashid bin ʿAli Al-ʿAsʿousi to ʿAbdulrahman bin Hussain Al-ʿAsʿousi (27 April 1918), ACK.

42. Interview with Muhammad Thunayyan Al-Ghanim on *Ṣafḥāt Min Tārīkh Al-Kuwayt*. Al-Hijji, *Nawākhithat Al-Safar Al-Shirāʿī*, 241–45.

43. Muhammad Thunayyan Al-Ghanim to Al-Matrook (26 September, 10 October, and 8 December 1921), MCK.

44. Shihab bin Muhammad to ʿAbdulrahman bin Hussain Al-ʿAsʿousi (13 January 1928), ACK.

45. See, for example, the letters from Jassim Muhammad Al-Ghanim and Ali Ibrahim Al-Kulaib to Muhammad Salem Al-Sudairawi, SCK.

46. Muhammad Thunayyan Al-Ghanim to Muhammad Salem Al-Sudairawi (22 December 1911), SCK.

47. Muhammad Thunayyan Al-Ghanim to Muhammad Salem Al-Sudairawi (24 November 1914), SCK.

48. Shihab bin Muhammad to ʿAbdulrahman bin Hussain Al-ʿAsʿousi (13 January 1928), ACK.

49. Al-Failakawi, Logbook 1, 119, AFC.

INSCRIPTION: LETTERS

1. Al-Khariji, *Al-Qawāʿid wal-Mīl*, 82.

2. See also Greif, "Reputation and Coalitions"; Aslanian, "The Salt in a Merchant's Letter"; Lydon, *On Trans-Saharan Trails*, 340–86. For critics of this approach, see Goldberg, "Choosing and Enforcing Business Relationships"; Edwards and Ogilvie, "Contract Enforcement, Institutions, and Social Capital." See also Greif, "The Maghribi Traders."

3. For similar claims on other merchant letters, see also Goldberg, "The Use and Abuse of Commercial Letters"; Sood, *India and the Islamic Heartland*; Trivellato, *The Familiarity of Strangers*, 177–93.

4. Trivellato, *The Familiarity of Strangers*, 191. On information and principal-agent relationships in the Indian Ocean specifically, see Mathew, "On Principals and Agents."

5. On useful information, see Goldberg, "The Use and Abuse of the Commercial Letter."

6. Saleh and Ibrahim ʿAbdullah Al-Fadhl to Al-Matrook (18 September 1921), MCK.

7. Muhammad Thunayyan Al-Ghanim to Al-Matrook (10 April 1922), MCK.

8. On letter-writing manuals, see Sood, *India and the Islamic Heartland*, 122–48; see also Trivellato, *The Familiarity of Strangers*, 184–89.

9. Muhammad and Jassem Al-Marzouq to Al-Matrook (16 October 1921), MCK.

10. Trivellato, *The Familiarity of Strangers*, 180.

11. Cables receipts, MCK.

12. Ibn Badran, *Al-ʿUqūd Al-Yāqūtiyya*, 257–82. I take up Ibn Badran's writings in much more detail in the chapter 6 inscription.

13. On Muslim jurists' skepticism regarding the place of written documents as evidence for commercial transactions, see also Lydon, "A Paper Economy of Faith."

14. Author unknown to 'Abdulrahman bin Hussain Al-'As'ousi (2 April 1922), ACK. The phrase is very similar to the one Gagan Sood found in letters from the eighteenth century; see Sood, *India and the Islamic Heartland*, 122.

CHAPTER SIX: BOMBAY

1. Al-Failakawi Logbook 1, 119, AFC.
2. *Annual Report on the Police of the Town and Island of Bombay* (1923): 30.
3. Al-Qitami, *Dalil*, 112–13.
4. Al-Failakawi, Logbook 1, 119, AFC.
5. 'Abdullah bin 'Abdulrahman Al-'As'ousi to 'Abdulrahman bin Hussain Al-'As'ousi (24 December 1921), ACK.
6. 'Abdullah bin Muhammad Al-Fozan to Al-Matrook (21 September and 26 September 1921), MCK; 'Abdulrazzaq Al-'Adwani to Al-Matrook (23 September and 26 October 1921), MCK.
7. 'Abdullah bin 'Abdulrahman Al-'As'ousi to 'Abdulrahman bin Hussain Al-'As'ousi (24 December 1921), ACK; his cousin ended up selling dates in Porbandar and Veravel. 'Abdullah bin 'Abdulrahman Al-'As'ousi to 'Abdulrahman bin Hussain Al-'As'ousi (1 February 1922), ACK.
8. Al-Fozan had experience with Gujaratis; his father had done business out of Surat during the nineteenth century, and they had been among the first to move to Bombay. Al-Ibrahim, *Min Al-Shirā'*, 83.
9. Al-Fozan to Al-Matrook (2, 21, and 26 September 1921), MCK.
10. See especially Stern, *The Company State*, 19–40, 83–99.
11. On the decline of Surat, see Das Gupta, *Indian Merchants and the Decline of Surat*; Subramanian, *Indigenous Capital and European Expansion*.
12. Maclean, *Guide to Bombay*, 90.
13. Subramanian, *Three Merchants of Bombay*, 88–143. On India's textile trade in the Indian Ocean and elsewhere, see Machado, *Ocean of Trade*; Parthasarathi, *Why Europe Grew Rich*, 23–46. On Gulf imports of textiles, see Pelly, "Remarks on the Tribes &c.," 44–45, 89.
14. Subramanian, *Three Merchants of Bombay*, 144–94.
15. Channeling Braudel's work on capitalism in Europe, K. N. Chaudhuri pointed this out out long ago in the last chapter of *Trade and Civilization in the Indian Ocean*.
16. Roy, *A Business History of India*, 64–65.
17. Colomb, *Slave-Catching in the Indian Ocean*, 86.
18. Siddiqui, *Bombay's People*, 84–141.
19. Chandavarkar, *The Origins of Industrial Capitalism in India*, 26, 247.
20. See also Bagchi, *The Presidency Banks and the Indian Economy*; Bagchi, "Transition from Indian to British Indian Systems." Roy, *A Business History of India*, 109–14; Siddiqui, *Bombay's People*, xvii.
21. Al-Ibrahim, *Min Al-Shirā'*, 81–84.

22. Fattah, *The Politics of Regional Trade*, 159–84.

23. Brucks, *Memoir Descriptive*, 575.

24. Benjamin, "Arab Merchants of Bombay and Surat," esp. 91.

25. See Mathew, "Khaliji Hindustan," 127.

26. McDow, *Buying Time*, 61–85; Mathew, "Khaliji Hindustan," 123–27. On Bahraini exiles in Bombay, Al-Baharna, *Legal Status of the Arabian Gulf States*, 34n1.

27. *A Review of the Administration of the Presidency (1923–24)*: 89. The report pointed out that the export of textiles had dropped during those years.

28. Al-Harbi, *Tārīkh Al-ʿIlāqāt*, 92.

29. Sharidah, *Merchants without Borders*.

30. Lorimer, *Gazetteer*, 1:2254. On Gulf pearl merchants in Bombay, see Albedwawi, "Pearl merchants of the Gulf"; Al-Shamlan, *Tārīkh Al-Ghawṣ*, 2:253–64.

31. Power of Attorney from ʿAbdulaziz bin ʿAli Al-Ibrahim to Jassem Al-Ibrahim (5 November 1906), "Wathāiq wa Asānīd."

32. US Department of Commerce, *Daily Consular and Trade Reports* 1 (1914): 202–3.

33. Al-Ibrahim, *Min Al-Shirāʿ*, 93–95.

34. *Report on the Trade of Maskat for the Year 1911–12*, 6. IOR/L/PS/10/647, 196.

35. Al-Ibrahim, *Min Al-Shirāʿ*; *Report on the Trade of Maskat for the Year 1914–15*, 4. IOR/L/PS/10/647, 124.

36. Al-Harbi, *Tārīkh Al-ʿIlāqāt*, 281; Political Agent, Kuwait, to Director of Persian Gulf Telegraphs, Karachi (1 July 1909), IOR R/15/5/87, 60.

37. Al-Harbi, *Tārīkh Al-ʿIlāqāt*, 279–86; Al-Madani, "Wakeel Shaikh Mubarak Al-Sabah Al-Kabir fi Bombay."

38. "Kuwait: Sheikh's Fao Properties, 1905–1910," IOR R/15/1/484, 19; "Koweit: Loans to Sheikh," IOR L/PS/10/48/2, 91.

39. Echenberg, *Plague Ports*, 47–78.

40. He had met the nakhoda ʿAbdullah bin ʿAbdulrahman Al-ʿAsʿousi many times before and had several dealings with both him and his father. ʿAbdulrahman bin Hussain Al-ʿAsʿousi to Al-Sudairawi (10 December 1912), SCK.

41. See letters between Nasser Al-Kharafi and Al-Sudairawi between 1913 and 1915, SCK.

42. Al-Sudairawi to Al-Matrook (4 February 1922), MCK.

43. See also Bishara, *A Sea of Debt*, 150–58; Wint, "From *Desh* to *Desh*"; Sharafi, "Bella's Case."

44. As one official in Bahrain explained to the Registrar of the Court of Small Causes in Bombay in 1918, "the Political Agent exercises jurisdiction but it has not yet been regulated, and he does not execute decrees other than his own." *Bahrain Order-in-Council*, IOR/R/15/2/948, 4.

45. "Messrs. Mahomed Ebrahim & Co.'s Claim of Recovery of Money from: (1) Saud bin Suleman (2) Abdulaziz bin Sultan Fleij (3) Mahomed bin Hassan, Kuwait Arabs," IOR/R/15/5/21, 163–224.

46. "Pearl Merchants' Claims Against Mohamed Al-Mushari," IOR/R/15/5/21, 15–34.

47. *Boggiano and Co v. Arab Steamers Co. Limited, Indian Law Reports* (hereafter, *ILR*) 40 Bom 529.

48. Al-Failakawi, Logbook 1, 118, AFC.

INSCRIPTION: TRANSFERS

1. The literature on the bill of exchange is enormous, but some useful entry points include Bolton and Boscoli, "'Your Flexible Friend'"; Braudel, *Civilization and Capitalism*, 1:470–78, 2:142–4, 3:66–67, 243–45; De Roover, *Business, Banking, and Economic Thought*, 48–75; North, "Institutions, Transaction Costs, and the Rise of Merchant Empires"; McCusker, *Money and Exchange*; Trivellato, *The Promise and Peril of Credit*.

2. On temples and finance, see also Rudner, *Caste and Capitalism*.

3. The literature on these instruments is scant compared with bills of exchange. See also Martin, *An Economic History of Hundi*; Habib, "Banking in Mughal India"; Lydon, "Paper Instruments in Early African Economies"; Ray, "Asian Capital."

4. Al-Duḥayyan was born in 1875 and spent his formative years studying the Quran with various scholars in Kuwait. After turning eighteen, he traveled north to Zubair, where he studied with the leading scholars of the town before returning to Kuwait. I draw much of Al-Duḥayyan's biography from Al-Rumi, *ʿUlamāʾ Al-Kuwayt*, 154–230.

5. Caeiro, "The Islamic Law of Pearling."

6. Ibn Badran, *Al-ʿUqūd*, 209–10.

7. See also Siegfried, "Concepts of Paper Money in Islamic Legal Thought."

8. Ibn Badran, *Al-ʿUqūd*, 216.

9. Ibn Badran, 218–20.

10. Ibn Badran, 222–23.

11. Ibn Badran, 223. It's worth pointing out that there were parallels between Ibn Badran's suspicions and those that thinkers in eighteenth- and nineteenth-century Europe raised. Even Adam Smith, whose *Wealth of Nations* laid out the many possible virtues of the market economy, voiced serious concerns surrounding the use of paper money and the excessive use of financial instruments. Smith's discussions of money in book 2, chapter 2, are generally instructive in this regard. Others voiced similar concerns: see also Poovey, *Genres of the Credit Economy*; Trivellato, *Promise and Peril of Credit*; Greenberg, *Bank Notes and Shinplasters*; Sklansky, *Sovereign of the Market*.

12. Ibn Badran, *Al-ʿUqūd*, 218–20.

13. See also Desan, *Making Money*; Stilt, *Islamic Law in Action*, 175–93.

14. Ibn Badran, *Al-ʿUqūd*, 216–17, 227–29.

15. Halevi, *Modern Things on Trial*, 107–13.

CHAPTER SEVEN: MALABAR

1. Al-Qitami, *Dalil*, 122–23.

2. Al-Qitami, *Dalil*, 124.

3. Al-Failakawi, Logbook 1, 117.

4. Yousef Al-Sager to ʿAbdullah bin ʿAbdulrahman Al-ʿAsʿousi (22 July 1922), ACK.

5. The nakhoda ʿAbdulwahhab Al-Qitami, ʿIsa Al-Qitami's son, said that when he visited Malindi during his own travels he traveled with a broker to his village, and was told that a Kuwaiti by the name of ʿAbdullah bin Yousef was buried there. Interview with Muhammad Al-Thunayyan Al-Ghanim, *Safḥāt min Tārīkh Al-Kuwayt* (Kuwait Television, 1966); Al-Wazzan, *Tijārat Al-Naql Al-Baḥrī*, 13–26.

6. Nainar, *Arab Geographers' Knowledge of Southern India*, 57–58.

7. Subrahmanyam and Alam, *Indo-Persian Travelers*, 64.

8. Prange, *Monsoon Islam*, 18.

9. Prange, *Monsoon Islam*, 110–12, 128–29; Prange, "The Mosque in a Land of Temples"; Kooria, *Islamic Law in Circulation*, 227–78.

10. Sood, "Through a Persian Looking Glass," 201–31.

11. Mikhail, *Nature and Empire in Ottoman Egypt*, 121–22; Wick, *The Red Sea*, 88–120.

12. Risso, "India and the Gulf," 194–95.

13. Shankar, "A Forest of Ships"; Mann, "Timber Trade on the Malabar Coast"; Rodrigues, "Commercialisation of Forests," esp. 810; Bennett, "The Origins of Timber Plantations," esp. 101–9.

14. Shankar, "A Forest of Ships," 692–93.

15. Al-Ghunaim, *Qirāʾāt fī Wathāʾiq Usrat Al-Nuṣuf*, 29–41, 47–67; see also appendixes on 121–41.

16. Dale, "The Hadrami Diaspora in Southwestern India," 181–82.

17. See Ali Barami Correspondence and Kunji Ahmad Correspondence, SCK. On Ahmad Kunji Ahmad, see "Tamarind for Bahrain & T.C.," IOR/R/15/2/789, 11.

18. Al-Duwaisan, 21–25.

19. Mann, "Timber Trade on the Malabar Coast," 408.

20. Al-Hijji, *Sināʿat Al-Sufun*, 41–47; Agius, *Seafaring in the Arabian Gulf*, 30–31.

21. Rammohan, "Coir in India."

22. Al-Wazzan, *Tijārat Al-Naql Al-Baḥrī*, 161–69; Bulley, *Bombay Country Ships*, 100.

23. Wilson, *A Glossary of Judicial and Revenue Terms*, 277–78.

24. Al-Wazzan, *Tijārat Al-Naql Al-Baḥrī*, 161–69.

25. Al-Hijji, *Nawākhithat Al-Safar*, 113–27; see Al-ʿAsʿousi letters to Al-Sudairawi from 1328 and 1332 AH, SCK.

26. Rashid bin ʿAli Al-ʿAsʿousi to ʿAbdulrahman bin Hussain Al-ʿAsʿousi (23 January 1924), ACK.

27. ʿAbdullah bin ʿAbdulrahman Al-ʿAsʿousi to ʿAbdulrahman bin Hussain Al-ʿAsʿousi (24 December 1921), ACK; Al-Sager to ʿAbdulrahman Al-ʿAsʿousi (13 January 1924), ACK; Al-Sager to ʿAbdullah Al-ʿAsʿousi (n.d.), ACK.

28. Undated list of purchases from Calicut, ACK. On *tiso* and *ʿank*, see Al-Hijji, *Ṣināʿat Al-Sufun*, 281, 289.

29. Al-Ghanim to Al-Matrook (8 December 1921), MCK. On the Mappila rebellion, see also Dhanagare, "Agrarian Conflict, Religion and Politics"; Dale, "The Mappilla Outbreaks."

30. ʿAbdullah bin ʿAbdulrahman Al-ʿAsʿousi to ʿAbdulrahman bin Hussain Al-ʿAsʿousi (23 February 1922), ACK.

31. Hussain bin ʿAbdulrahman to ʿAbdullah bin ʿAbdulrahman (10 January 1922), ACK.

32. ʿAbdullah bin ʿAbdulrahman Al-ʿAsʿousi to ʿAbdulrahman bin Hussain Al-ʿAsʿousi (23 February 1922), ACK.

33. See, for example, Palgrave, *Narrative of a Year's Journey.*

34. Ferguson, *Planting Directory for India and Ceylon*, 41.

35. Langley, *Century in Malabar*, 37. I am grateful to Wilson Chacko Jacob for this reference.

36. On at least one occasion, in 1921, Mohideen issued ʿAbdullah a money order that the latter could take to a Nandramdass Hiranand to collect nearly Rs 1,150, and on another occasion Mohideen's relative and partner sent another note for Rs 1,000 with a Bahraini merchant that was passing through on his way to Calicut. Rashid bin ʿAli to ʿAbdullah bin ʿAbdulrahman (25 January 1922), ACK; Hundi from Mohideen to ʿAsousi (27 January 1921), ACK; ʿAbdulqadir Abubaker Mohidin to ʿAbdullah bin ʿAbdulrahman (31 January 1922), ACK.

37. Mahmood bin Mohideen to ʿAbdullah bin ʿAbdulrahman (10 February 1922), ACK.

38. Al-Hijji, *Kuwait and the Sea*, 20.

39. Hussain bin ʿAbdulrahman to ʿAbdulrahman bin Hussain (1 March 1924), ACK.

40. Hussain bin ʿAbdulrahman to ʿAbdullah bin ʿAbdulrahman (25 November 1937), ACK.

41. Rashid bin ʿAli to ʿAbdullah bin ʿAbdulrahman (n.d.), ACK.

INSCRIPTION: CONVERSIONS

1. Al-Failakawi Logbook 3, 121.

2. Crosby, *The Measure of Reality*, 17.

3. See also Baptist, *The Half Has Never Been Told*, 75–144; Rosenthal, *Accounting for Slavery*, 85–120; Thompson, "Time, Work-Discipline, and Industrial Capitalism."

4. Cronon, *Nature's Metropolis*, esp. 97–147.

5. The discussion that follows draws from my article "Circulation and Capitalism in a Maritime Bazaar."

6. Al-Qitami, *Al-Khāliṣ Min Kulli ʿAyb*. I rely on the version republished by the Center for Research and Studies on Kuwait in 2007.

7. Tajirian, *The Early Modern Global Trade of Diamonds and Gems*; Jetha Dayal and Jamnadas Dayal Rawji, *Motina Hishabni* (1911), Sharjah Museums Department, SM1996–3916.

8. One major exception was the Mashhad *mithqal*, which weighted at 29 grams and was used for the valuation of lesser-quality pearls, which were sold en masse. Carter, *Sea of Pearls*, 245, 247.

9. Carter, *Sea of Pearls*, 244.

10. "Note on the Weights and Measures Employed in the Pearl Trade of the Persian Gulf," IOR V/23/49, 111–12.

11. Carter, *Sea of Pearls*, 249.

12. See also Wise, *The Values of Precision*; Hahn, *Making Tobacco Bright*.

CHAPTER EIGHT: CROSSINGS

1. Al-Failakawi, Logbook 1, 114.

2. Al-Hijji, *Nawākhithat Al-Safar*, 54–56.

3. Al-Failakawi, Logbook 1, 114.

4. Al-Qitami tells the prospective navigator to take measurements "at third hour of the day, or ninth hour firangi." Al-Qitami, *Dalil*, 91.

5. Of course, it wasn't always possible to catch the sun precisely at nine. Sometimes the nakhoda had to measure its angle earlier or later in the day. In those situations, he had recourse to diagrams that Al-Khariji left him, which instructed him on how to correct his sextant's readings based on the sun's position relative to the ship. Al-Failakawi, Logbook 3, 114–15.

6. Shihab, *Al-Muʿjam*, 63–65.

7. Al-Failakawi, Logbook 3, 122–28. Many of these are surveyed and explained in Al-Hijji, "Arab Navigational Methods," 38–43.

8. Al-Qitami, *Dalil*, 90–93.

9. Fisher, *The Makers of the Blueback Charts*. They also occasionally referred to the book as *Al-Nōd*.

10. Schotte, *Sailing School*; Reidy and Rozwadowski, "The Spaces in Between"; Steinberg, *The Social Construction of the Ocean*, 110–58.

11. Al-Failakawi, Logbook 1, 113–14.

12. "Slave Trade (East African Courts) Act, 1873," *The Law Journal Reports for the Year 1873* (London, 1873): 177–79; Hertslet, *A Complete Collection of the Treaties and Conventions*, 1–2.

13. Colomb, *Slave-Catching*, 261–63.

14. Colomb, 32–33, 66–67.

15. William Francis Prideaux to the Secretary to the Government of India (25 Aug 1874), ZNA AA 2/14.

16. Sullivan, *Dhow Chasing*, 65–66.

17. Colomb, *Slave-Catching*, 107.

18. William Francis Prideaux to the Secretary to the Government of India (25 Aug 1874), ZNA AA 2/14.

19. Colomb, *Slave-Catching*, 213–15, 217.

20. Colomb, 152–53.

21. Colomb, 74.

22. "Complaint of Kuwait Nakhodas at Search of their Ships by British Men of War," IOR/R/15/5/21, 227–52.

23. Hunter, *An Account of the British Settlement*, 90.

24. On Aden and global trade, see Khalili, *Sinews of War and Trade*, 69–75; Gavin, *Aden Under British Rule*, 174–94.

25. There is a growing literature on Parsis in Aden. See also Bhattacharyya, *Ocean Bombay*, 171–225; Toussia Cohen, "Parsi Capital and Imperial Infrastructure."

26. *Annuario d'Italia per l'Esportazione e l'Importazione* (1911–1912): 1326.

27. See Hasanali letters to Al-Sager reproduced in Al-Wazzan, *Al-Tijāra Wal-Naql*, 112–23. Al-Sudairawi's collection includes twenty-seven letters from Hasanali & Bros, spanning the early 1910s.

28. For most of Al-Hamad's biographical information, I rely on television interviews conducted with him on *Ṣafḥāt Min Tārīkh Al-Kuwayt*. On the coffee trade in Aden, see Hunter, *An Account of the British Settlement*, 100–103.

29. Al-Wazzan, *Tijārat Al-Naql Al-Baḥrī*, 128, 130.

30. See the Aden Civil and Criminal Justice Act, 1864, IOR L/PS/12/1463.

31. *Abdul Karim Fateh Mahomed vs The Municipal Officer, Aden*, *ILR* 27 Bom 374.

32. *The Municipal Officer, Aden vs Hajee Ismail Hajee Allana and Others* (1905) 3 *Calcutta Law Journal*, 5.

33. *Bhimbai Jamalbhoy vs Mariam Binte Abdul Rasool and Others* (1909), *ILR* 34 Bom 267.

34. Robbins, "The Legal Status of Aden Colony," 702.

35. See Knox-Mawer, *Law Reports of Aden*.

36. On Islamic law and Aden, see Reese, *Imperial Muslims*, 79–108.

37. I draw this from the nakhoda ʿIsa Al-ʿUthman's interview on *Ṣafḥāt Min Tārīkh Al-Kuwayt*.

38. Al-Khariji, *Al-Qawāʿid wal-Mīl*, 54.

39. *Queen Empress vs Mangal Tekchand* (1886), *ILR* 10 Bom 258, *ILR* 10 Bom 263, *ILR* 10 Bom 274.

40. Al-Failakawi, Logbook 1, 111. On birds and land sighting, see Al-Hijji, *Kuwait and the Sea*, 176.

INSCRIPTION: MAPS

1. See also Wigen, *A Malleable Map*; Edney, *Mapping an Empire*; Edelson, *A New Map of Empire*; Steinberg, *The Social Construction of the Ocean*, 99–109; Winichakul, *Siam Mapped*; Tagliacozzo, *In Asian Waters*, 344–68.
2. Wick, *The Red Sea*, 121–54; Raj, *Relocating Modern Science*, 60–94; Fernando, "Mapping Oysters and Making Oceans"; Khalili, *Sinews of War and Trade*, 16–22.
3. Owen, *Narrative of Voyages*, 1:vi–vii.
4. Appadurai, *The Social Life of Things*, 5.
5. Schotte, *Sailing School*, 46–47; Edelson, *A New Map of Empire*, 21–64.
6. Webb, *The Expansion of British Naval Hydrographic Administration*, 228–74.
7. Fisher, *The Makers of the Blueback Charts*; Ritchie, *The Admiralty Chart*.
8. See, for example, Villiers, *Sons of Sindbad*, 75, 143, 149, 186–87.
9. Al-Falah, *Ruznāmat*, 51.
10. Villiers, *Sons of Sindbad*, 143.

CHAPTER NINE: MUSCAT

1. The most common sighting after a crossing was Ras Al-Hadd, the headland that formed the easternmost tip of the Arabian Peninsula, or Jabal Abu Dawood, a hill on the stretch of coast between Ras Al-Hadd and Muscat.
2. Al-Failakawi, Logbook 1:111, AFC.
3. Villiers, *Sons of Sindbad*, 285–86.
4. "Report on the Trade of Muscat for the Year 1924–25," *The Persian Gulf Trade Reports*, Vol. 1 (Muscat).
5. Villiers, *Sons of Sindbad*, 286.
6. Alavi, *Sovereigns of the Sea*, 199–254.
7. Al-ʿAraimi, "Taʿarruf ʿalā Sayyid Yusuf Al-Zawawi."
8. See, for example, Al-Zawawi to Al-Sudairawi (19 Rajab 1336), SCK.
9. Al-Zawawi to Al-Sudairawi (14 Rabiʿ Al-Awwal 1334 and 18 Rabiʿ Al-Thani 1334), SCK.
10. Green, *Bombay Islam*.
11. Al-Loghani, "Wathīqa lahā tārikh"; Al-Manar Library to Al-Sudairawi (15 July 1914), SCK; Jirji Zidan, Al-Hilal, to Al-Sudairawi (31 July 1913), SCK. Nasser bin Mubarak is an interesting figure: though blind, he kept the largest collection of books and manuscripts in Kuwait. Al-Shaybani, *Kaʿb Al-Akhbār*.
12. Al-ʿAraimi, "Taʿarruf ʿalā Sayyid Yusuf Al-Zawawi."
13. ʿAbdulmasih Antaki to Sudairawi (1 February 1911), SCK.
14. Rida, "Riḥlatuna Al-Hindiyya Al-ʿArabiyya," 396–99.
15. "Personalities. 'Iràq (Exclusive of Baghdad and Kàdhimain)," (1919), IOR/L/PS/20/221, 144.

16. The touchstone work on this is Hourani, *Arabic Thought in the Liberal Age*, esp. 260–323. For a more recent set of takes, see also Hanssen and Weiss, *Arabic Thought Beyond the Liberal Age.*

17. ʿAbdulmasih Al-Antaki, *Al-Ayāt al-ṣabāḥ fī madāʾiḥ mawlānā ṣāḥib al-sumū amīr al-Kuwayt Shaikh Mubarak Bāshā bin Ṣabāḥ* (1907); *Al-riyāḍ al-muzhara bayn al-Kuwayt wal-Muḥammarah* (1907); *Al-durr al-ḥusān fī ʾimārat ʿArabistān* (1908); *Al-Riyāḍ Al-Khazʿaliyyah fī Al-Siyāsa Al-Insāniyya* (1911).

18. See also Choudhry, *Hajj across Empires*, 269–98; Morgenstein Fuerst, *Indian Muslim Minorities*; Stephens, "The Phantom Wahhabi"; Minault, *The Khilafat Movement*; Jacob, *For God or Empire.*

19. "Personalities. 'Iràq (Exclusive of Baghdad and Kàdhimain)," (1919), p. 128, IOR/L/PS/20/221, 144.

20. For recent discussions, see also Alavi, *Muslim Cosmopolitanism*; Laffan, *Under Empire*, 296–315.

21. Gilbert, *Dhows and the Colonial Economy*, 59–109; Miller, *The Lunatic Express.*

22. Chatterjee, "Tracing the History and Legacy of Mangalore Tiles."

23. Al-Failakawi, Logbook 2:49–58.

24. "Ports Decrees," ZNA AB 44/3; Martin and Martin, *Cargoes of the East*, 41–47.

25. His family members recall learning that he had a wife in "Zanzibar"—though given that there is no record of him making it as far as that island, they may well have conflated Zanzibar with Mombasa. Muhammad, *Al-Nawkhithā*, 23. In lots of different writings, "Zanzibar" is used as a metonym for the entire Swahili coast.

26. Al-ʿAsʿousi, *Ruznāmat*, 356–63.

27. Villiers, *Sons of Sindbad*, 214.

28. Burton, *The Lake Regions*, 447. See also Gilbert, *Dhows and the Colonial Economy of Zanzibar*, 112–14; Sunseri, *Wielding the Ax*, 30–46.

29. Gilbert, *Dhows and the Colonial Economy*, 121–23; Sunseri, *Wielding the Ax*, 91–92.

30. Villiers, *Sons of Sindbad*, 229; see also Curtin, "African Enterprise"; Gilbert, "Sailing from Lamu and Back."

31. Al-ʿAsʿousi, *Ruznāmat*, 363–64.

32. Al-Falah's *ruznamah* records his voyages to the Rufiji delta between 1944 and 1948. Al-Falah, *Ruznāmat*, 57–59; Villiers, *Sons of Sindbad*, 231.

33. Al-Falah, *Ruznāmat*, 59. At different times, there were different numbers of dhow agents in Zanzibar; one document from 1957 lists nine of them. The majority of these were of Yemeni extraction: the names Ba-Harun, Ba-Hurmuz, Bin Braik, Al-Yafiʿi, and Al-Suwaydi all speak to roots in Yemen (and more specifically in Hadramawt), though these would have been individuals whose families had long settled on the East African coast. Minutes of Dhow Agents' Meeting (19 December 1957), ZNA AK 18/13, 6A. Bin Gurnah was the father of the celebrated novelist Abdulrazak Gurnah.

34. Al-Khariji, *Al-Qawāʿid wal-Mīl*, 84.

35. Glassman, *War of Words*, 198–201. On Manga Arabs, their recent immigration, and their unsavory reputation more generally, see McMahon, *From Honor to Respectability*, 163–65.

36. I chart much of this out in *A Sea of Debt*, 24–106.

37. Glassman, *War of Words*, 23–61.

38. Villiers, *Sons of Sindbad*, 169–74.

39. Bang, *Sufis and Scholars*, 136; Ghazal, "The Other Frontiers of Arab Nationalism," 114–15; Prestholdt, "From Zanzibar to Beirut," 212–13.

40. Al-Kindi, *Al-Ṣahāfa Al-ʿUmāniyya*.

41. See esp. *Al-Falaq* issues 484–490; Ghazal, *Islamic Reform*, 104–5; Al-Kindi, *Al-Ṣahāfa Al-ʿUmāniyya*, 126–29.

42. Ghazal, *Islamic Reform*, 101.

43. On at least one instance for which we have reportage, *Al-Najāḥ* published a piece criticizing the sultan of Muscat's appointment of certain counselors. IOR R/15/6/43, 124.

44. "Wājib ʿArabi," *Al-Falaq*, no. 489 (May 28, 1938): 2.

45. "Al-Wiḥda Al-ʿArabiyya," *Al-Falaq*, no. 482 (April 9, 1938): 2.

46. "Laytak Tadri," *Al-Falaq*, no. 588 (April 20, 1940): 1–2.

47. *Al-Falaq* nos. 579–591, dating February 17 to May 11, 1940.

48. "Al-ʿArab wa Al-Muslimūn," *Al-Falaq*, no. 591 (May 11, 1940): 1.

49. Bishara, *A Sea of Debt*, 217–45.

50. Sheriff, "Race and Class in the Politics of Zanzibar," 308.

51. See also Green, "Waves of Heterotopia"; Green, "The Languages of Indian Ocean Studies"; Green, "Introduction: Arabic as a South Asian Language"; Zaman, "Afterword: Reassessing Arabic in South Asia."

52. See Al-Maskari's correspondence in Al-Kindi, *Al-Ṣahāfa Al-ʿUmāniyya*, 188–237. "Muscat Intelligence Summary No. 5 for the Period 1–15 March 1948," IOR R/15/6/361, 12; "Muscat Intelligence Summary No. 23 for the Period 16th–31st December 1947," IOR L/PS/12/2973A, 18. On Sudairawi, see Al-Ghunaim, *Quṭūf Tārīkhiyya*, 173–96.

53. See also Aiyar, *Indians in Kenya*; Ghazal, *Islamic Reform*; Green, "Forgotten Futures"; Laffan, *Under Empire*, 231–333.

54. For this in the Gulf, see also "Proscribed Newspapers and Seditious Press Articles," IOR R/15/2/126; "Vernacular Newspapers & Periodicals, Extracts from Pan-Islamism," IOR R/15/5/62. For a parallel example, see Hofmeyr, *Dockside Reading*.

55. Al-Failakawi, Logbook 1:108–10.

INSCRIPTION: POEMS

1. The *jāh* is the point on the compass rose pointing north; *qutb* points south. The two *farqads* point slightly northeast and northwest, and the two *silbārs* lie

opposite them. Likewise, the two *naʿsh* (north by northeast and north by northwest) are opposite the two *suhayls*.

2. Ibn Majid, *Al-Nūniya Al-Kubrā*; Tibbetts, *Arab Navigation*; Ferrand, *Le Pilote des Mers de l'Inde*; Khoury, "Les poèmes nautiques"; Staples, "Navigating the Gulf."

3. Al-ʿAbdulmughni, "Jassem Al-Nasrallah."

4. Al-Khariji, *Al-Qawāʿid wal-Mīl*, 14. The sailor ʿAbdullah Al-Duwaish, who worked aboard both pearling and trading dhows, wrote a long poem that mapped out the voyage from Kuwait to Calicut and back, replete with shorthand references to the coastal landmarks that nakhodas relied upon to give shape to their mental map of the voyage—the mountains, headlands, port cities, and other markers that Al-Failakawi and Al-Khariji would have immediately recognized. Al-Fulayḥ, *Ḥayyahalā*, 54–55. I am indebted to Tareq Al-Rabei for alerting me to this source.

5. On *madīḥ* poetry in West Africa, see Ogunnaike, *Poetry in Praise of Prophetic Perfection*. For other examples of *madīḥ* poetry from the Gulf, see Al-Roumi, *Fahad Rāshid Burisly*, 175–80.

6. Al-Failakawi, Logbook 3, pages not numbered. The poems include verses by Abi Bakr bin Zahir Al-Andalusi, Bahaʾ Al-Din Al-Juyushi, Al-Waʾwaʾ Al-Dimashqi, to name a few. It's unclear whether the poems were inscribed by Al-Khariji or Al-Failakawi; some appear to be in the latter's hand, but others align more with Al-Khariji's handwriting.

7. Al-Khariji, *Al-Qawāʿid wal-Mīl*, 2.

8. Al-Khariji, 1–2. I'm grateful to Ahmad Al-Maazmi for help with these.

9. I am grateful to Gabriel Lavin for his insights on the collection.

10. Cohen, *The Social Life of Poems*, 10.

11. See crew list in Al-Failakawi, *Ruznāmat*, 503. On *nahhāms* and performance, see also Ulaby, "On the Decks of Dhows"; Al-Rifai, *Al-Nehmah wal-Nahham*, 121–255.

12. Villiers, *Sons of Sindbad*, 19.

13. On poetic recitation and alteration in Yemen, see Caton, *Peaks of Yemen I Summon*. For a similar phenomenon in nineteenth-century American newspaper poetry, see also Cordell and Mullen, "Fugitive Verses."

14. Lavin, "Arabian Passings," 165–66; Al-Salhi, "The Recordings of ʿAbdullatif al-Kuwaiti."

15. For a comparable phenomenon, see also Schofield, *Music and Musicians*.

16. Cohen, *The Social Life of Poems*, 19.

CHAPTER TEN: BAHRAIN

1. Lorimer, *Gazetteer*, 2:234–35.

2. As he recorded his stay on the island in his logbook, he seemed to have dropped a month out of his English calendar; his entries go from April 31 to June 1, leaving out the entire month of May. That was not because he didn't bother to record

anything for the month of May, though; the entries for the Nairuz and Hijri calendars proceed with no interruption, so we can assume that the missing month was due to a clerical error.

3. Lorimer, *Gazetteer*, 2:245–47.

4. "Report on the Trade of the Bahrein Islands for the year ending 31st March 1925," in *The Persian Gulf Trade Reports*, Vol. 3.

5. See also Lienhardt, *Shaikhdoms*, 143–44; Al-Hijji, *Kuwait and the Sea*, 93–98.

6. See also Hopper, *Slaves of One Master*, 80–104; Suzuki, "Baluchi Experiences"; Morse, "Pearling and the Language of Freedom."

7. Al-Zayyani, *Al-Ghawṣ Wal-Ṭawāshah*, 56. By contrast, haulers only took two shares.

8. Abu Hakima, *History of Eastern Arabia*, 91–124; Farah, *Protection and Politics in Bahrain*, 7–8, 12. The standard Arabic reference for Bahrain's history remains Al-Nabhani, *Al-Tuḥfa Al-Nabhāniyya*.

9. Lorimer, *Gazetteer*, 1:866–83; Al-Nabhani, *Al-Tuḥfa Al-Nabhāniyya*, 87–120.

10. Political Agent, Zanzibar, to Chief Secretary of the Government at Bombay (27 March 1849), ZNA AA 3/8, 121–26.

11. MSA Political Department 1853 Vol. 78, Comp. 313; MSA 1868 Vol. 97, Comp. 64; Al-Nabhani, *Al-Tuḥfa Al-Nabhāniyya*, 121–29.

12. Lorimer, *Gazetteer*, 1:898–99; Bhacker, "Family Strife and Foreign Intervention."

13. Lorimer, *Gazetteer*, 1:858 and 2:248–49; Farah, *Protection and Politics*, 11–12; Fuccaro, *Histories of City and State*, 57.

14. Bu-Safwan, *Al-Baḥrayn 1923*, 113–32.

15. Farah, *Protection and Politics*, 66, 84. On the Ottomans, see Anscombe, *The Ottoman Gulf.*

16. Fuccaro, *Histories of City and State*, 80, 120–26; Stephenson, *Rerouting the Persian Gulf*, 22–28; Onley, *The Arabian Frontier of the British Raj*, 119–26; Al-Baharna, *British Extra-Territorial Jurisdiction*, 10–20.

17. E. C. Ross, "Report on the Administration of the Persian Gulf Political Residency and Muscat Political Agency for the year 1877–78," IOR/V/23/32, No. 152, 39–40; Hopper, *Slaves of One Master*, 82.

18. He later turned his attention to other consumer goods: tobacco, perfumes, and even automobiles. Kanoo, *The House of Kanoo.*

19. Kanoo, 22–23.

20. See also Onley, "Transnational Merchants"; Fuccaro, *Histories of City and State*, 55–70; Stephenson, "Rerouting the Persian Gulf," 45–78.

21. Fuccaro, *Histories of City and State*, 55–60, 76–82, 87–90, 95–104. Gulf Resident to Secretary to the Government of India (17 December 1875), MSA Political Department 1876, Vol. 197, Comp. 358.

22. Here, I draw from Laura Edwards's insights into legal culture in the American South; Edwards, *The People and Their Peace*, 27–28, 44.

23. *Government of Bahrain Administrative Reports for the Years 1926–1937*, 21–22; Caeiro, "The Islamic Law of Pearling."

24. IOR R/15/1/336, 221.

25. Daly to Gulf Resident (1 August 1923), IOR R/15/1/336, 45.

26. IOR R/15/1/336, 178–83.

27. Daly to Gulf Resident (1 August 1923), IOR R/15/1/336, 36. For the history of intellectual movements in Bahrain during this time, see also Alshehabi, *Contested Modernity*, 91–130.

28. Daly to Gulf Resident (6 October 1923), IOR R/15/1/336, 193–94.

29. Immediately after Daly suggested taxing pearling dhows, one individual produced a chit issued by Shaikh ʿIsa in 1895 that exempted him and all his dependents from any pearling-related taxes, along with a note from Shaikh Hamad in 1897 declaring that he would respect his father's decision.

30. IOR R/15/2/132, 38–39, 81–82.

31. Daly to Gulf Resident (6 October 1923), IOR R/15/1/336, 190–91.

32. Daly to Gulf Resident (17 March 1923), IOR R/15/1/336, 75.

33. Daly to Gulf Resident (11 February and 17 March 1923), IOR/R/15/1/336, 62–67, 72–76.

34. IOR R/15/1/336, 47–48.

35. IOR R/15/1/336, 105–109.

36. IOR/R/15/2/132, 91–93.

37. IOR R/15/2/132, 120–27.

38. IOR/R/15/1/349, 20.

39. Secretary to the Gulf Resident to Political Agent, Bahrain (1 Jan 1927), IOR R/15/2/132, 152.

40. ʿAli bin Ibrahim Al-Zayyani, Bahrain, to Fahad Al-Khalid and his brothers, Kuwait (4 January 1927), published in Al-Ghunaim, *Wathāʾiq Min ʿAṣr Al-Luʾluʾ*, 61–62.

41. Secretary to the Gulf Resident to Political Agent, Bahrain (1 Jan 1927), IOR R/15/2/132, 149–153.

INSCRIPTION: ACCOUNTS

1. See the appendix to ʿIsa Al-Qitami's *Dalil*, in which his son ʿAbdulwahhab (also a nakhoda) describes the maritime economy of the Gulf. Al-Qitami, *Dalil*, 225–27. We see similar arrangements in other parts of the world, from the caravans of the Muslim Sahara to the whaling ships of eighteenth-century New England. See also Lydon, *On Trans-Saharan Trails*; Hohman, "Wages, Risk, and Profit."

2. The literature on inheritance and shares is too vast to list. On inheritance and commerce in the Indian Ocean, see Bishara, *A Sea of Debt*, 88–89. On math and inheritance in another setting, see Trouillot, *Mathematics in the Desert*.

3. Yamey, "Accounting and the Rise of Capitalism."

4. For examples of this sort of narrative, see also Crosby, *The Measure of Reality*; Gleeson-White, *Double Entry*.

5. See also Porter, *Trust in Numbers*; Poovey, *A History of the Modern Fact*; Alborn, *Regulated Lives*; Soll, *The Reckoning*; Deringer, *Calculated Values*.

6. See also Johnson, *River of Dark Dreams*, 176–208; Rosenthal, *Accounting for Slavery*; Hill Edwards, *Unfree Markets*, 125–52. For a more general reflection on accounting and managerial power, see Miller, "Accounting as Social and Institutional Practice."

7. There is, by contrast, very little to cite on the history of accounting and bookkeeping in the Islamic world—this, despite the widespread acknowledgment that accounting must have arrived in Europe by way of the Middle East. Chris Bayly's brief discussion of the topic suggests that Indian merchants engaged in rational bookkeeping practices, but he offers little by way of detail. Bayly, "Pre-Colonial Indian Merchants and Rationality." Another article observes bookkeeping practices in early Muslim society; see Zaid, "Accounting Systems and Recording Procedures in the Early Islamic State."

8. On crossing out entries in the act of copying them into a separate ledger, see Poovey, *A History of the Modern Fact*, 56. In her work on Saharan traders, however, Ghislaine Lydon suggests that merchants crossed out copies of loan contracts after debts were repaid. Lydon, *On Trans-Saharan Trails*, 331.

CHAPTER ELEVEN: RETURNS

1. Villiers, *Sons of Sindbad*, 320.

2. I draw this from Violet Dickson's account of dhow berthing in Dickson, *Kuwait and Her Neighbors*, 467–68.

3. "Report on the Trade of Kuwait for the Year 1925–26," *The Persian Gulf Trade Reports*, 6:11–18; Villiers, *Sons of Sindbad*, 330–31.

4. Al-Hijji, *Ṣināʿat Al-Sufun*, 71–182.

5. Al-Hijji, *Kuwait and the Sea*, 93–98.

6. There is still an enormous amount of work left to be done on the caravan trade. For enlightening discussions, see Pétriat, "Caravan Trade in the Late Ottoman Empire"; Pétriat, "The Uneven Age of Speed"; Altorki and Cole, *Arabian Oasis City*, 67–82.

7. It is surprisingly difficult to speak with much confidence about what women did while the town's men were away; the archival and oral-historical material is limited by a strong male bias. My information here is pieced together from various hints, including Villiers, *Sons of Sindbad*, 335; Al-Nakib, *Kuwait Transformed*, 84, 88; Al-Qinaʿi, *Safḥāt*, 86; Jamal, *Aswāq Al-Kuwayt*, 184–85, 276–77.

8. For a similar phenomenon in East Africa, see Prestholdt, *Domesticating the World*, 59–87.

9. Muhammad, *Al-Nawkhithā*, 21.

10. Stephenson, *Rerouting the Persian Gulf*, 69–77; Floor, *The Persian Gulf*, 101–3; Al-ʿAbbasi, *Tārīkh Lingah*, 24–29.

11. Al-Khariji, *Al-Qawāʿid wal-Mīl*, 75.

12. Strunk, *The Reign of Shaykh Khazʿal*, 337–468.

13. Al-Rasheed, *A History of Saudi Arabia*, 45–68.

14. See also Copland, *The Princes of India in the Endgame of Empire, 1917–1947*; Menon, *The Story of the Integration of the Indian States*; Ramusack, *The Indian Princes and Their States*, 245–75.

15. See Al-Rashoud, *Modern Education and Arab Nationalism*, 44–85; Al-Rushaid, *Tārīkh Al-Kuwayt*, 253–345; Al-Qinaʿi, *Ṣafḥāt*, 43–46.

16. Crystal, *Oil and Politics*, 41; Al-Rashoud, *Modern Education and Arab Nationalism*, 104; Alebrahim, *Kuwait's Politics*, 45–46.

17. Al-Rashoud, *Modern Education and Arab Nationalism*, 111; Crystal, *Oil and Politics*, 47–55; Alebrahim, *Kuwait's Politics*, 66–68.

18. Al-Rashoud, *Modern Education and Arab Nationalism*, 100–136; Al-Wazzan, *ʿAbdullah Ḥamad Al-Ṣager*; Alebrahim, *Kuwait's Politics*, 59–64.

19. Al-Qitami himself had left Kuwait for Jabal Al-Akhdar in Oman in the 1920s, out of dissatisfaction with the political climate there. See the twenty-one-part interview with Lulwah Al-Qitami on Al-Qabas Television in May 2020, especially part 7.

20. IOR/L/PS/12/3894A, 256, 259, 261–62.

21. Crystal, *Oil and Politics*, 48–49.

22. On the *salifa* in Kuwait, see Al-Hijji, *Kuwait and the Sea*, 54; Saleh, "Ṭawāriʾ wa Mashākil."

23. ʿAbdullatif bin ʿAbdulaziz Al-Mishari to Salman bin Hussain bin Matar (23 Ramadan 1339). I am grateful to Fidaa Al-Zaidani at the Bahrain National Museum for letting me quote this letter.

24. *Government of Bahrain Administrative Report for the Years 1926–1937*, IOR/R/15/1/750/1, 50.

25. Hilal bin Fajhan Al-Mutairi to ʿAbdulrahman bin Hussain Al-ʿAsʿousi (7 May 1937), ACK.

26. "Correspondence Regarding Mortgages," IOR/R/15/2/1908; IOR/R/15/1/352, 290. On litigation involving property, see Morse, *Crude Legalities*.

27. Crystal, *Oil and Politics*, 39; Young, *Losing Ground*.

28. Villiers, *Sons of Sindbad*, xx.

29. ʿIsa Bishara interview with Rida Al-Fayli on Kuwait Television.

30. Al-Hijji, *Min Al-Fulklūr*, 152–57.

31. Some of Al-Failakawi's crew members did die during their voyages, but it is unclear whether those deaths had anything to do with the risks he may have taken; they were most likely for unrelated reasons.

32. Mathew, *Margins of the Market*, 113–72; Segal, "Merchants' Networks in Kuwait," 713–14; Mathew, "At the Crossroads of Empire and Nation-State."

33. For a good, brief overview of this history, see Yergin, *The Prize*, 280–302.

34. "Status of Kuwait Bay and Extent of Territorial Waters in Kuwait," IOR R/15/1/708; Khalili, *Sinews of War and Trade*, 97–105; Erich, *Extraction, Property, and Rights*, 143–83.

35. Al-Failakawi, *Ruznāmat*, 411, 503–7. Bishara, *Ruznāmat*, 408, 470, 498, 529, 598, 619; Al-ʿUthman, *Ruznāmat*, 67, 200, 219, 243, 270, 296, 331, 353, 383, 403, 426. Most sailors are impossible to locate geographically; nakhodas frequently only identified them as "X, son of Y"—Saʿad bin Farhan or Jumʿa bin Faraj, for example. In one of ʿIsa Bishara's crew lists, one mariner is listed only as "Yunus." Bishara, *Ruznāmat*, 408.

36. Boodrookas, *The Making of a Migrant Working Class*, 193–213.

37. Bahrain Government Proclamation No. 2 of 1360, IOR R/15/2/1230, 15–18.

38. See, for example, the attachment order in IOR R/15/3/9344.

39. *Government of Bahrain, Annual Report for the Year 1359 AH (1940–41)*: 12–13, and *Government of Bahrain, Annual Report for the Year 1360 AH (1941–42)*: 16, IOR R/15/1/750.

40. This arrangement, called the *khammās*, grew increasingly popular during the late 1930s and 1940s. Fuccaro, *Histories of City and State*, 162–63.

41. Thesiger, *Arabian Sands*, 277–79.

42. IOR/R/15/6/107, 19.

EPILOGUE: TRIUMPH AND LOSS

1. There are many possible translations for *Fateḥ Al-Khair*; in a previous article, I translated it as "The Opening of Good Fortune." *Fateḥ* can be translated as opening, but also as victory, conquest, or triumph. Meanwhile, *khair* has multiple meanings, all of which relate to "that which is good"—blessing, fortune, wealth, and bounty.

2. Bishara, *Ruznāmat*, 563–97.

3. Al-Hijji, *Fateḥ Al-Khair*, 12–21.

4. Al-Rashoud, *Modern Education and Arab Nationalism*, 111–12, 122, 126, 129, 249.

5. Interview with ʿAbdulrahman Al-Bahar, Mishʿan Al-Khudhair, and Khaled ʿAbdullatif Al-Hamad on *Ṣafḥāt Min Tārīkh Al-Kuwayt*.

6. The series ran for sixteen years, from 1964 to 1980, and then included sporadic episodes throughout the 1980s and 1990s. It still serves as one of the richest repositories of oral narratives on Kuwait's past, and in its online recirculations it continues to educate its viewers about Kuwait's history.

7. On the issue of lost language, it is interesting to note that this is just after the British lexicographers Johnstone and Muir came to Kuwait to undertake their study of the Gulf maritime vernacular—a project in which they relied on the assistance of the nakhoda Hussain bin ʿAbdulrahman Al-ʿAsʿousi. Johnstone and Muir, "Some Nautical Terms in the Kuwaiti Dialect of Arabic."

8. Al-Shamlan, *Min Tārīkh Al-Kuwayt*.

9. Interview with ʿIsa ʿAbdullah Al-ʿUthman on *Ṣafḥāt Min Tārīkh Al-Kuwayt*.

10. This kind of maritime history, I should be clear, is rather old-fashioned these days. For more recent takes on the field, see Manning, "Maritime History and Global History"; Fusaro and Polónia, *Maritime History as Global History*.

11. Interview with Al-Bahar, Al-Khudhair, and Al-Hamad on *Ṣafḥāt Min Tārīkh Al-Kuwayt*.

12. Al-Hijji, *Fateḥ Al-Khair*, 12–13.

13. Conversation with ʿAbdulrahman Al-Mulla Al-Failakawi, June 2023.

14. Ricciardo, *The Voyage of the Mir El-Lah*; Parsa, *Bādbānhā-i Junūb*.

15. "Nasser Al-Kharafi & Family," *Forbes*, March 11, 2011, https://www.forbes.com/profile/nasser-al-kharafi/?sh=65a935fc188d.

16. See also Field, *The Merchants*.

17. Al-Hijji, *Nawākhithat Al-Safar*, 113–33, 205; Muhammad, *Al-Nawkhitha ʿAbdulmajeed*, 25.

18. Alhabib, "Muḥammad ʿAbdullah Al-Matrūk."

19. Al-Khariji, *Al-Qawāʿid wal-Mīl*, 11–12.

20. Al-Falah, *Ruznāmat*, 104.

BIBLIOGRAPHY

PRIMARY SOURCES

Archives

Al-ʿAsʿousi Collection, Kuwait; ACK
Al-Failakawi Collection, Kuwait; AFC
Ali Akbar Bushihri Collection, NYU Abu Dhabi; ABC
Ali Raʾis Collection, Kuwait; ARC
Al-Matrook Collection, Kuwait; MCK
Al-Sudairawi Collection, Kuwait; SCK
Foreign Office Records, National Archives, London, United Kingdom.
India Office Records, British Library, London, United Kingdom; IOR
Maharashtra State Archives, Mumbai, India; MSA
Ministère des Affaires Étrangères, Centre des Archives Diplomatiques, La Courneuve, Paris, France; MAE
Sharjah Museums Department
Zanzibar National Archives, Zanzibar, Tanzania; ZNA

Interviews

Al-Ghanim, Muhammad Thunayyan. Interview with Saif Al-Shamlan. *Ṣafḥāt Min Tārīkh Al-Kuwayt.* June 23, 1966. www.youtube.com/watch?si=6mSodBFR_Amgm40J&v=SToGXNy-cxo&feature=youtube.

Al-Hamad, Khaled ʿAbdullatif. Interview with Rida Al-Fayli. *Ṣafḥāt Min Tārīkh Al-Kuwayt.* November 26, 1985. https://youtu.be/KZU7qbHlV8Y?si=uDDMtsKy5TIYUHpf.

Al-Khudair, Mishʿan, ʿAbdulrahman Al-Bahar, and Khaled ʿAbdullatif Al-Hamad. Interview with Rida Al-Fayli. *Ṣafḥāt Min Tārīkh Al-Kuwayt.* www.youtube.com/watch?v=TDJ5SvoQ4Us&t=1787s.

Al-Qitami, Loulwah. Twenty-one-part interview with ʿAmmar Taqi. Al-Qabas Television. May 2020. www.youtube.com/watch?v=sxeiJTcsQDc&t=1396s.

Al-ʿUthman, ʿIsa ʿAbdullah. Interview with Saif Al-Shamlan. *Ṣafḥāt Min Tārīkh Al-Kuwayt*. December 16, 1985. www.youtube.com/watch?v=unBVZeRPv_o&t=3805s.

Bishara, ʿIsa. Interview with Rida Al-Fayli. Kuwait Television. Mid-1990s. www.youtube.com/watch?v=4WB8qqPFmZ4&t=192s.

Unpublished

Loch, Francis Erskine. *Diary in Persian Gulf and Sketch Books*. National Records of Scotland, Edinburgh, [1835?].

Published

Aitchison, C. U. *A Collection Of Treaties, Engagements, and Sanads*, Vol. 7. Office of the Superintendent of Government Printing, India, 1892.

ʿAbdulmasih Al-Antaki. *Al-Ayāt Al-Ṣabāḥ fī Madāʾiḥ Mawlānā Ṣāḥib Al-Sumū Amīr Al-Kuwayt Shaikh Mubarak Bāshā bin Ṣabāḥ*. Maṭbaʿat Al-ʿArab, 1907.

ʿAbdulmasih Al-Antaki. *Al-Durr al-Ḥusān fī Imārat ʿArabistān*. Maṭbaʿat Al-ʿArab, 1908.

ʿAbdulmasih Al-Antaki. *Al-Riyāḍ Al-Khazʿaliyyah fī Al-Siyāsa Al-Insāniyya*. Maṭbaʿat Al-ʿArab, 1911.

ʿAbdulmasih Al-Antaki. *Al-Riyāḍ Al-Muzhara Bayn Al-Kuwayt Wal-Muḥammarah*. Maṭbaʿat Al-ʿArab, 1907.

Al-ʿAsʿousi, Hussain bin ʿAbdulrahman. *Ruznāmat Al-Nawkhithā Ḥussain bin ʿAbdulraḥmān Al-ʿAsʿūsī*. Edited by Yacoub Yousef Al-Hijji. Self-Published, 2019.

Al-Azkawi, Sirhan. *Kashf Al-Ghumma Al-Jāmiʿ li-Akhbār Al-Aʾimma*. Edited by ʿAbdulmajeed Al-Qaysi. 4th ed. Oman Ministry of Heritage and Culture, 2005.

Al-Failakawi, ʿAbdulmajeed. *Ruznāmat Al-Nawkhitha ʿAbdulmajīd Al-Mulla Aḥmad Al-Faylakāwī*. Edited by Yacoub Yousef Al-Hijji. Center for Research and Studies on Kuwait, 2001.

Al-Falah, Mufleh Saleh. *Ruznāmat Al-Nawkitha Mufleḥ Ṣāleḥ Al-Falāḥ*. Edited by Yacoub Yousef Al-Hijji. Center for Research and Studies on Kuwait, 2000.

Al-Falaq. Oman Ministry of Endowments and Religious Affairs Electronic Library. https://elibrary.mara.gov.om/en/zanzibar-library/newspapers/al-falaq/.

Al-Ghunaim, ʿAbdullah, ed. *Ḥadīth Al-Wathāʾiq: Ṣafḥāt Min Wathāʾiq Usrat Al-ʿAbduljalīl*. Center for Research and Studies on Kuwait, 2014.

Al-Ghunaim, ʿAbdullah, ed. *Qirāʾāt fī Wathāʾiq Usrat Al-Nuṣuf*. Center for Research and Studies on Kuwait, 2016.

Al-Ghunaim, ʿAbdullah, ed. *Quṭūf Tārīkhiyya min Al-Wathāʾiq Al-Ahliyya al-Kuwaytiyya*. Center for Research and Studies on Kuwait, 2022.

Al-Ghunaim, ʿAbdullah, ed. *Wathāʾiq Min ʿAṣr Al-Luʾluʾ*. Center for Research and Studies on Kuwait, 2017.

Al-Khaduri, Nasser bin ʿAli. *Maʿdan al-Asrār fī ʿIlm Al-Biḥār.* Oman Ministry of Heritage and Culture, 2015.
Al-Khariji, Mansur Ibrahim. *Al-Qawāʿid wal-Mīl wal-Natīja fī ʿIlm Al-Baḥar.* Edited by Hassan Salih Shihab. Center for Research and Studies on Kuwait, 2006.
Al-Kindi, Muhammad bin Ibrahim. *Bayān Al-Sharʿ*. Oman Ministry of Heritage and Culture, 1988.
Al-Nabhani, Muhammad bin Khalifa. *Al-Tuḥfa Al-Nabhāniyya fī Tārīkh Al-Jazīra Al-ʿArabiyya*. 2nd ed. Bahrain National Library, 1999.
Al-Qinaʿi, Yousef bin ʿIsa. *Ṣafḥāt Min Tārīkh Al-Kuwayt*. 5th ed. Thāt es-Salāsil, 1987.
Al-Qitami, ʿIsa bin ʿAbdulwahhab. *Al-Khāliṣ Min Kulli ʿAyb fī Waḍʿ al-Jayb*. Reissue. Center for Research and Studies on Kuwait, 2007.
Al-Qitami, ʿIsa bin ʿAbdulwahhab. *Al-Mukhtaṣar al-Khāṣṣ lil-Musāfir wal-Baḥḥār wal-Ghawwāṣ*. 2nd ed. Maṭbaʿat al-Kuwayt, 1924.
Al-Qitami, ʿIsa bin ʿAbdulwahhab. *Dalīl Al-Muḥtār fī ʿIlm Al-Biḥār.* 3rd ed. Dār Al-Salām, 1963.
Al-Reshaid, ʿAbdulaziz. *Tārīkh Al-Kuwayt.* Dar Maktabat Al-Ḥayāt, 1962.
Al-Shamlan, Saif Marzouq. *Min Tārīkh Al-Kuwayt.* Thāt es-Salāsil, 1988.
Al-Shaqsi, Khamis bin Saʿid. *Manhaj Al-Ṭālibīn wa Balāgh Al-Rāghibīn*. Ministry of Heritage and Culture, 1998.
Al-ʿUthman, ʿIsa. *Ruznāmat Al-Nawkhitā ʿIsa ʿAbdullah Al-ʿUthman*. Edited by Yacoub Yousef Al-Hijji. Center for Research and Studies on Kuwait, 1999.
Al-Wazzan, Faisal. *ʿAbdullah Ḥamad Al-Ṣager: Aḍwāʾ ʿalā Sīratuh wa Dawruh al-Siyāsī wal-Iqtiṣādī*. Center for Research and Studies on Kuwait, 2019.
Al-Wazzan, Faisal. *Tijārat Al-Naql Al-Baḥrī fī Al-Kuwayt min Khilāl Sīrat Ḥamad ʿAbdullah Al-Sager.* Center for Research and Studies on Kuwait, 2019.
Annual Report on the Police of the Town and Island of Bombay. Bombay Police Department, 1923.
Annuario d'Italia per l'Esportazione e l'Importazione. 10th ed. Tipografia Nazionale di G. Bertero & C., 1911.
A Review of the Administration of the Presidency (1923–24). Bombay: Government Central Press, 1925.
Bishara, ʿIsa Yacoub. *Ruznāmat Al-Nawkhitā ʿIsa Yaʿqub Bishara*. Edited by Yacoub Yousef Al-Hijji. Center for Research and Studies on Kuwait, 2003.
Brucks, George Barnes. *Memoir Descriptive of the Navigation of the Gulf of Persia; with Brief Notices of the . . . People Inhabiting Its Shores and Islands*. Bombay Education Society's Press, 1856.
Buckingham, James Silk. *Travels in Assyria, Media, and Persia: Including a Journey from Bagdad by Mount Zagros . . . to Bombay.* 2 vols. Henry Colburn and Richard Bentley, 1830.
Burton, Richard F. *Sind Revisited.* 2 vols. Richard Bentley and Son, 1877.
Burton, Richard F. *The Lake Regions of Central Equatorial Africa*. Longman, Green, Longman, and Roberts, 1860.

Calcutta Law Journal, beginning 1905.
Colomb, Philip. *Slave-Catching in the Indian Ocean: A Record of Naval Experiences*. Longmans, Green, and Co., 1873.
Curzon, George N. *Persia and the Persian Question*. 2 vols. Longmans, Green, 1892.
De Barros, João. *Décadas da Ásia*. 3 vols. Na Regina Officina Typografica, 1778.
Dickson, H. R. P. *Kuwait and Her Neighbors*. George Allen & Unwin, 1956.
Dods, Marcus, ed. *The Works of Aurelius Augustine, Bishop of Hippo*. Vol. 1, *The City of God*. T & T Clark, 1884.
Dowson, V. H. W. *Dates and Date Cultivation of the ʿIraq*. W. Heffer & Sons, 1921.
Ferguson, Alastair, and John Ferguson. *The Planting Directory for India and Ceylon*. Colombo, 1878.
Ferrand, Gabriel. *Le Pilote des Mers de l'Inde, de la Chine, et de l'Indonesie Par Sulayman Al-Mahri et Šihāb ad-Dīn Aḥmad Bin Mājid*. Librarie Orientaliste, 1925.
Gazetteer of Bombay Presidency. 27 vols. Bombay: Government Central Press, 1884.
Grotius, Hugo. *The Free Sea*. Translated by Richard Hakluyt. Edited and with an introduction by David Armitage. Liberty Fund, 2004.
Hertslet, Edward. *A Complete Collection of the Treaties and Conventions and Reciprocal Regulations at Present Subsisting between Great Britain and Foreign Powers*. Butterworths, 1885.
Hunter, F. M. *An Account of the British Settlement of Aden in Arabia*. Trübner & Co, 1877.
Ibn Badran, ʿAbdulqadir bin Ahmad Al-Dumi. *Al-ʿUqūd Al-Yāqūtiyya fī Jayd Al-Asʾila Al-Kuwaytiyya*. Edited by ʿAbdulsattar Abu-Ghudda. Al-Sadawa Library, 1984.
Ibn Bishr, ʿUthman bin ʿAbdullah. *ʿUnwān Al-Majd fī Tārīkh Najd*. 2 vols. 4th ed. Dārat Al-Malik ʿAbd Al-ʿAziz, 1982.
Ibn Majid, Shihab Al-Din Ahmad. *Al-Nūniya Al-Kubrā Maʿa Sitt Qaṣāʾid Ukhrā*. Edited by Hassan Saleh Shihab. Oman Ministry of Heritage and Culture, 2016.
Ibn Ruzayq, Ḥamīd. *Al-Fatḥ Al-Mubīn fī Sīrat Al-Sāda Al-Busaīdiyīn*. 2 vols. 6th ed. Edited by Muhammad Habib Saleh and Mahmoud bin Mubarak Al-Sulaymi. Ministry of Heritage and Culture, 2016.
Indian Law Reports, Bombay Series. Thacker, Spink and Co., beginning 1876.
Ives, Edward. *A Voyage from England to India in the Year 1754*. Charles and Dilly, 1773.
Karachi Handbook and Directory for 1922–1923. Karachi: Daily Gazette Press, 1922.
Knox-Mawer, R., ed. *Law Reports of Aden, Containing Cases Determined by the Supreme Court of the Colony of Aden*. 4 vols. Government Press, 1955–64.
Langley, W. K. M., ed. *Century in Malabar: The History of Peirce Leslie & Co., Ltd., 1862–1962*. Madras Advertising Company, 1962.
Lorimer, John G. *Gazetteer of the Persian Gulf, Oman, and Central Arabia*. 2 vols in multiple parts. Superintendent Government Printing, 1915.
Maclean, James Mackenzie. *A Guide to Bombay*. Steam Press, 1889.
Owen, W. F. W. *Narrative of Voyages to Explore the Shores of Africa, Arabia, and Madagascar*. 2 vols. J. & J. Harper, 1833.

Palgrave, William. *Narrative of a Year's Journey through Central and Eastern Arabia, 1862–63*. Macmillan, 1908.
Pearson, T. A. *The Law of Agency in British India*. W. H. Allen, 1890.
Pelly, Lewis. "Remarks on the Tribes &c. around the Shores of the Persian Gulf." *Transactions of the Bombay Geographical Society* 17 (1865): 32–112.
Persian Gulf Trade Reports, 1905–1940. 8 vols. Archive Editions, 1987.
Quarterly Journal of the Department of Agriculture, Bengal 4. Bengal Secretariat Press, 1911.
Rida, Rashid. "Riḥlatuna Al-Hindiyya Al-ʿArabiyya: Shukr ʿAlani li-Ahl ʿUman wa al-Kuwayt." *Al-Manar* 16, no. 5 (May 7, 1913): 396–99.
Selden, John. *Mare Clausum: Of the Dominion, or Ownership, of the Sea*. William Du-Gard, 1652.
"Slave Trade (East African Courts) Act, 1873." *Law Journal Reports for the Year 1873*. London, 1873.
Stocqueler, J. H. *Fifteen Months Pilgrimage through Untrodden Tracts in Khuzistan and Persia . . . in the Years 1831 and 1832*. Saunders & Otley, 1832.
Sullivan, G. L. *Dhow Chasing in Zanzibar Waters and on the Eastern Coast of Africa*. 2nd ed. Sampson Low, Marston, Low, & Searle, 1873.
Persian Gulf Trade Reports, 1905–1940. 8 vols. Archive Editions, 1987.
Thesiger, Wilfred. *Arabian Sands*. Penguin Classics Edition. Penguin, 2007.
Tibbetts, G. R. *Arab Navigation in the Indian Ocean Before the Portuguese*. Routledge, 2004.
US Department of Commerce. *Daily Consular and Trade Reports*. Vol. 1. US Government Printing Office, 1914.
US Navy Hydrographic Office. *Sailing Directions for the West Coast of India from Point Calimere to Cape Monze Including Ceylon, Pamban Pass, and Palk Gulf*. 2nd ed. US Government Printing Office, 1930.
Villiers, Alan. *Sons of Sindbad*. Charles Scribner's Sons, 1969.
"Wathāʾiq wa Asānīd." Bayt Bin Ibrahim. 2020. www.al-ibrahim.org/category/general/wathaeq/.
Watson, J. W. *Statistical Account of Porbandar*. Education Society's Press, 1879.
Wilson, Horace-Hayman. *A Glossary of Judicial and Revenue Terms and of Useful Words Occurring in Official Documents Relating to the Administration of British India*. W. H. Allen, 1855.

Secondary Sources

Abdullah, Thabit. *Merchants, Mamluks, and Murder: The Political Economy of Trade in Eighteenth-Century Basra*. SUNY Press, 2001.
Abu Hakima, Ahmad Mustafa. *History of Eastern Arabia: The Rise and Development of Bahrain, Kuwait, and Wahhabi Saudi Arabia*. Probsthain, 1988.
Abu Hakima, Ahmad Mustafa. *The Modern History of Kuwait*. International Book Center, 1983.

Acevedo, Juan. "A New Arabic Nautical Manuscript in Lisbon." *Comparative Oriental Manuscript Studies Bulletin* 7 (2021): 9–36.

Agius, Dionisius. *Seafaring in the Arabian Gulf and Oman: The People of the Dhow.* Routledge, 2009.

Aiyar, Sana. *Indians in Kenya: The Politics of Diaspora.* Harvard University Press, 2015.

Al-ʿAbbasi, Hussain bin ʿAli Al-Wahaydi Al-Khunji. *Tārīkh Lingah: Ḥāḍirat Al-ʿArab ʿalā Al-Sāḥil Al-Sharqī lil-Khalīj.* Dar Al-Umma, 1988.

Al-ʿAbdulmughni, ʿAdel. "Jassim Al-Nasrallah . . . thaqāfa wāsiʿa wa qāmat adabiyya Kuwaitiyya." *Al-Jarida*, September 2, 2020. www.aljarida.com/articles/1598975795167168700.

Al-ʿAraimi, Muhammad bin Hamad. "Taʿarruf ʿalā Sayyid Yusuf Al-Zawawi, Iḥdā' Al-Shakhṣiyyāt al-ʿUmāniyya al-Bāriza." *Atheer*, December 29, 2019. www.atheer.om/archives/513908/.

Alavi, Seema. *Sovereigns of the Sea: Oman Ambition in the Age of Empire.* Allen Lane, 2023.

Alavi, Seema. *Muslim Cosmopolitanism in the Age of Empire.* Harvard University Press, 2015.

Al-Baharna, Hussain. *British Extra-Territorial Jurisdiction in the Gulf, 1913–1971: An Analysis of the System of the British Protected States of the Gulf during the Pre-independence Era.* Brill, 1998.

Al-Baharna, Hussain. *The Legal Status of the Arabian Gulf States: A Study of Their Treaty Relations and Their International Problems.* University of Manchester Press, 1968.

Albedwawi, Saif. "Pearl Merchants of the Gulf and Their Life in Bombay." *Proceedings of the Seminar for Arabian Studies* 47 (2017): 1–7.

Alborn, Timothy. *Regulated Lives: Life Insurance and British Society, 1800–1914.* University of Toronto Press, 2009.

Al-Duwaisan, Rasha. "How Far Did Kuwaitis Integrate into Indian Society?" Unpublished master's thesis, Harvard University, 2008.

Alebrahim, Abdulrahman. *Kuwait's Politics before Independence: The Role of the Balancing Powers.* Gerlach Press, 2019.

Alexandrowicz, C. H. *An Introduction to the History of the Law of Nations in the East Indies.* Clarendon Press, 1967.

Alexandrowicz, C. H. "Freitas versus Grotius." *British Yearbook of International Law* 35 (1959): 162–182.

Al-Fulayḥ, Sulaiman. *Ḥayyahalā: Qiṣas wa Ashʿār min Al-Bādiya.* Maktabat Sulayman Al-Fulayḥ, 2021.

Al-Ghaylani, Hamoud. *Asyād Al-Biḥār.* Self-published, 2016.

Al-Habad, Hamad ʿAbdullah. *Fann Al-Sangani.* Kuwait National Council for Culture, Arts, and Letters, 2014.

Alhabib, Mohammed. "Muḥammad ʿAbdullah Al-Matrūk: Sīrat Tājir Tanāsāhu Al-Tārīkh." *Al-Qabas*, August 19, 2012.

Al-Harbi, Hussa. *Tārīkh Al-ʿIlāqāt Al-Kuwaytiyya Al-Hindiyya, 1895–1965: Al-Mawsūʿa Al-Kāmila wal-Muṣawwara.* Self-published, 2017.

Al-Hijji, Yacoub Y. *Al-Nashāṭāt al-Baḥriyya al-Qadīma fi Al-Kuwait*. Cener for Research and Studies on Kuwait, 2007.
Al-Hijji, Yacoub Y. "Arab Navigational Methods after the 19th Century." In *The Principles of Arab Navigation*, edited by Anthony R. Constable and William Facey, 35–45. Arabian Publishing: 2013.
Al-Hijji, Yacoub Y. *Fateḥ Al-Khair*. Kuwait Foundation for the Advancement of Sciences, n.d.
Al-Hijji, Yacoub Y. *Kuwait and the Sea: A Brief Economic and Social History*. Arabian Publishing, 2010.
Al-Hijji, Yacoub Y. *Min al-Fulklūr al-Baḥrī al-Kuwaytī*. Center for Research and Studies on Kuwait, 2009.
Al-Hijji, Yacoub Y. *Nawākhithat Al-Safar Al-Shirā'ī fi Al-Kuwayt*. Al-Rubay'ān Publishing, 1993.
Al-Hijji, Yacoub Y. *Nawākhithat Al-Safar Al-Shirā'ī fī Al-Kuwayt*. Center for Research and Studies on Kuwait, 2005.
Al-Hijji, Yacoub Y. *Ṣinā'at Al-Sufun Al-Shirā'iyya fī Al-Kuwayt*. Markaz Al-Turath Al-Sha'bi, 1988.
Al-Hulw, 'Ali Ni'ma. *Al-Aḥwāz, 'Arabistān*. 4 vols. Maktabat Al-Qurī Al-Hadītha, 1969–1970.
Ali, Tariq Omar. *A Local History of Global Capital: Jute and Peasant Life in the Bengal Delta*. Princeton University Press, 2018.
Al-Ibrahim, Yacoub Yousef. *Min Al-Shirā' Ilā Al-Bukhār: Qiṣṣat Awwalu Sharikat Malāḥa Baḥriyya 'Arabiyya*. Al-Rubay'ān Publishing, 2002.
Al-Kindi, Muhsin. *Al-Ṣaḥāfa Al-'Umāniyya Al-Muhājira: Ṣaḥīfat Al-Falaq wa Shakhṣiyyātuhā*. Riad El-Rayyes Books, 2001.
Al-Shaikh Jassim bin Muhammad Al-Ibrahim. Self-published.
Allen, Calvin H. "Ratansi Purshottam, the Gujaratis of Muscat, and the Global Connectivity of Indian Ocean Transregional Markets." In *Transregional Trade and Traders: Situating Gujarat in the Indian Ocean from Early Times to 1900*, edited by Edward Alpers and Chhaya Goswami, 269–84. Oxford University Press, 2019.
Al-Loghani, Bassem. "Wathiqa Lahā Tārikh: Al-Shaykh Nasser Ṭalaba Shirā' Riwāyāt Bonsūn al-Faransi wa Kitāb 'Naḍrāt' lil-Manfalūṭī 'ām 1913." *Al-Jarida*, February 5, 2016. www.aljarida.com/articles/1463611857603193000/.
Al-Madani, 'Abdullah. "Wakeel Shaikh Mubarak Al-Sabah Al-Kabir fī Bombay." *Al-Ayyam*, August 22, 2014.
Al-Matrook, Muhammad bin Dhargham. *Min Tijārat Al-Māḍī*. Self-published, 2015.
Al-Nakib, Farah. *Kuwait Transformed: A History of Oil and Urban Life*. Stanford University Press, 2016.
Al-Najjar, Mustafa 'Abdulqadir. *'Arabistān Khilāla Ḥukm Al-Shaikh Khaz'al Al-Ka'bi, 1897–1925*. Al-Dār Al-'Arabiyya lil-Mawsū'āt, 2009.
Al-Naqeeb, Khaldoun. *State and Society in the Gulf and Arab Peninsula: A Different Perspective*. Routledge, 1990.

Alpers, Edward. *The Indian Ocean in World History*. Oxford University Press, 2014.
Al-Qasimi, Sultan bin Muhammad. *Power Struggles and Trade in the Gulf, 1620–1820*. Forest Row, 1999.
Al-Rasheed, Madawi. *A History of Saudi Arabia*. Cambridge University Press, 2010.
Al-Rashoud, Talal. "Modern Education and Arab Nationalism in Kuwait, 1911–1961." PhD diss., School of Oriental and African Studies, 2017.
Al-Rifai, Hissa. "Al-Nehmah wal-Nahham: A Structural, Functional, Musical, and Aesthetic Study of Kuwaiti Sea Songs." PhD diss., Indiana University, 1982.
Al-Roumi, Ahmad Al-Bishr. *Fahad Rāshid Burisly, 1918–1960: Shāhid ʿAṣr al-Taḥwīlāt*. Thāt al-Salāsil, 2013.
Al-Rumi, ʿAdnan bin Salem. *ʿUlamāʾ Al-Kuwayt wa Aʿlāmuhā Khilāl Thalāthat Qurūn*. Al-Manar, 1999.
Al-Salhi, Ahmad. "The Recordings of ʿAbdullatif al-Kuwaiti: 1927–1947." In *Music in Arabia: Perspectives on Heritage, Mobility, and Nation*, edited by Issa Boulos, Virginia Danielson, and Anne Rasmussen, 105–24. Indiana University Press, 2021.
Al-Shamlan, Saif. *Tārīkh Al-Ghawṣ ʿala al-Luʾluʾ fi Al-Kuwayt wa Al-Khalīj Al-ʿArabī*. 2 vols. That es-Salāsil, 1989.
Al-Shaybani, Muhammad bin Ibrahim. *Kaʿb Al-Akhbār Al-Kuwayti: Al-Shaykh Nasser bin Mubarak Al-Sabah*. Markaz al-Makhṭūṭāṭ wal-Turāth wal-Wathāʾiq, 2011.
Alshehabi, Omar. *Contested Modernity: Sectarianism, Nationalism, and Colonialism in Bahrain*. Oneworld Publications, 2019.
Altorki, Soraya, and Donald P. Cole. *Arabian Oasis City: The Transformation of ʿUnayzah*. University of Texas Press, 1989.
Al-Zayyani, Rashid. *Al-Ghawṣ Wal-Ṭawāshah*. Al-Ayyam Publishing, 1998.
Amrith, Sunil. *Crossing the Bay of Bengal: The Furies of Nature and the Fortunes of Migrants*. Harvard University Press, 2015.
Anscombe, Frederick. *The Ottoman Gulf: The Creation of Kuwait, Saudi Arabia, and Qatar*. Cambridge University Press, 1997.
Appadurai, Arjun, ed. *The Social Life of Things: Commodities in Cultural Perspective*. Cambridge University Press, 1986.
Armitage, David. *The Ideological Origins of the British Empire*. Cambridge University Press, 2000.
Ashkanani, Z. A. M. "Middle-Aged Women in Kuwait: Victims of Change." PhD diss., Durham University, 1988.
Aslanian, Sebouh. *From the Indian Ocean to the Mediterranean: The Global World of Armenian Merchants from New Julfa*. University of California Press, 2011.
Aslanian, Sebouh. "The Salt in a Merchant's Letters: The Culture of Julfan Correspondence in the Indian Ocean and the Mediterranean." *Journal of World History* 19, no. 2 (2008): 127–88.
Avery, P. W., Gavin Hambly, and Charles Melville, eds. *The Cambridge History of Iran*, Vol. 7, *From Nadir Shah to the Islamic Republic*. Cambridge University Press, 1991.

Axworthy, Michael. "Nader Shah and Persian Naval Expansion in the Persian Gulf, 1700–1747." *Journal of the Royal Asiatic Society* 21, no. 1 (2011): 31–39.
Bagchi, Amiya. *The Presidency Banks and the Indian Economy, 1876–1914*. Oxford University Press, 1990.
Bagchi, Amiya. "Transition from Indian to British Indian Systems of Money and Banking, 1800–1850." *Modern Asian Studies* 19, no. 3 (1985): 501–19.
Bang, Anne. *Sufis and Scholars of the Sea: Family Networks in East Africa, 1860–1925*. Routledge, 2004.
Banga, Indu. "Karachi and Its Hinterland under Colonial Rule." In *Ports and Their Hinterlands in India (1700–1950)*, edited by Indu Banga, 337–58. Manohar, 1992.
Baptist, Edward. *The Half Has Never Been Told: Slavery and the Making of American Capitalism*. Basic Books, 2016.
Barendse, R. N. *Arabian Seas, 1700–1763*. 4 vols. Brill, 2009.
Bathurst, Roland D. "The Ya'rubi Dynasty of Oman." PhD diss., Oxford University, 1967.
Bayly, Christopher. *Imperial Meridian: The British Empire and the World, 1780–1830*. Longman Group, 1989.
Bayly, Christopher. "Pre-Colonial Indian Merchants and Rationality." In *India's Colonial Encounter: Essays in Memory of Eric Stokes*, edited by Mushirul Hasan and Narayani Gupta, 3–24. Manohar Publishers, 1993.
Benjamin, N. "Arab Merchants of Bombay and Surat (c. 1800–1840)." *Indian Economic and Social History Review* 13, no. 1 (1976): 85–95.
Bennett, Brett M. "The Origins of Timber Plantations in India." *Agricultural History Review* 62, no. 1 (2014): 98–118.
Benton, Lauren. *A Search for Sovereignty: Law and Geography in European Empires, 1400–1900*. Cambridge University Press, 2009.
Bhacker, Reda M. "Family Strife and Foreign Intervention: Causes in the Separation of Zanzibar from Oman; A Reappraisal." *Bulletin of the School of Oriental and African Studies, University of London* 54, no. 2 (1991): 269–80.
Bhacker, Reda M. *Trade and Empire in Muscat and Zanzibar: Roots of British Dominion*. Routledge, 1992.
Bhattacharyya, Tania. "Ocean Bombay: Space, Itinerancy and Community in an Imperial Port City, 1839–1937." PhD diss., Columbia University, 2019.
Bishara, Fahad Ahmad. "Circulation and Capitalism in a Maritime Bazaar: Notes from a Pearl Merchant's Chest." *Comparative Studies of South Asia, Africa, and the Middle East* 42, no. 1 (2022): 107–17.
Bishara, Fahad Ahmad. "No Country but the Ocean: Reading International Law on the Deck of an Indian Ocean Dhow, ca. 1900." *Comparative Studies in Society and History* 60, no. 2 (2018): 338–66.
Bishara, Fahad Ahmad. *A Sea of Debt: Law and Economic Life in the Western Indian Ocean, 1780–1950*. Cambridge University Press, 2017.
Bishara, Fahad Ahmad, and Hollian Wint. "Into the Bazaar: Indian Ocean Vernaculars in the Age of Global Capitalism." *Journal of Global History* 16, no. 1 (2021): 44–64.

Bolton, James and Guidi Boscoli, "'Your Flexible Friend': The Bill of Exchange in Theory and Practice in the Fifteenth Century." *Economic History Review* 74 (2021): 873–91.

Boodrookas, Alex C. "The Making of a Migrant Working Class: Contesting Citizenship in Kuwait and the Persian Gulf, 1925–1975." PhD diss., New York University, 2020.

Bose, Sugata. *A Hundred Horizons: The Indian Ocean in the Age of Global Empire.* Harvard University Press, 2006.

Braudel, Fernand. *Civilization and Capitalism, 15th–18th Century.* 3 vols. University of California Press, 1992.

Braudel, Fernand. *The Mediterranean and the Mediterranean World in the Age of Philip II.* 2 vols. Harper and Row, 1972.

Bulley, Anne. *The Bombay Country Ships, 1790–1833.* Routledge, 2000.

Bu-Safwan, ʿAbbas. *Al-Baḥrayn 1923: Naqd Manhajiyyun lil-Sardiyya Al-Sāʾida ʿan ʿIshrīniyyāt Al-Qarn Al-ʿIshrīn.* London Printing and Publishing: 2024.

Caeiro, Alexandre. "The Islamic Law of Pearling: Ritual Obligation and Economic Practice in the Arabian Gulf, ca. 1910–1940." *Islamic Law and Society* 29, no. 4 (2022): 457–94.

Carter, Robert. *Sea of Pearls: The History of Pearl Fishing in Bahrain and the Gulf.* Arabian Publishing, 2012.

Casale, Giancarlo. *The Ottoman Age of Exploration.* Oxford University Press, 2010.

Caton, Steven C. *Peaks of Yemen I Summon: Poetry as Cultural Practice in a North Yemeni Tribe.* University of California Press, 1993.

Cattelan, Steffano. "Iberian Expansion over the Oceans: Law and Politics of Mare Clausum on the Threshold of Modernity." *Historia et Ius*, no. 18 (2020).

Çetinsaya, Gökhan. *Ottoman Administration of Iraq, 1890–1908.* Routledge, 2006.

Chakrabarty, Dipesh. *Provincializing Europe: Postcolonial Thought and Historical Difference.* Princeton University Press, 2000.

Chand, Vijaya, Abhishek Mishra, and Kriti Bole. *Maritime Trade of Gujarat's Princely States: Nawanagar and Porbandar.* IIMA, 2023.

Chandavarkar, Rajnarayan. *The Origins of Industrial Capitalism in India: Business Strategies and the Working Classes in Bombay, 1900–1940.* Cambridge University Press, 1994.

Chandler, Alfred D. *The Visible Hand: The Managerial Revolution in American Business.* Belknap, 1977.

Chatterjee, Arijit. "Tracing the History and Legacy of Mangalore Tiles." *Architectural Digest*, May 22, 2022. www.architecturaldigest.in/story/tracing-the-history-and-legacy-of-mangalore-tiles/.

Chaudhuri, K. N. *Trade and Civilization in the Indian Ocean: An Economic History from the Rise of Islam to 1750.* Cambridge University Press, 1985.

Choudhry, Rishad. *Hajj across Empires: Pilgrimage and Political Culture after the Mughals, 1739–1857.* Cambridge University Press, 2024.

Cohen, Michael C. *The Social Life of Poems in Nineteenth-Century America.* University of Pennsylvania Press, 2015.

Cole, Camille. "Empire on Edge: Land, Law, and Capital in Gilded Age Basra." PhD diss., Yale University, 2020.
Cole, Camille. "Precarious Empires: A Social and Environmental History of Steam Navigation on the Tigris." *Journal of Social History* 50, no. 1 (2016): 74–101.
Copland, Ian. *The Princes of India in the Endgame of Empire, 1917–1947*. Cambridge University Press, 1997.
Cordell, Ryan, and Abby Mullen. "'Fugitive Verses': The Circulation of Poems in Nineteenth-Century American Newspapers." *American Periodicals* 27, no. 1 (2017): 29–52.
Cronon, William. *Nature's Metropolis: Chicago and the Great West*. W. W. Norton, 1991.
Crosby, Alfred. *The Measure of Reality: Quantification and Western Society, 1250–1600*. Cambridge University Press, 1997.
Crawford, Harriet E. W. *Dilmun and Its Gulf Neighbors*. Cambridge University Press, 1998.
Crystal, Jill. *Oil and Politics in the Gulf: Rulers and Merchants in Kuwait and Qatar*. Cambridge University Press, 1994.
Cunha, João Teles e. "Oman and Omanis in Portuguese Sources in the Early Modern Period (ca. 1500–1750)." In *Oman and Overseas*, edited by Michaela Hoffman-Ruf and Abdulrahman Al-Salimi, 227–63. Georg Olms Verlag, 2013.
Curtin, Philip D. "African Enterprise in the Mangrove Trade: The Case of Lamu." *African Economic History*, no. 10 (1981): 23–33.
Dale, Stephen. "The Hadhrami Diaspora in Southwestern India: The Role of the Sayyids of the Malabar Coast." In *Hadhrami Traders, Scholars, and Statesmen in the Indian Ocean, 1750s–1960s*, edited by Ulrike Freitag and W. G. Clarence-Smith, 175–84. Brill, 1997.
Dale, Stephen. *The Muslim Empires of the Ottomans, Safavids, and Mughals*. Cambridge University Press, 2010.
Dale, Stephen. "The Mappilla Outbreaks: Ideology and Social Conflict in Nineteenth-Century Kerala." *Journal of Asian Studies* 35, no. 1 (1975): 85–97.
Das Gupta, Ashin. *Indian Merchants and the Decline of Surat, c. 1700–1750*. Franz Steiner Verlag, 1979.
Das Gupta, Ashin. *The World of the Indian Ocean Merchant, 1500–1800: Collected Essays of Ashin Das Gupta*. Oxford University Press, 2004.
Davies, Charles. *The Blood-Red Arab Flag: An Investigation into Qasimi Piracy, 1797–1820*. University of Exeter Press, 1997.
Davis, Natalie Zemon. *The Return of Martin Guerre*. Harvard University Press, 1984.
Deringer, William. *Calculated Values: Finance, Politics, and the Quantitative Age*. Harvard University Press, 2018.
De Roover, Raymond. *Business, Banking, and Economic Thought in Late Medieval and Early Modern Europe*. University of Chicago Press, 1976.
Desan, Christine. *Making Money: Coin, Currency, and the Coming of Capitalism*. Oxford University Press, 2014.

Dewey, Clive. *Steamboats on the Indus: The Limits of Western Technological Superiority in South Asia*. Oxford University Press, 2014.

Dhanagare, D. N. "Agrarian Conflict, Religion and Politics: The Moplah Rebellions in Malabar in the Nineteenth and Early Twentieth Centuries." *Past and Present*, no. 74 (1977): 112–41.

Dodge, Toby. *Inventing Iraq: The Failure of Nation Building and a History Denied*. Columbia University Press, 2003.

Doerflinger, Thomas. *A Vigorous Spirit of Enterprise: Merchants and Economic Development in Revolutionary Philadelphia*. University of North Carolina Press, 2001.

Doumani, Beshara. *Family Life in the Ottoman Mediterranean: A Social History*. Cambridge University Press, 2017.

Echenberg, Myron. *Plague Ports: The Global Urban Impact of Bubonic Plague, 1894–1901*. New York University Press, 2007.

Edelson, S. Max. *The New Map of Empire: How Britain Imagined America before Independence*. Harvard University Press, 2017.

Edney, Matthew. *Mapping an Empire: The Geographical Construction of British India, 1765–1843*. University of Chicago Press, 1997.

Edwards, Jeremy, and Sheilagh Ogilvie. "Contract Enforcement, Institutions and Social Capital: The Maghribi Traders Reappraised." *Economic History Review* 65, no. 2 (2012): 421–44.

Edwards, Laura F. *The People and Their Peace: Legal Culture and the Transformation of Inequality in the Post-Revolutionary South*. University of North Carolina Press, 2009.

Elliott, Derek, and Sebastian Prange. "Beyond Piracy: Maritime Violence and Colonial Encounters in Indian History." In *Beyond the Line: Cultural Narratives of the Southern Oceans*, edited by Michael Mann and Ineke Phaf-Rheinberger, 95–120. Neofelis Verlag, 2009.

Erich, Scott. "A Deeper History of the 'World's Largest Dead Zone' in the Gulf of Oman." *International Review of Environmental History* 9, no. 1 (2023): 121–34.

Erich, Scott. "Extraction, Property, and Rights in Southeast Arabian Seascapes, c. 1820 to Present." PhD diss., CUNY Graduate Center, 2023.

Erman, Sam. *Almost Citizens: Puerto Rico, the U.S. Constitution, and Empire*. Cambridge University Press, 2018.

Farah, Talal Toufic. *Protection and Politics in Bahrain, 1869–1915*. American University of Beirut Press, 1985.

Fattah, Hala. *The Politics of Regional Trade in Iraq, Arabia, and the Gulf, 1750–1900*. SUNY Press, 1997.

Fernando, Tamara. "Mapping Oysters and Making Oceans in the Northern Indian Ocean, 1886–1906." *Comparative Studies in Society and History* 65, no. 1 (2023): 53–80.

Field, Michael. *The Merchants: The Big Business Families of Saudi Arabia and the Gulf States*. Overlook Press, 1984.

Fisher, Suzannah. *The Makers of the Blueback Charts: A History of Imray, Laurie, Norie and Wilson Ltd*. Regatta Press, 2001.

Floor, Willem. *The Persian Gulf: A Political and Economic History of Five Port Cities, 1500–1730*. Mage Publishers, 2006.
Floor, Willem. *The Rise and Fall of Bandar-e Lengeh, the Distribution Center for the Arabian Coast, 1750–1930*. Mage Publishers, 2010.
Floor, Willem. "The Rise and Fall of the Banū Kaʿb. A Borderer State in Southern Khuzestan." *Iran: Journal of the British Institute of Persian Studies* 44, no. 1 (2006): 277–315.
Frank, Andrew Gunder. *ReOrient: The Global Economy in the Asian Age*. University of California Press, 1997.
Fuccaro, Nelida. *Histories of City and State in the Persian Gulf: Manama since 1800*. Cambridge University Press, 2009.
Fusaro, Maria, and Amélia Polónia, eds. *Maritime History as Global History*. International Maritime History Association, 2010.
Gavin, R. J. *Aden under British Rule, 1839–1967*. Hurst Publishers, 1975.
Geertz, Clifford. "The Bazaar Economy: Information and Search in Peasant Marketing." *American Economic Review* 68, no. 2 (1978): 28–32.
Ghazal, Amal. *Islamic Reform and Arab Nationalism: Expanding the Crescent from the Mediterranean to the Indian Ocean (1880s–1930s)*. Routledge, 2010.
Ghazal, Amal. "The Other Frontiers of Arab Nationalism: Ibadis, Berbers, and the Arabist-Salafist Press in the Interwar Period." *International Journal of Middle East Studies* 42, no. 1 (2010): 105–22.
Ghobrial, John Paul. "Introduction: Seeing the World Like a Microhistorian." Supplement, *Past and Present* 242, S14 (2019): S1–22.
Gilbert, Erik. *Dhows and the Colonial Economy of Zanzibar, 1860–1970*. Ohio University Press, 2004.
Gilbert, Erik. "Sailing from Lamu and Back: Labor Migration and Regional Trade in Colonial East Africa." *Comparative Studies of South Asia, Africa and the Middle East* 19, no. 2 (1999): 9–15.
Ginzburg, Carlo. *The Cheese and the Worms: The Cosmos of a Sixteenth-Century Miller*. Johns Hopkins University Press, 1980.
Glassman, Jonathon. "The Bondsman's New Clothes: The Contradictory Consciousness of Slave Resistance on the Swahili Coast." *Journal of African History* 32, no. 2 (1991): 277–312.
Glassman, Jonathon. *War of Words, War of Stones: Racial Thought and Violence in Colonial Zanzibar*. Indiana University Press, 2011.
Gleeson-White, Jane. *Double Entry: How the Merchants of Venice Created Modern Finance*. W. W. Norton, 2012.
Goldberg, Jessica. "Choosing and Enforcing Business Relationships in the Eleventh-Century Mediterranean: Re-Examining the 'Maghribī Traders.'" *Past and Present* 215, no. 2 (2012): 3–40.
Goldberg, Jessica. "The Use and Abuse of Commercial Letters from the Cairo Geniza." *Journal of Medieval History* 38, no. 2 (2012): 127–54.
Goswami, Chhaya, *Globalization before Its Time: The Gujarati Merchants from Kachchh*. Portfolio, 2016.

Green, Nile. *Bombay Islam: The Religious Economy of the West Indian Ocean, 1840–1915*. Cambridge University Press, 2011.

Green, Nile. "Forgotten Futures: Indian Muslims in the Trans-Islamic Turn to Japan." *Journal of Asian Studies* 72, no. 3 (2013): 611–31.

Green, Nile. "Introduction: Arabic as a South Asian Language." *International Journal of Middle East Studies* 55, no. 1 (2023): 106–21.

Green, Nile. "The Languages of Indian Ocean Studies: Models, Methods, and Sources." *History Compass* 20, no. 7 (2021).

Green, Nile. "Waves of Heterotopia: Toward a Vernacular Intellectual History of the Indian Ocean." *American Historical Review* 123, no. 3 (2018): 846–74.

Greenberg, Joshua. *Bank Notes and Shinplasters: The Rage for Paper Money in the Early Republic*. University of Pennsylvania Press, 2020.

Greif, Avner. "The Maghribi Traders: A Reappraisal?" *Economic History Review* 65, no. 2 (2012): 445–69.

Greif, Avner. "Reputation and Coalitions in Medieval Trade: Evidence on the Maghribi Traders." *Journal of Economic History* 49, no. 4 (1989): 857–82.

Guyer, Jane. "Wealth in People, Wealth in Things—Introduction." *Journal of African History* 36, no. 1 (1995): 83–90.

Guyer, Jane, and Samuel M. Eno Belinga, "Wealth in People as Wealth in Knowledge: Accumulation and Composition in Equatorial Africa." *Journal of African History* 36, no. 1 (1995): 91–120.

Habib, Irfan. "Banking in Mughal India." In *Contributions to Indian Economic History*, edited by Tapan Raychaudhuri, 1:1–7. K. L. Mukhopadhyay, 1960.

Hahn, Barbara. *Making Tobacco Bright: Creating an American Commodity, 1617–1937*. Johns Hopkins University Press, 2011.

Halevi, Leor. "Arabians for Guns: Wahhabi Matchlocks, World Trade, and the Rise of the First Saudi State." *Journal of the Royal Asiatic Society* 33, no. 2 (2023): 401–42.

Halevi, Leor. *Modern Things on Trial: Islam's Global and Material Reformation in the Age of Rida, 1865–1935*. Columbia University Press, 2019.

Hall, Richard. *Empires of the Monsoon: A History of the Indian Ocean and Its Invaders*. HarperCollins, 1998.

Hancock, David. *Citizens of the World: London Merchants and the Integration of the British Atlantic Community, 1735–1785*. Cambridge University Press, 1997.

Hanna, Nelly. *Making Big Money in 1600: The Life and Times of Ismaʿil Abu Taqiyya, Egyptian Merchant*. Syracuse University Press, 1998.

Hanssen, Jens, and Max Weiss, eds. *Arabic Thought beyond the Liberal Age: Towards an Intellectual History of the Nahda*. Cambridge University Press, 2016.

Hill Edwards, Justene. *Unfree Markets: The Slaves' Economy and the Rise of Capitalism in South Carolina*. Columbia University Press, 2021.

Ho, Engseng. "Empire through Diasporic Eyes: A View from the Other Boat." *Comparative Studies in Society and History* 46, no. 2 (2004): 210–46.

Ho, Engseng. "Inter-Asian Concepts for Mobile Societies." *Journal of Asian Studies* 76, no. 4 (2017): 907–28.

Hodgson, Marshall. *The Venture of Islam: Conscience and History in a World Civilization*. 3 vols. University of Chicago Press, 1974.
Hofmeyr, Isabel. *Dockside Reading: Hydrocolonialism and the Custom House*. Duke University Press, 2022.
Hohman, Elmo P. "Wages, Risk, and Profits in the Whaling Industry." *Quarterly Journal of Economics* 40, no. 4 (1926): 644–71.
Hopper, Matthew. "The Globalization of Dried Fruit: Transformation in the Eastern Arabian Economy, 1860s–1920s." In *Global Muslims in the Age of Steam and Print*, edited by James L. Gelvin and Nile Green, 158–82. University of California Press, 2014.
Hopper, Matthew. *Slaves of One Master: Globalization and Slavery in Arabia in the Age of Empire*. Yale University Press, 2015.
Horden, Peregrine, and Nicholas Purcell. *The Forbidding Sea: A Study in Mediterranean History*. Wiley-Blackwell, 2000.
Hourani, Albert. *Arabic Thought in the Liberal Age, 1798–1939*. Cambridge University Press, 1962.
Hourani, George. *Arab Seafaring in the Indian Ocean in Ancient and Early Medieval Times*. Princeton University Press, 1995.
Huber, Valeska. *Channeling Mobilities: Migration and Globalisation in the Suez Canal Region and Beyond, 1869–1914*. Cambridge University Press, 2013.
Jacob, Wilson Chacko. *For God or Empire: Sayyid Fadl and the Indian Ocean World*. Stanford University Press, 2019.
Jakes, Aaron. *Egypt's Occupation: Colonial Economism and the Crises of Capitalism*. Stanford University Press, 2020.
Jamal, Muhammad ʿAbdulhadi. *Aswāq Al-Kuwayt Al-Qadīma*. Center for Research and Studies on Kuwait, 2001.
Johnson, L. G. *General T. Perronet Thompson, 1783–1869: His Military, Literary, and Political Campaigns*. Routledge, 1957.
Johnson, Walter. *River of Dark Dreams: Slavery and Empire in the Cotton Kingdom*. Harvard University Press, 2013.
Johnstone, T. M., and J. Muir. "Some Nautical Terms in the Kuwaiti Dialect of Arabic." *Bulletin of the School of Oriental and African Studies, University of London* 27, no. 2 (1964): 299–332.
Jones, Stephanie. *Two Centuries of Overseas Trading: The Origins and Growth of the Inchcape Group*. Palgrave Macmillan, 1986.
Kanoo, Khalid. *The House of Kanoo: A Century of an Arabian Family Business*. London Centre of Arab Studies, 1997.
Kelly, J. B. *Britain and the Persian Gulf, 1795–1880*. Clarendon Press, 1968.
Keshavarzian, Arang. *Bazaar and State in Iran: The Politics of the Tehran Marketplace*. Cambridge University Press, 2007.
Khalifoh, Bashar Muhammad. *Sūr Al-Kuwait Al-Thālith wa Tārīkh Bawwābātuhu*. Center for Research and Studies on Kuwait, 2009.
Khalili, Laleh. *Sinews of War and Trade: Shipping and Capitalism in the Arabian Peninsula*. Verso, 2020.

Khazʿal, Hussain Khalaf Al-Shaikh. *Tārīkh Al-Kuwayt Al-Siyāsī.* Al-Hilal, 1962.

Khoury, Ibrahim. "Les poèmes nautiques d'Ahmad Ibn Magid, 2ème partie. Les poèmes à rime unique: Al-Qaṣāʾid. Texte arabe établi avec introduction et analyse en français." *Bulletin d'études Orientales* 37/38 (1985/86): 226–30.

Knight, William S. M. "Seraphin de Freitas: Critic of *Mare liberum.*" *Transactions of the Grotius Society* 11, no. 1 (1926).

Kooria, Mahmood. *Islamic Law in Circulation: Shafiʿi Texts across the Indian Ocean and Mediterranean.* Cambridge University Press, 2022.

Kuran, Timur. *The Long Divergence: How Islamic Law Held Back the Middle East.* Princeton University Press, 2011.

Laffan, Michael. *Under Empire: Muslim Lives and Loyalties across the Indian Ocean World, 1775–1945.* Columbia University Press, 2022.

Lamoreaux, Naomi, Daniel M. G. Raff, and Peter Temin. "Beyond Markets and Hierarchies: Toward and New Synthesis of American Business History." *American Historical Review* 108, no. 2 (2003): 404–33.

Lari, Yasmeen. *The Dual City: Karachi during the Raj.* Oxford University Press, 1996.

Lavin, Gabriel. "Arabian Passings in Indian Ocean History: Troubadours, Technology, and the Longue Durée, 1656–1963." In *Sounding the Indian Ocean: Musical Circulations in the Afro-Asiatic Seascape*, edited by Jim Sykes and Julia Suzanne Byl, 158–78. University of California Press, 2023.

Layton, Simon. "Hydras and Leviathans in the Indian Ocean World." *International Journal of Maritime History* 25, no. 2 (2013): 213–25.

Lepore, Jill. "Historians Who Love Too Much: Reflections on Microhistory and Biography." *Journal of American History* 88, no. 1 (2001): 129–44.

Levi, Giovanni. "On Microhistory." In *New Perspectives on Historical Writing*, edited by Peter Burke, 97–119. Pennsylvania State University Press, 1992.

Lienhardt, Peter. *Shaikhdoms of Eastern Arabia.* Palgrave, 2001.

Lipartito, Kenneth. "Reassembling the Economic: New Departures in Historical Materialism." *American Historical Review* 121, no. 1 (2016): 101–39.

Lydon, Ghislaine. *On Trans-Saharan Trails: Islamic Law, Trade Networks, and Cross-Cultural Exchange in Nineteenth-Century West Africa.* Cambridge University Press, 2009.

Lydon, Ghislaine. "A Paper Economy of Faith without Faith in Paper: A Reflection on Islamic Institutional History." *Journal of Economic Behavior and Organization* 71 (2009): 647–59.

Lydon, Ghislaine. "Paper Instruments in Early African Economies and the Debated Role of the *Suftaja.*" *Cahiers d'Études Africaines*, no. 263 (2019): 1091–118.

Machado, Pedro. *Ocean of Trade: South Asian Merchants, Africa, and the Indian Ocean c. 1750–1850.* Cambridge University Press, 2015.

Maloni, Ruby. "Control of the Seas: The Historical Exegesis of the Portuguese *Cartaz.*" *Proceedings of the Indian History Congress* 72, pt. 1 (2011): 476–84.

Mandaville, Jon. "The Ottoman Province of al-Hasā in the Sixteenth and Seventeenth Centuries." *Journal of the American Oriental Society* 90, no. 3 (1970): 486–513.

Mann, Michael. "Timber Trade on the Malabar Coast, c. 1740–1840." *Environment and History* 7, no. 4 (2001): 403–25.

Manning, Patrick. "Global History and Maritime History." *International Journal of Maritime History* 25, no. 1 (2013): 1–22.

Marcocci, Giuseppe. "Saltwater Conversion: Trans-Oceanic Sailing and Religious Transformation in the Iberian World." In *Space and Conversion in Global Perspective*, edited by Giuseppe Marcocci, Wietse de Boer, Aliocha Maldavsky, and Ilaria Pavan, 233–59. Brill, 2015.

Margariti, Roxani. "Mercantile Networks, Port Cities, and 'Pirate' States: Conflict and Competition in the Indian Ocean World of Trade before the Sixteenth Century." *Journal of the Economic and Social History of the Orient* 51, no. 1 (2008): 543–77.

Markovits, Claude. *The Global World of Indian Merchants, 1750–1947: Traders of Sind from Bukhara to Panama*. Cambridge University Press, 2000.

Martin, Esmond Bradley, and Chryssee Perry Martin. *Cargoes of the East: The Ports, Trade, and Culture of the Arabian Seas and Western Indian Ocean*. Elm Tree Books, 1978.

Martin, Marina. "An Economic History of Hundi, 1858–1978." PhD diss., London School of Economics, 2012.

Mathew, Johan. "Khaliji Hindustan: Towards a Diasporic History of Khalijis in South Asia from the 1780s to the 1960s." In *The Gulf in World History: Arabia at the Crossroads*, edited by Allen Fromherz, 120–36. Edinburgh University Press, 2018.

Mathew, Johan. *Margins of the Market: Trafficking and Capitalism in the Arabian Sea*. University of California Press, 2016.

Mathew, Johan. "On Principals and Agency: Reassembling Trust in Indian Ocean Commerce." *Comparative Studies in Society and History* 61, no. 2 (2019): 242–68.

Mathew, Nisha. "At the Crossroads of Empire and Nation-State: Partition, Gold Smuggling, and Port Cities in the Western Indian Ocean." *Modern Asian Studies* 54, no. 3 (2019): 898–929.

McCusker, John J. *Money and Exchange in Europe and America, 1600–1755*. University of North Carolina Press, 1978.

McDow, Thomas F. *Buying Time: Debt and Mobility in the Western Indian Ocean*. Ohio University Press, 2018.

McMahon, Elisabeth. *Slavery and Emancipation in East Africa: From Honor to Respectability*. Cambridge University Press, 2013.

Menon, V. P. *The Story of the Integration of the Indian States*. MacMillan, 1956.

Metcalf, Thomas. *Imperial Connections: India in the Indian Ocean, 1860–1920*. University of California Press, 2006.

Miller, Charles. *The Lunatic Express: An Entertainment in Imperialism*. Macmillan, 1972.

Miller, Peter. "Accounting as Social and Institutional Practice: An Introduction." In *Accounting as Social and Institutional Practice*, edited by Anthony Hopwood and Peter Miller, 1–39. Cambridge University Press, 1994.

Mikhail, Alan. *Nature and Empire in Ottoman Egypt: An Environmental History.* Cambridge University Press, 2011.

Minault, Gail. *The Khilafat Movement: Religious Symbolism and Political Mobilization in India.* Columbia University Press, 1982.

Morgenstein Fuerst, Ilyse R. *Indian Muslim Minorities and the 1857 Rebellion: Religion, Rebels and Jihad.* I. B. Tauris, 2017.

Morse, Robyn. "Crude Legalities: Law, Society, and Oil in Bahrain, 1920–1950." PhD diss., University of Virginia, 2024.

Morse, Robyn. "Pearling and the Language of Freedom: Navigating the Manumission System in the 1930s Persian Gulf." *Journal of Global Slavery* 7, no. 3 (2022): 317–46.

Muhammad, Khalid Salem. *Al-Nawkhithā ʿAbd Al-Majeed Al-Mulla, Min Nawākhithat Jazīrat Faylaka Al-Bārizīn.* Self-published, 1999.

Muir, Edward, and Guido Ruggiero, eds. *Microhistory and the Lost Peoples of Europe: Selections from Quaderni Storici.* Johns Hopkins University Press, 1991.

Nainar, Syed Muhammad Husayn. *Arab Geographers' Knowledge of Southern India.* Other Books, 2011.

Nicholls, C. S. *The Swahili Coast: Politics, Diplomacy and Trade on the East African Littoral, 1798–1856.* George Allen & Unwin, 1971.

North, Douglass. "Institutions, Transaction Costs, and the Rise of Merchant Empires." In *The Political Economy of Merchant Empires*, edited by James D. Tracy, 22–40. Cambridge University Press, 1991.

Ogunnaike, Oludamini. *Poetry in Praise of Prophetic Perfection: A Study of West African Arabic Madih Poetry and its Precedents.* Islamic Texts Society, 2020.

Onley, James. *The Arabian Frontier of the British Raj: Merchants, Rulers, and the British in the Nineteenth-Century Gulf.* Oxford University Press, 2007.

Onley, James. "Transnational Merchants in the Nineteenth-Century Gulf: The Case of the Safar Family." In *Transnational Connections and the Arab Gulf,* edited by Madawi Al-Rasheed, 59–89. Routledge, 2004.

Özbaran, Salih. *Ottoman Expansion Towards the Indian Ocean in the 16th Century.* Bilgi Üniversitesi, 2009.

Pannikar, K. M. *Asia and Western Dominance: A Survey of the Vasco da Gama Epoch of Asian History, 1498–1945.* George Allen & Unwin, 1953.

Parsa, Ali. *Bādbānhā-i Junūb: Daryānūr-i Bādbān-i dar Khalīj Fāris va Aqyānūs Hind.* Farhang Nashr Now, 2013.

Parthasarathi, Prasannan. *Why Europe Grew Rich and Asia Did Not: Global Economic Divergence 1600–1850.* Cambridge University Press, 2011.

Pearson, Michael N. "Littoral Society: The Case for the Coast." *The Great Circle* 7, no. 1 (1985): 1–8.

Perry, John. "The Zand Dynasty." In *The Cambridge History of Iran*, edited by P. Avery, G. R. G. Hambly, and C. Melville, 63–103. Cambridge University Press, 1991.

Peterson, J. E. "Britain and the Gulf: At the Periphery of Empire." In *The Persian Gulf in History*, edited by Lawrence Potter, 277–93. Palgrave Macmillan, 2009.

Pétriat, Philippe. "Caravan Trade in the Late Ottoman Empire: The *'Aqīl* Network and the Institutionalization of Overland Trade." *Journal of the Economic and Social History of the Orient* 63 (2020): 38–72.
Pétriat, Philippe. "The Uneven Age of Speed: Caravans, Technology, and Mobility in the Late Ottoman and Post-Ottoman Middle East." *International Journal of Middle East Studies* 53, no. 2 (2021): 273–90.
Polanyi, Karl. "Ports of Trade in Early Societies." *Journal of Economic History* 23, no. 1 (1963): 30–45.
Pomeranz, Kenneth. *The Great Divergence: China, Europe, and the Making of the Modern World*. Princeton University Press, 2000.
Poovey, Mary. *Genres of the Credit Economy: Mediating Value in Eighteenth- and Nineteenth-Century Britain*. University of Chicago Press, 2008.
Poovey, Mary. *A History of the Modern Fact: Problems of Knowledge in the Sciences of Wealth and Society*. University of Chicago Press, 1998.
Porter, Theodore. *Trust in Numbers: The Pursuit of Objectivity in Science and Public Life*. Princeton University Press, 1995.
Potter, Lawrence, ed. *The Persian Gulf in History*. Palgrave Macmillan, 2009.
Potts, Daniel. *The Arabian Gulf in Antiquity*. 2 vols. Oxford University Press, 1991.
Prakash, Gyan. *Mumbai Fables*. Princeton University Press, 2010.
Prange, Sebastian. "The Contested Sea: Regimes of Maritime Violence in the Pre-Modern Indian Ocean." *Journal of Early Modern History*, 17 (2003): 9–33.
Prange, Sebastian. *Monsoon Islam: Trade and Faith on the Medieval Malabar Coast*. Oxford University Press, 2018.
Prange, Sebastian. "The Mosque in a Land of Temples: Reading Malabar's Muslim Monuments." In *Malabar in the Indian Ocean: Cosmopolitanism in a Maritime Historical Region*, edited by Mahmood Kooria and M. N. Pearson, 338–54. Oxford University Press, 2018.
Prestholdt, Jeremy. *Domesticating the World: East African Consumerism and the Genealogies of Globalization*. University of California Press, 2008.
Prestholdt, Jeremy. "From Zanzibar to Beirut: Sayyida Salme bint Said and the Tensions of Cosmopolitanism." In *Global Muslims in the Age of Steam and Print*, edited by James L. Gelvin and Nile Green, 204–26. University of California Press, 2013.
Prins, A. H. J. *Sailing from Lamu: A Study of Maritime Culture in East Africa*. Van Gorcum, 1965.
Raj, Kapil. *Relocating Modern Science: Circulation and the Construction of Knowledge in South Asia and Europe, 1650–1900*. Palgrave Macmillan, 2007.
Rammohan, K. T. "Coir in India: History of Technology." In *Encyclopaedia of the History of Science, Technology, and Medicine in Non-Western Cultures*, edited by Helaine Selin, 596–600. Springer, 2008.
Ramusack, Barbara. *The Indian Princes and Their States*. Cambridge University Press, 2004.
Ray, Rajat Kanta. "Asian Capital in the Age of European Domination: The Rise of the Bazaar, 1800–1914." *Modern Asian Studies* 29, no. 3 (1995): 445–554.

Rediker, Marcus. *Between the Devil and the Deep Blue Sea: Merchant Seamen, Pirates, and the Anglo-American Maritime World, 1700–1750.* Cambridge University Press, 1989.

Reese, Scott. *Imperial Muslims: Islam, Community and Authority in the Indian Ocean, 1839–1937.* Edinburgh University Press, 2018.

Reidy, Michael S., and Helen M. Rozwadowski. "The Spaces in Between: Science, Ocean, Empire." *Isis* 105, no. 2 (2014): 338–51.

Ricciardo, Lorenzo. *The Voyage of the Mir El-Lah.* Collins, 1980.

Risso, Patricia. "India and the Gulf: Encounters from the Mid-Sixteenth to the Mid-Twentieth Centuries." In *The Persian Gulf in History*, edited by Lawrence Potter, 189–203. Palgrave, 2009.

Ritchie, G. S. *The Admiralty Chart: British Hydrography in the Nineteenth Century.* American Elsevier, 1967.

Roberts, Nicholas. "Oceanic Wahhabism." *Journal of World History* 36, no. 1 (2025): 21–49.

Roberts, Nicholas. "A Tolerated Terror: Rahmah bin Jabir and the Age of Revolutions in the Gulf, 1760–1830." *Itinerario* (2025): 1–15.

Robbins, Robert R. "The Legal Status of Aden Colony and the Aden Protectorate." *American Journal of International Law* 33, no. 4 (1939): 700–715.

Rodrigues, Louiza. "Commercialisation of Forests, Timber Extraction, and Deforestation of Malabar: Early Nineteenth Century." *Proceedings of the Indian History Congress* 73 (2012): 809–19.

Rosaldo, Renato. *Ilongot Head-Hunting, 1883–1974: A Study in Society and History.* Stanford University Press, 1980.

Rosenthal, Caitlin. *Accounting for Slavery: Masters and Management.* Harvard University Press, 2018.

Roy, Tirthankar. *A Business History of India: Enterprise and the Emergence of Capitalism from 1700.* Cambridge University Press, 2018.

Rudner, David West. *Caste and Capitalism in Colonial India: The Nattukottai Chettiars.* University of California Press, 1994.

Saleh, Sharif. "Ṭawāri' wa Mashākil fī 'Arḍ al-Baḥar: Shāhīn Al-Ghānim Akhiru man Tawallā Muhimmat 'Rā'ī al-Sālifa.'" *Annahar*, December 7, 2014. www.annaharkw.com/Article.aspx?id=505717&date=07122014.

Schofield, Katherine Butler. *Music and Musicians in Late Mughal India: Histories of the Ephemeral, 1748–1858.* Cambridge University Press, 2023.

Schotte, Margaret. *Sailing School: Navigational Science and Skill, 1500–1800.* Johns Hopkins University Press, 2019.

Segal, Eran. "Merchants' Networks in Kuwait: The Story of Yusuf Al-Marzuk." *Middle Eastern Studies* 45, no. 5 (2009): 709–19.

Shahvanaz, Shahbaz. *Britain and the Opening Up of South-West Persia, 1880–1914: A Study in Imperialism and Economic Dependence.* Routledge, 2005.

Shankar, Devika. "A Forest of Ships: Malabar's State Forests and Bombay's Dockyards, 1795–1822." *South Asia: A Journal of South Asian Studies* 46, no. 3 (2023): 682–96.

Sharafi, Mitra. "Bella's Case: Parsi Identity and the Law in Colonial Rangoon, Bombay and London, 1887–1925." PhD diss., Princeton University, 2006.

Sharidah, Mansour. "Merchants without Borders: Qusman Traders in the Arabian Gulf and Indian Ocean, c. 1850–1950." PhD diss., University of Arkansas, 2020.

Sheriff, Abdul. "Race and Class in the Politics of Zanzibar." *Africa Spectrum* 36, no. 3 (2001): 301–18.

Sheriff, Abdul. *Slaves, Spices, and Ivory in Zanzibar: Integration of an East African Commercial Empire into the World Economy, 1770–1873*. Ohio University Press, 1987.

Shihab, Hassan. *Al-Muʿjam al-Mufaṣṣal fī Muṣṭlaḥāt al-Malāḥa al-ʿArabiyya al-Qadīma wal-Ḥadītha fī al-Muḥīṭ al-Hindī*. Center for Research and Studies on Kuwait, 2010.

Siddiqui, Asiya. *Bombay's People, 1860–98: Insolvents in the City*. Oxford University Press, 2017.

Siegfried, Nikolaus. "Concepts of Paper Money in Islamic Legal Thought." *Arab Law Quarterly* 16, no. 4 (2001): 319–32.

Simpson, Edward. *Muslim Society and the Western Indian Ocean: The Seafarers of Kachchh*. Routledge, 2006.

Sklansky, Jeffrey. *Sovereign of the Market: The Money Question in Early America*. University of Chicago Press, 2017.

Soll, Jacob. *The Reckoning: Financial Accountability and the Rise and Fall of Nations*. Basic Books, 2014.

Sood, Gagan D. S. *India and the Islamic Heartland: An Eighteenth-Century World of Circulation and Exchange*. Cambridge University Press, 2016.

Sood, Gagan D. S. "Through a Persian Looking Glass: Malabar's World in the Middle of the Eighteenth Century." In *Malabar in the Indian Ocean: Cosmopolitanism in a Maritime Historical Region*, edited by Mahmood Kooria and Michael Pearson, 201–31. Oxford University Press, 2018.

Staples, Eric. "Navigating the Gulf: Aḥmad b. Mājid's Poem on Gulf Navigation." *Proceedings of the Seminar for Arabian Studies* 51 (2022): 371–78.

Steinberg, Philip. *The Social Construction of the Ocean*. Cambridge University Press, 2001.

Stephens, Julia. "The Phantom Wahhabi: Liberalism and the Muslim Fanatic in Mid-Victorian India." *Modern Asian Studies* 47, no. 1 (2013): 22–52.

Stephenson, Lindsey. "Rerouting the Persian Gulf: The Transnationalization of Iranian Migrant Networks." PhD diss., Princeton University, 2018.

Stern, Philip. *The Company State: Corporate Sovereignty and the Early Modern Foundations of the British Empire in India*. Oxford University Press, 2011.

Stilt, Kristen. *Islamic Law in Action: Authority, Discretion, and Everyday Experiences in Mamluk Egypt*. Oxford University Press, 2011.

Strunk, William T. "The Reign of Shaykh Khazʿal ibn Jabir and the Suppression of the Principality of ʿArabistān: A Study of British Imperialism in Southwestern Iran, 1897–1925." PhD diss., Indiana University, 1977.

Subrahmanyam, Sanjay. *The Career and Legend of Vasco da Gama*. Cambridge University Press, 1997.

Subrahmanyam, Sanjay, and Muzaffar Alam. *Indo-Persian Travelers in the Age of Discoveries, 1400–1800*. Cambridge University Press, 2007.

Subramanian, Lakshmi. *Indigenous Capital and European Expansion: Bombay, Surat, and the West Coast*. Oxford University Press, 1996.

Subramanian, Lakshmi. *The Sovereign and the Pirate: Ordering Maritime Subjects in India's Western Littoral*. Oxford University Press, 2016.

Subramanian, Lakshmi. *Three Merchants of Bombay: Business Pioneers of the Nineteenth Century*. Penguin, 2016.

Sunseri, Thaddeus. *Wielding the Ax: State Forestry and Social Conflict in Tanzania, 1820–2000*. Ohio University Press, 2009.

Suzuki, Hideaki. "Baluchi Experiences under Slavery and the Slave Trade of the Gulf of Oman and the Persian Gulf, 1921–1950." *Journal of the Middle East and Africa*, no. 4 (2013): 205–23.

Tagliacozzo, Eric. *In Asian Waters: Oceanic Worlds from Yemen to Yokohama*. Columbia University Press, 2023.

Tajirian, Sona. "The Early Modern Global Trade of Diamonds and Gems: An Armenian Family Firm on the Crossroads of Caravan and Maritime Trade (ca. 1670–1730)." PhD diss., University of California Los Angeles, 2020.

Tanaka, Stefan. "History without Chronology." *Public Culture* 28, no. 1 (2016): 161–86.

Thomaz, Luis. "Precedents and Parallels of the Portuguese Cartaz System." In *The Portuguese, Indian Ocean, and European Bridgeheads, 1500–1800: Festschrift in Honour of Prof. K. S. Mathew*, edited by Pius Malekandathil and Jamal Mohammed, 67–85. Institute for Research in Social Sciences and Humanities of MESHAR, 2001.

Thompson, E. P. "Time, Work-Discipline, and Industrial Capitalism." *Past and Present*, no. 38 (1967): 56–97.

Tomlinson, B. R. *The Economy of Modern India: From 1860 to the Twenty-First Century*. Cambridge University Press, 2013.

Toth, Anthony. "Tribes and Tribulations: Bedouin Losses in the Saudi and Iraqi Struggles over Kuwait's Frontiers, 1921–1943." *British Journal of Middle Eastern Studies* 32, no. 2 (2005): 150–55.

Toussia Cohen, Itamar. "Parsi Capital and Imperial Infrastructure: Shipping and Shopping in the Port of Aden, 1840–1888." *Journal of Global History* 19, no. 2 (2023): 260–80.

Tripp, Charles. *A History of Iraq*. Cambridge University Press, 2000.

Trivellato, Francesca. *The Familiarity of Strangers: The Sephardic Diaspora, Livorno, and Cross-Cultural Trade in the Early Modern Period*. Yale University Press, 2009.

Trivellato, Francesca. *The Promise and Peril of Credit: What a Forgotten Legend about Jews and Finance Tells Us about the Making of European Commercial Society*. Princeton University Press, 2019.

Trouillot, Alexis. "Mathematics in the Desert: Computing Texts and Intellectual Authority in 19th Century Sahara." PhD diss., Universite de Paris, 2023.

Ulaby, Laith. "On the Decks of Dhows: Musical Traditions of Oman and the Indian Ocean World." *World of Music* 1, no. 2 (2012): 43–62.
Varisco, Daniel. *Seasonal Knowledge and the Almanac Tradition in the Arab Gulf.* Palgrave Macmillan, 2022.
Vieira, Mónica Brito. "Mare Liberum vs. Mare Clausum: Grotius, Freitas, and Selden's Debate on Dominion over the Seas." *Journal of the History of Ideas* 64, no. 3 (2003): 361–77.
Webb, Adrian. "The Expansion of British Naval Hydrographic Administration, 1808–1829." PhD diss., University of Exeter, 2010.
Wick, Alexis. *The Red Sea: In Search of Lost Space.* University of California Press, 2016.
Wigen, Karen. *A Malleable Map: Geographies of Restoration in Central Japan, 1600–1912.* University of California Press, 2012.
Wilkinson, John C. *The Arabs and the Scramble for Africa.* Equinox, 2015.
Wilkinson, John C. *The Imamate Tradition of Oman.* Cambridge University Press, 1985.
Winichakul, Thongchai. *Siam Mapped: A History of the Geo-Body of a Nation.* University of Hawai'i Press, 1994.
Wint, Hollian. "Credible Relations: Indian Finance and East African Society in the Indian Ocean, c. 1860–1940." PhD diss., New York University, 2016.
Wint, Hollian. "From *Desh* to *Desh*: The Family Firm as Trans-Local Household in the Nineteenth-Century Western Indian Ocean." *Journal of World History* 34, no. 2 (2023): 187–216.
Wise, M. Norton, ed. *The Values of Precision.* Princeton University Press, 1997.
Yahaya, Nurfadzilah. *Fluid Jurisdictions: Colonial Law and Arabs in Southeast Asia.* Cornell University Press, 2020.
Yamey, Basil S. "Accounting and the Rise of Capitalism: Further Notes on a Theme by Sombart." *Journal of Accounting Research* 2, no. 2 (1964): 117–36.
Yergin, Daniel. *The Prize: The Epic Quest for Oil, Money, and Power.* Simon & Schuster, 1991.
Young, Gabriel. "Losing Ground: State, Land, and Oil in an Iraqi Periphery, 1920–1963." PhD diss., New York University, 2024.
Zaid, Omar Abdullah. "Accounting Systems and Recording Procedures in the Early Islamic State." *Accounting History Journal* 31, no. 2 (2004): 149–70.
Zaman, Muhammad Qasim. "Afterword: Reassessing Arabic in South Asia." *International Journal of Middle East Studies* 55, no. 1 (2023): 165–70.

INDEX

ʿAbdullah ʿAbdulrasul, Hassan and ʿAli, 292
Abdulqadir, Ahmad, 180
Abu Al-Khasib, town of, 61
Abu Dhabi, 14
Abyssinia, 118
accounting/bookkeeping, 266–67, 333n7; Al-Matrook's ledgers, 269–70, 271*fig.*, 272; capitalism and, 267–68; managerial power and, 268–69
Aden, 5, 53, 98, 204; in Bombay Presidency, 160–61, 164, 209; British efforts to suppress slave trade and, 204, 205; captured by East India Company, 110; Court of the Resident, 210–11; Kuwaiti merchants in, 11, 209; Portuguese attempt to capture, 101; strategic importance for British Empire, 208–9; Supreme Court, 211
Adulteration of Produce Decree, 233
Afrasiyab, 78
Africans, 24, 25, 207
agency contract (*wakāla*), 130
Ahmad, Kunji, 180
Al-ʿAbdulkarim, ʿAbdulmuhsin bin Nasser, 310n46
Al-ʿAbdulkarim, Fatima bint ʿIsa, 61
Al-ʿAbdulkarim, Lulwah bint ʿIsa, 310n46
Al-Antaki, ʿAbdulmasih, 227, 228
Al-ʿAsʿousi, ʿAbdullah bin ʿAbdulrahman, 151, 187, 321n44, 324n36
Al-ʿAsʿousi, ʿAbdulrahman bin Hussain, 33–34, 185–86
Al-ʿAsʿousi, ʿAbdulwahhab, 33–34
Al-ʿAsʿousi, Hussain bin ʿAbdulrahman, 203, 231, 232–33, 236, 237; *Al-Mohallab* dhow of, 300; transregional public sphere and, 238
Al-ʿAsʿousi, Hussain bin Rashid, 185, 187, 189
Al-ʿAsʿousi, Hussain Muhammad, 95
Al-ʿAsʿousi, Rashid bin ʿAli, 33, 63–65, 140, 185, 189–90
Al-ʿAsʿousi family, 297–98
Al-Bahar, ʿAbdulrahman, 293, 294
Al-Balushi, Hussain, 45
Al-Barʿi, ʿAbdulrahim, 243
Al-Bassam family, 157
Albuquerque, Afonso de, 101–2
Al-Busaʿidi, Sultan Faisal bin Turki, 112, 225, 301–2
Al-Busaʿidi, Saʿid bin Sultan, 82, 83, 87, 95, 110, 179, 224–25, 252
Al-Busaʿidi, Sultan bin Ahmad, 82
Al-Busaʿidi dynasty, 83, 84, 85, 177, 206, 234
Al-Dhakir, Muqbil bin ʿAbdulrahman, 173–74
Al-Dora, town of, 61, 62
Al-Doukhi, ʿAwadh, 245
Al-Duḥayyan, ʿAbdullah, 170–71, 173, 174, 322n4
Alexander the Great, 88
Alexandria (Egypt), city of, 125, 153

Al-Fadaghiya, date gardens of, 63
Al-Failakawi, ʿAbdulmajeed Al-Mulla Ahmad, 1, 11, 27, 116, 120, 283; Al-Ghanim as first employer of, 35, 36; Al-Khariji's logbook and, 15; in Bahrain, 247–49, 261, 264–65, 272; in Bombay, 150–51, 156, 167–68; in Calicut, 176, 188–89; contracts for transport of dates, 67; on converting units of measure, 191, 192*fig.*, 197; crew assembled by, 37–40; in East Africa, 230; family background, 33; final voyage (1945), 280, 286; as free agent, 35; in Goa, 189–90; in Gujarat, 140, 141–42; Malabar teak chest of, 1, 2, 15–16, 214, 220, 283; maps used by, 215*map*, 216–17, 219, 220–21; as microhistory protagonist, 7–9; money transfer templates from Al-Khariji, 169; in Muscat, 239; ordinariness of, 8, 303n11; as pilot of oil ship, 298; poems inscribed in notebook of, 241, 242, 243, 246; portrait of, 8*fig.*; *qawls* (safe-passage documents) and, 93–94, 97; relations with crew, 75; on return to Kuwait, 274–75; time spent with family and friends, 28; trail of documentation, 71–72; transregional public sphere and, 238; turnaround time for voyages, 290; on voyage home over open sea, 198–99, 200, 203–4, 212; writings of, 12, 17. See also *Crooked* [*Al-ʿAway* or *Al-Aʿwaj*]; *Fateḥ Al-Khair* [*Triumph of Good Fortune*]
Al-Failakawi, ʿAbdulrahman, 15–16, 65, 214, 306n33
Al-Failakawi, Mulla Ahmad, 301
Al-Failakawi, *ruznamahs* (logbooks) of, 1–3, 2*fig.*, 6, 14, 75, 213, 301; bought in Calicut, 176; debt contracts copied into, 41–42, 44, 46; first full-year voyage (1924), 18, 138, 304n25; pass from Kuwaiti emir, 93; on setting sail from Bombay, 168; on stopping to take water, 98
Al-Falah, Mufleh, 232, 233
Al-Falaq (newspaper), 235–39
Al-Faraj, ʿAbdullah, 245
Al-Faw, 47, 48, 54; Al-Sabah family and, 57–58; Al-Sager family and, 62; British telegraph station at, 66, 108n11
Al-Fayli, Redha, 293–96
Al-Fozan, ʿAbdullah, 151, 320n8
Al-Ghanim, Marzouq, 297
Al-Ghanim, Muhammad Shaheen, 33, 35, 40, 284
Al-Ghanim, Muhammad Thunayyan, 63, 140, 286, 297, 307n50; letter to Muhammad Al-Matrook, 145*fig.*, 146–48; as owner of Al-Failakawi's dhow, 242
Al-Ghanim, Shaheen, 284
Al-Ghanim, Thunayyan, 63
Al-Ghibba (deep gulf of water), 198, 199
Al-Ghunaim, Khalid, 131
Al-Hamad, Khaled ʿAbdullatif, 209–10, 296, 297
Al-Ḥarbi [the *Warship*] (dhow of Al-Failakawi), 280
Al-Hasa rulers, 31
Al-Hashemi, Faisal bin Hussain, 59
Al-Hijji, Yacoub, 3, 292, 316n3
Al-Hilāl (journal), 235, 238
Al-Hindi, ʿAbdullah bin Saḥāq, 69
Ali, Mehmet, 86
Al-Ibrahim, ʿAbdullah bin ʿIsa, 60–61
Al-Ibrahim, ʿAbdulrahman bin ʿAbdulaziz, 159
Al-Ibrahim, Jassem bin Muhammad, 155–56, 158–60, 165
Al-Ibrahim, Yousef, 61, 62, 311n53
Al-Ibrahim family, 60–62, 63, 310n46
Al-Iqbāl (newspaper), 238
Al-Jaber, Shaikh Ahmad, 287
Al-Jalahmah, Rahmah bin Jaber, 120
Al-Jalahmi family, 80
Al-Jibla (quarter of Kuwait), 27
Al-Kaʿbi, Jaber bin Mirdaw, 57, 309n30
Al-Kaʿbi, Muhammad Jaʿfar, 51
Al-Khaḍuri, Nasser bin ʿAli, 117
Al-Khalifa, Ahmad bin Muhammad, 251
Al-Khalifa, ʿIsa bin ʿAli, 252–53, 257, 260–61, 332n29
Al-Khalifa, Muhammad, 251–52
Al-Khalifa, Shaikh ʿAbdullah, 251
Al-Khalifa, Shaikh Hamad, 261, 263, 265, 332n29
Al-Khalifah tribe, 28, 80
Al-Khalil, Mansur bin Al-Hajj Ibrahim. *See* Al-Khariji, Mansur

Al-Khāliṣ Min Kulli ʿAyb fi Waḍʿ Al-Jayb [*The Perfect Pocketbook*] (Al-Qitami, 1924), 194
Al-Kharafi, ʿAbdulmuhsin, 51, 131, 132, 140, 163, 297, 318n40
Al-Kharafi, Ahmad, 35, 133
Al-Kharafi, Nasser bin ʿAbdulmuhsin, 51, 131, 132, 133, 140; Al-Sudairawi and, 163; Gujarati merchants and, 135
Al-Kharafi family, 297
Al-Khariji, ʿAbdulmuhsin bin Nasser, 282
Al-Khariji, Mansur, 14–16, 15*fig.*, 46, 70–71, 213; accounting formulas of, 266–67, 272–73; advice for nakhodas, 115–16; Al-Failakawi as student of, 16–17, 33, 66, 272; on Al-Faw under British attack, 58; on clash between British and Omani Arabs, 233, 239; contracts for transport of dates, 67–71, 72; on converting units of measure, 191, 193, 197; handwriting of, 41, 201, 218, 330n6; letter template for Eid greetings, 143, 149; money transfer templates of, 169; notebook transcription published (2007), 298; in old age, 289; poems inscribed in notebook of, 241, 242, 246, 330n6; Porbandar (Khormiyan) sketched by, 138, 139*fig.*; *qawls* (safe-passage documents) and, 93, 94, 96, 97; *ruznamah* (logbook) of, 68; transregional public sphere and, 238; writings of, 17
Al-Khiḍr (patron saint of mariners), 28, 306n22
Al-Khudair, Mishʿan, 293, 294
Al-Kuwaiti, ʿAbdullatif, 245
Alliance Bank of Simla, 158
Al-Mahri, Sulaiman, 241
almanacs, 16, 117, 203, 239
Al-Manar (journal), 174, 226
Al-Marzouq, Fahad, 131
Al-Marzouq, Jassim, 131
Al-Marzouq, Marzouq bin Muhammad, 128, 131
Al-Marzouq, Muhammad, 131
Al-Marzouq, Yousef, 131
Al-Maskari, Hashil, 237
Al-Matrook, ʿAbdullah, 298
Al-Matrook, ʿAbdullah bin Yousef, 177
Al-Matrook, Aslan, 131
Al-Matrook, Hajji Muhammad bin ʿAbdullah, 11, 50–54, 51*fig.*, 69, 177–78; account ledgers of, 269–70, 271*fig.*, 272; archive of, 68; death of, 298; Gujarati merchants and, 135; letter from Al-Ghanim to, 145*fig.*, 146–48; merchants' letters to, 151; partnership with Al-Kharafi, 131, 132, 282; Political Agent in Kuwait and, 136; wartime embargo on wheat exports from India and, 133
Al-Matrook family, 63
Al-Maymani, ʿAbdulkarim bin ʿAbdullatif, 141
Al-Mishari, ʿAbdulwahhab, 159
Al-Mishari, Muhammad, 165–67
Al-Mohallab (dhow of ʿAbdulrahman Al-ʿAsʿousi), 300
Al-Muʾayyad (newspaper), 238
Al-Muhaini, ʿAbdullah, 38
Al-Mukhtaṣar (Al-Qitami), 119
Al-Muqaṭṭam (newspaper), 238
Al-Muwaṭṭa (Malik ibn Anas), 173
Al-Najāḥ (newsletter in Zanzibar), 235, 329n43
Al-Najdi, ʿAli, 245
Al-Natīja Al-Kuwaytiyya (ʿAsfour and Al-ʿAsʿousi, 1933), 203
Al-Qawāʿid wal-Mīl wal-Natīja fī ʿIlm Al-Baḥar [*The Principles, Declinations, and Almanac in the Science of the Seas*] (Al-Khariji, logbook, 2006), 14, 15, 139*fig.*
Al-Qitami, ʿAbdulwahhab, 119–20, 198, 282, 323n5, 334n19
Al-Qitami, ʿIsa bin ʿAbdulwahhab, 116–21, 175, 282, 298, 316n8, 323n5; *chau* manual authored by, 194; navigation almanacs used by, 202; on route to Calicut, 175–76
Al-Rushaid, ʿAbdulaziz, 28–30, 309n33
Al-Sabah, Ahmad Al-Jaber, 29, 93
Al-Sabah, Jaber, 292, 309n30, 309n33, 317n11
Al-Sabah, Mubarak, 26, 29–30, 62, 162, 226; agents of, 225; conscription policies of, 295; date gardens of Al-Fadaghiya

Al-Sabah, Mubarak (*continued*)
owned by, 63; half brothers of, 62, 311n53; land estates owned by, 57–58, 309n31; in legal dispute over pearls, 166; relations with British and Ottomans, 57; as shareholder in Arab Steamers Limited, 159
Al-Sabah, Nasser bin Mubarak, 226, 327n11
Al-Sabah family, 28, 32, 61, 280, 282–83
Al-Sager, ʿAbdullah, 282
Al-Sager, ʿAbdullah bin Yousef, 95, 177, 323n5
Al-Sager, Bilal, 286
Al-Sager, Hamad bin ʿAbdullah, 67, 69, 70–71, 135, 140, 177; commission agents and, 129, 317n16; debt claims pursued in Aden courts, 210; merchant-led advisory council and, 281; Political Agent in Kuwait and, 136; residence in Calicut, 176–78
Al-Sager, Yousef, 176–82, 181*fig.*, 183, 184–85, 282
Al-Sager family, 62–63, 209
Al-Saraji, village of, 61, 310n45
Al-Saʿud, ʿAbdulaziz, 85
Al-Sbaghah, Jassim, 21
Al-Shamlan, Saif Marzouq, 295, 296
Al-Shaqsi, Saʿid bin Khamis, 106–9, 110, 315n29, 315–16n33
Al-Sharq (quarter of Kuwait), 27
Al-Shirwani, Ahmad Al-Ansari, 244
Al-Sudairawi, Muhammad Salem, 11, 128, 160–64, 161*fig.*, 185, 254; Al-Zawawi correspondence with, 225, 226; business partner in Calicut, 180; coffee merchants and, 188; newspapers and journals subscribed to, 237
Al-Sudairawi, Salem, 162
Al-ʿUqūd Al-Yāqūtiyya fī Jayd Al-Asʾila Al-Kuwaytiyya [*The Ruby Necklaces in the Excellence of the Kuwaiti Questions*] (Ibn Badran), 171
Al-ʿUthman, ʿIsa, 38
Al-Wasat (quarter of Kuwait), 27
Al-Yaʿrubi, Nasser bin Murshid, 105, 106, 107
Al-Zawawi, Yousef bin Ahmad, 224–29
ʿamarah (warehouses), 26, 33
Anglo-Persian Oil Company, 287
ʿAntar bin Yaqut, 39
Arab Association, 237
Arabia/Arabian Peninsula, 4, 83, 134, 153; caravan routes through, 25; detritus of history on coastline of, 9; Hajar Mountains, 222; immigrants in East Africa, 234–35; pan-Arab/pan-Islamist currents in, 235
Arabic language, 51, 69, 117, 141, 298; annotations on maps, 220; *chau* manuals in, 195; interpreters hired by British anti–slave-trade patrols and, 206–7; as lingua franca, 11; nautical terms in, 2; newspapers and journals in, 235–39, 281
Arab Steamers Limited, 159, 167
Armenians, 24, 153
Asanand, Haridass, 135–36

baggarah (type of smaller boat), 277
Baghdad, 29, 33, 245
baghlah (type of boat), 32, 47, 141, 198, 304n1, 306n33
Bahrain, 5, 15, 37, 52, 79, 118, 160; Bombay court in, 164; British Political Agent in, 256, 260, 263, 264; court proceedings between nakhodas and mariners, 288–89; Daly's reform of pearl trade, 256–61; Joint Court, 260, 263; marketplaces of, 247–48; navigation route to, 247; oil concessions in, 265, 287, 288; pearl market (*Sūq Al-Ṭawāwīsh*), 248–50, 249*fig.*, 253; pearl merchants and bankers in, 254–56; Shiʿa population, 257; strategic importance, 250; Trucial Coast and, 90; ʿUtub raids on, 80
balam (type of boat), 63
Balochistan, 38, 98, 243, 280
Bandar ʿAbbas, 68, 76*map*, 77, 83
Bandar Lingeh, 118
Bani Kaʿb, 79, 312n11
Bani Khaled tribe, 29–31
Bank of Bombay, 154, 158
banks, 13, 85, 158, 161, 162; Al-Sudairawi and, 161; bills of exchange and, 170; in Bombay, 128, 150, 152, 158, 162, 164; in Europe, 172; merchants' need for, 154;

pearl merchants and, 159; state and private, 155; Western, 170
Barros, João de, 102
barwa (debt accounting), 44–45, 46
Basra, 1, 4, 5, 153, 180; Al-ʿAshar neighborhood, 50; British Agent in, 96; under British occupation, 122; dates produced in, 49, 308n4; "Gilded Age" of, 60; imperial rivalries and, 55–56; as independent port town, 31; under Kingdom of Iraq, 281; maps of, 220; merchants' ties to Bombay, 156; Ottomans' loss of direct control over, 78–79; printers in, 10; prominence as port city, 47; steamships from India in, 52; strategic location of, 99
Bay of Bengal, 126, 316n8
bazaars, 11, 12, 13, 277, 303–4n15
Bedouin merchants, 277
Beirut, 238
bills of exchange, 170
bills of lading, 128, 135
Bin ʿAli Jinaʿ, Hussain, 298
Bin Ghaith, ʿAbdulkarim, 286
Bin Gurnah, Salim, 233
Bin ʿIsa, Hussain, 298
Bin Moosa, Saleh, 206–7
Bin Qais, ʿAzzan, 225
Bin Saʿid, Jumʿa, 302
Bin Salamah, Saʿid, 219
Bin Sultan, ʿAbdulrazzaq bin Salim, 165, 166
Bishara, Fatima, 298–99
Bishara, ʿIsa Yacoub, 3, 35, 38, 286, 291, 292
Bishara, Yousef, 34
blacksmiths, 24
Bombay, 18, 32, 50, 53, 68, 123*map*, 226, 276; Aden's commercial connections to, 209; Al-Sudairawi's connections in, 160–64, 161*fig.*; Arabic-language journals from, 242; Arab Lane, 156, 159, 164; Arsenal Castle, 150; banks in, 161, 162; bubonic plague in, 163; cotton trade and, 44, 154–55, 156–57; Gujarati merchants in, 151; HMV Studios, 245; information on markets in, 129; Karachi as twin city to, 124–25, 126; *kotiya* boats from, 47; Kuwaiti bankers in, 128, 134; legal institutions of, 164–67; mercantile elite of, 61; as metropolis, 150, 151; pearl market in, 158–59; printers in, 10; rise of, 153–54; as textile exporter, 138, 279
Bombay High Court, 164, 166, 167, 210–13
Bombay Presidency, 110, 137, 155, 161; Aden as district of, 160–61, 164, 209, 213; expansion of, 164. *See also* India, British-ruled
Boodai, Jassim, 129–30
boom (type of boat), 33, 47, 63, 67, 189, 304n1; in convoy, 198; as small light boat, 306n33
Braudel, Fernand, 121, 170
Briefe Instruction and Maner how to Keepe Bookes of Accompts, A (Mellis, 1588), 268
British Empire, 7, 202; colonial capitalist elite and, 160; emergence of, 9; impact on *nakhodas*' everyday life, 10; Kuwait and, 30; loss of American colonies, 179; maps and, 216; Pax Britannica, 92, 95, 251; piracy and, 86–92. *See also* East India Company, English; India, British-ruled
British India Steam Navigation Company (BISN), 52, 136, 159, 180
British Petroleum, 287
brokers, 11, 18, 129, 153, 176
Brussels Act (1890), 111
Buckingham, James Silk, 84
Bukhit bin Khamis, 39
Burton, Richard, 125
Bushehr, 1, 79, 156, 280

Cairo, 161, 173, 227; Arabic-language publications based in, 226, 235, 238; letters from Cairo Geniza, 147
Calcutta, 126, 154, 161, 177, 229, 244
Calicut, 5, 123*map*, 129, 160, 277; Kuwaiti traders in, 177; lumber yards around, 183; route to, 175–76; as source of shipbuilding materials, 179; timber trade in, 18, 181, 181*fig.*, 189; Yousef Al-Sager in, 176–82, 181*fig.*; Zamorins of, 178
Canning, Charles, 110
capitalism, 137, 155; bazaar as site of, 12, 304n15; bills of exchange in history of, 169–70; bookkeeping and, 267–68;

capitalism *(continued)*
European colonial, 11, 124, 125; history of banknotes, 172; math of commodification and, 193; steamships and, 12; translocal business practices and, 13
caravan routes, 25, 277
carpenters (*gallāfs*), 22, 24
cartels, 12
Chabahar, port of, 98
Charles II, king of England, 152
Chartered Bank of India, 158
chau manuals, 193–95, 197
Chevron oil company, 287
China, 77, 208
chip log (*bāṭili*), 201
Christianity, 101, 103
Civil War, American, 153
Cochin, 189
coffee, 186–88, 210, 248, 249, 276
coffeehouses, 28, 36*fig.*, 223, 239
coir rope, 183–84, 277
Colomb, Philip Howard, 204–8
colonialism, 229, 233, 238
commission agency, 129–30
commodification, 193–94, 196, 300
Comoros, 95, 111, 118
copra coconut, 233–34
coral reefs, 230
cotton boom, 153–54
Cowasji Dinshaw Adenwalla, firm of, 209
Cox, Sir Percy, 229
credit, 13, 23, 158, 162, 163, 166, 277; debt acknowledgment and, 43; expansion of, 44; local ties and, 40; long-term credit arrangements, 25; money transfers and, 170; provision of goods on credit, 36, 39; regional credit market, 44, 134; rice as credit, 42
Crooked [*Al-ʿAway* or *Al-Aʿwaj*] (dhow of Al-Failakawi), 18, 25, 32, 188, 280; beached at Ras Al-Khaimah, 296, 297*fig.*; in Bombay, 150; crew of, 37, 40, 204, 212, 245, 275, 334n31; dates as cargo, 66, 73; lumber as cargo, 189; origin of nickname, 21–22; prepared in shipyards, 33; repairs to, 168; return voyage to Kuwait, 274–75; routes taken by, 20*map*, 98, 113, 121, 123*map*, 136, 175, 222–23, 247. See also *Fateḥ Al-Khair* [*Triumph of Good Fortune*]
customs houses, 9, 66; British Political Agency in Kuwait and, 26; Daly reforms in Bahrain and, 258, 260; of Ibn Saud, 25; in Karachi, 122, 124; red tape of imperial power and, 97

Da Gama, Vasco, 100–101, 314n5 (ch. 4)
Dalīl Al-Muḥtār fī ʿIlm Al-Biḥār [*The Perplexed's Guide to the Science of the Seas*] (Al-Qitami, 1916), 116, 118–19, 194, 202, 316n3
Daly, Clive Kirkpatrick, 256–61, 272, 332n29
dates, 33, 43, 52, 54, 249, 261; date plantations of Mubarak Al-Sabah, 58; exports of, 59, 310n39; groves damaged by flooding, 53; markets for, 49, 59, 69, 70, 122, 128, 137, 209; plantations of Qasba, 63–64; prices for, 128, 151, 285; of Shatt Al-ʿArab, 47–49, 62, 67
debt contracts, 16, 43; claims pursued in courts, 210; model contracts (templates), 67–69, 72; uncertainties inscribed into, 69
De Indis De Jure Belli (Vittoria, 1557), 315n20
dhow crews, 37–40, 74*fig.*; division of labor among, 73–75; mutinous, 77–78; tasks onboard, 73
dhows, 11, 298–99, 304n1; dates as cargo, 64, 64*fig.*; formal names and nicknames of, 21; human capital and, 39; illicit activity associated with, 92; *nakhodas* overpowered by crew or passengers, 77–78; naming of, 37; outfitted with motors, 296; as pirate vessels, 82; prepared in shipyards, 22–23, 23*fig.*; provisioning of, 42; repair of, 168; as security for borrowing money, 34–35; *sinyār* (convoy) of, 198; slave trade and, 204, 205*fig.*, 206, 207, 302; types of, 47, 63; unloading of cargo after return home, 276
dhow trade, 7, 12, 31, 120, 230, 246, 259, 285; agricultural products, 27; decline of, 289; efflorescence of, 286; expansion of,

180; financial and legal obligations in, 284; growth of Kuwait and, 31–32; itinerary of, 141; oil economy and, 290; al-safar (oceangoing trade), 283; survival and persistence of, 12–13
dinkiya (type of boat), 140
dīri dates, 49, 128
Diu, island of, 120, 142
Djibouti, 111
Dryad, HMS, 204, 206, 207–8
Dubai, 37, 76*map*, 219, 236, 286–87, 292, 296
Dutch empire, 3, 7, 55, 122; Gulf rulers and, 91; Portuguese conflict with, 108. *See also* East India Company, Dutch (VOC)
Dwarkadhish Temple, 137

East Africa, 2, 3, 134, 153; detritus of history on coastline of, 9; East India Company protectorate in, 111; enslaved mariners from, 39; European empires in, 111; littorals of, 5; mangrove pole trade, 18, 46, 223, 229–33, 247, 276; maps of, 218, 220
Eastern Bank Ltd., 52, 255, 258, 308n12
East India Company, Dutch (VOC), 103, 108, 178. *See also* Dutch empire
East India Company, English, 32, 81, 91, 110, 153, 244; Bombay as headquarters of, 152; Hydrographic Office charts, 217, 218, 219, 221; military blows against Maratha Empire, 153; Portuguese conflict with, 105–6; raids on EIC vessels, 85; Zamorins and, 178. *See also* British Empire
Ebrahim, Muhammad, 165, 167
economic crises, 154
economic depression (1930s), 44, 285
Eden (East India Company brig), 82
Edward VII, king of England, 225
Egypt, 235, 236, 258; Arabic newspapers published in, 281; Bahrainis in, 257; Egyptian merchants in Bombay, 156; Napoleonic invasion of, 30, 56
Elijah's Mountain (*Al-Yāsh*), 239
English language, 51, 141, 220
Ethiopia, 210
Euphrates River, 47, 48, 55
European empires, 3, 4–5, 7, 9, 203

Failaka island, 1, 14, 27–28, 76*map*, 306nn18–19
"family firm," 60
Fateḥ Al-Khair [*Triumph of Good Fortune*], 18, 21, 67, 93, 242, 291–92, 301; brief second wind of, 296; moored at Kuwait Scientific Center, 292–93, 292*fig.*; as severed artifact, 300
first nature and second nature, 193–94
forts, 9, 10, 81, 92, 251, 252; founding of Kuwait and, 28; Portuguese, 101, 103, 106
France, 7, 110, 111, 112
freightage, 65, 67, 70, 71, 168, 266, 286, 290; fees for, 230; freightage contracts, 69, 70, 72; partnerships among nakhodas and, 266; regulatory regimes and, 287
Freitas, Seraphim de, 104, 105, 108, 109
Fulan bin Fulan, 41–42, 93
Fuller, Melville, 110, 112
furḍah (government wharf), 26

General Treaty (1820), 88–92, 95, 110
Germany, colony in East Africa, 111
Ghandi, Mahatma, 237
Goa (India), port of, 22, 118, 120, 123*map*, 189–90; boat repairs in, 190; lighthouse of, 175
Gregorian calendar, 6–7
Grotius, Hugo, 103, 104, 108, 110, 315n20
Gujarat, 18, 22, 63, 120, 136; information on markets in, 129; maps of, 220; Memon community in, 141; textile exports from, 153
Gujarati language, 11, 51, 141, 195
Gulf of Aden, 118
Gulf of Cambay, 142
Gulf of Kachchh, 137, 215*map*, 218, 219
Gulf Oil Company, 287
Gwadar, 113, 116n41

Hadhrami middlemen, 11
Hadramawt, 180
Haider Ali, king of Mysore, 178–79
ḥalāwi dates, 49, 50, 151

Hamza, Hajji ʿAbdullah, 218
Hanseatic League, 172
ḥarīm (protected areas), 108, 109, 316n33
Hassan bin ʿAli, 41
Hassan bin Rahmah, 86
Hassan bin Yaʿqub, 37–38
Hawalli (dhow), 34
Hengam Island, 118
Hijri calendar, 6
Hill Bros firm, 59
Hindi language, 69
Holmes, Frank, 265
Hormuz, port of, 77
Hormuz island, 83

Ibn ʿAbdulwahhab, Muhammad, 83, 84, 85
Ibn Badran Al-Dumi, ʿAbdulqadir, 170–73, 322n11
Ibn Majid, Shihab Al-Din Ahmad, 99, 100, 109, 110, 203, 214, 298; advice for nakhodas attributed to, 115–16; Al-Qitami compared with, 116, 117, 316n3; myth connected to Vasco da Gama, 100–101, 314n5 (Ch. 4); poetic verse of, 241
Ibn Saud, ʿAbdulaziz, 25, 27, 29, 84, 236, 277, 281
Ibn Saʿud, Muhammad, 84
Ibrahim Pasha, 86, 88
India, 2, 77, 279; detritus of history on coastline of, 9; littorals of, 5; market for dates in, 49, 59; Mughal Empire and successor states, 30; steamship trade and, 52
India, British-ruled, 10, 30, 110, 122, 124, 129, 208; Aden and administration of, 209; financial capital circuits and, 162; Indian Rebellion (1857), 228; Princely States in, 137–38; Punjab as principal agrarian province, 125–26; railroads in, 123*map*, 125; wartime embargo on wheat exports, 132–33. *See also* Bombay Presidency; East India Company, English
Indian Contract Act, 167
Indian Ocean world, 11, 13, 20*map*, 100, 152, 251, 300; banking infrastructure of, 163; British regulatory efforts in, 95; demand for luxury items in, 44; economic geography of, 10; economic violence in, 77; empires around, 4; European expansion in, 55; global imperial contests in, 9; Indian Ocean history, 4, 8, 295; intellectual labor behind, 17; maps of, 216, 220; Portuguese presence in, 101
Indian Specie Bank, 158
Indus Steam Flotilla, 125
Industrial Revolution, 12
Ingrizi (English) calendar, 6, 7
Iraq, 11, 25, 27, 63; Al-Hashemi as ruler of, 59; British invasion during First World War, 52; in Ottoman Empire, 31; Shiʿite shrines in, 84

jalbut (type of smaller boat), 277
Jews, 24, 153
jihad, right to traverse the sea and, 107–8, 113
Juju, x
Jumʿa bin Saleh (Al-Suri), 38
Junagarh, Princely State of, 140

Kachchh, 69, 71, 137
Ka-Ibrahim, Ahmad bin Hussain, 218
Ka-Ibrahim, Hajji Hassan, 1
"Kaki Island" (Velliyamkallu), 175–76
Kannada language (Karnataka, India), 11
Kanoo, Yousef bin Ahmad, 254, 255
Karachi, 19*map*, 50, 63, 68, 114, 123*map*, 138, 276; Al-Failakawi in, 73, 124–25, 126; bazaars of, 124, 127; Bombay as twin city to, 124–25, 126; in Bombay Presidency, 164; date market in, 5, 128, 129; as export-oriented colonial city, 124; Gujarati merchants in, 134; Kharadar neighborhood, 126–27; Kiamari Anchorage, 122; kitchenwares sold out of, 157; Manora Lighthouse, 122, 126; Napier Mole wharf, 122; rice exported from, 42, 127; as shipping hub, 10; society of brokers in, 129–30; wheat exported from, 127–28, 186
Karam, Mohammad Ibrahim, 218
Karwar, 123*map*, 175
Kathiawar Peninsula, 72, 120, 142
kaws (gale-force winds), 64, 65
Kharj (*Khārī*) island, 1, 14, 27, 33
kharjiah (loan), 249, 278
Khazʿal, Shaikh, 57, 78, 228, 280

Kitāb Al-Fawāʾid fī Uṣūl ʿIlm Al-Biḥār wa Al-Qawāʿid [*The Book of Useful Information on the Principles and Rules of the Science of the Seas*] (Ibn Majid), 99–100, 314n5 (Ch. 4)
Konkan coast, 150, 175
kotiya (type of boat), 47, 53, 68, 140, 151
Kundapura, 123*map*
Kut Al-Shaikh, village of, 60
Kuwait, 1, 5, 14, 39, 57; advantageous location of, 25, 305n10; British Political Agent in, 26, 96, 97, 136, 164, 165, 207–8; dhow voyages from port of, 21; diverse inhabitants of, 24–25; founding as fort, 28–29, 80; Ibn Saud's blockade on, 247, 277, 278; Iraqi invasion of (1990), 300; Majlis movement, 282–83; oil deposits in, 289–90; Ottoman customs administration and, 31; political history, 29–31; quarters of, 27; relations with the interior, 25–26; social life in, 24, 305n7; *sūqs* in, 305n5; ties to Bombay, 162, 164; trade with Najdi interior, 277, 278; Trucial Coast and, 90
Kuwait Oil Company, 287

Lakhmidas, Lalji, 134, 136
Lammasch, Heinrich, 109–10
Lamu, port of, 206, 219, 229, 276
landowners (*fellāḥs*), 53–54, 308n17
ledger books, 21, 39, 40
letters, 11, 50, 51, 52, 140–41, 188; Al-Sudairawi's correspondence, 160–61; formulaic structure of, 147–48; formulas for writing, 144–46; mercantile letter (*khaṭṭ*), 143, 147, 149; as technologies of coordination, 143–44; telegraph communication compared with, 148–49
Lingeh, 51, 68, 70, 87, 118, 160, 206; burned by Anglo-Omani forces, 86; erosion of autonomy of, 280
Loch, Francis Erskine, 82, 83, 86, 87, 88, 110
Lynch Bros firm, 60

Madagascar, 95, 111, 219
Maʿdan Al-Asrār fī ʿIlm Biḥār (Al-Khaḍuri), 117
Madras, 126, 154, 187
Majid bin Saʿid, Sultan, 95
Makran coast, 83, 99, 113, 175
Malabar, 23, 31, 90, 100, 118, 176; Al-Naibar (Malabar coast), 65; Muslim community in, 178; shipbuilding timber from forests of, 179, 184, 189, 193; spices in, 186
Malik ibn Anas, 173
mʿallim (navigator), 38, 74
manātikh (coastal landmarks), 119, 120
Mangalore, 65, 118, 123*map*, 160, 175; coffee beans exported from, 187–88; roofing tiles bound for East Africa from, 230, 231
mangrove poles, 46, 223, 229–33, 247, 276
manifests, 7, 50, 67, 69, 231
Mappila rebellion, 186
maps, 214–15; "blueback charts," 217; British Admiralty charts, 214, 216, 218; market for uncorrected maps, 217–18; Mercator projection, 215–16; *nakhodas*' inscriptions on, 219–20; territories rendered legible to empires by, 216
Maratha Empire, 153
Mare Clausum (Selden), 104
Mare Liberum [*The Free Sea*] (Grotius, 1609), 103–4
mariners/sailors, 5, 7, 18, 287–88, 335n35; advances on credit from *nakhodas*, 39; debts and expenses of, 22, 43, 45; duties under law, 284; families at homecoming of, 275, 278; *faṣl* (severance) pay for, 288; half-share sailor (*baḥḥār nuṣṣ glāṭa*), 74; mainsail maneuvered by, 74*fig.*; *tabbāb* (boy sailor), 73–74; temporary bonds with *nakhodas*, 45
Marzouq (*boom* dhow), 66
masāj (distance traveled by a ship), 200, 201
Maskin bin Fairuz, 38, 39
Mauritius, 111
Mayotte, 111
Mecca, 25, 86, 161, 228
Mecca, Sharif of, 227
Mellis, John, 268
merchants, 5, 11, 24, 166, 176; Arab, 11, 134, 135, 147; Armenian, 147; debt contracts of, 43, 46; financial capital circuits and,

merchants *(continued)*
161; "Gilded Age" of Basra and, 60; global fashion trends and, 158; Gujarati, 53, 134, 142, 151; Indian, 11, 44, 81, 110, 131, 156, 169, 234, 253; *nakhodas'* relationship with, 35–37, 36*fig.*; oil economy and, 297; Persian, 147, 178; wharfs (*niga*ʿ) owned by, 26, 33
microhistory, 7–9, 303n8
middlemen, 11, 54, 69, 129
Min Tārīkh Al-Kuwayt (Al-Shamlan, 1958), 294
Mistri, Jamshetji, 160
mjaddimi (first mate), 37, 38–39, 40, 73, 267
mjaddimi al-ṣgheer (assistant first mate), 74
Mohammerah, emirate of, 33, 51, 56, 160, 180, 280; Al-Sager family and, 62; annexed by Persian empire, 281
Mohideen, Mahmood bin, 188, 324n36
Mombasa, 106, 111, 216, 218, 234, 276
money transfers (*ḥawālas*, *ṣakks*, or *hundis*), 16, 169–74, 188, 209, 225–26
monsoon winds, 100, 123*map*, 146
mosques, 24, 26, 227; endowed by merchants, 281, 298; Muslim community in India and, 178
Mozambique, 83, 106
Mughal Empire, 30, 153
Muhammad ʿAbdullah Hasanali & Brothers, 209
Muḥayya, 243
Muscat, 44, 56, 208; division of Omani empire and, 110; Gujarati merchants in, 134, 135; maps of, 218; merchants' ties to Bombay, 156; Mutrah harbor, 223, 224*fig.*; Portuguese-Ottoman conflict and, 103; as Qawasim power base, 83; Sultan of, 86, 88, 120, 225, 236
Muscat Dhows case (1905), 230
Muslim bin Al-Hajjaj, 173
Mysore, Kingdom of, 178

Nabhanis dynasty, 105
Nafḥat Al-Yemen (anthology, Al-Shirwani, ed.), 243–44
nahhām (singer), 38, 39, 40, 245
Najdi mariners, 99, 157
nakhodas (dhow captains), 5, 7, 11, 33; accounting and ledger books of, 266–67; biographical dictionary of, 303n11; coffeehouses and, 224*fig.*; consumer preferences and, 50; Daly reforms in Bahrain and, 261; date harvests and, 50; debt contracts of, 43, 45; *faṣl* (severance) pay for mariners and, 288; handwriting of, 1–2; *jaʿdi* ("sitting") nakhodas, 35; letters written by, 140–41; as "lost people" of Indian Ocean history, 8; maps used by, 215, 218, 219; mathematics of measurement and, 196; navigation manuals by, 116, 117, 203, 239; oil economy and, 297–98; open-sea crossings of, 199, 200, 201; pearl trade and, 254, 260; persistence in age of industrial capitalism, 12; poetry as medium of expression for, 241; prayers onboard led by, 73; process of becoming a *nakhoda*, 16; protégés of, 16; *qawls* (safe-passage documents) and, 94–97; response to pearl trade reforms, 261–63; rights and responsibilities under law, 283; summers spent with families, 22; as thinkers, 17; weights and measures of marketplace and, 191
Napoleon I, 30
Nasrallah, Jassim bin ʿIsa bin, 242–43
National Bank of Kuwait, 297
navigation, 14, 16, 33, 34, 45, 72; coastal landmarks, 119; decline of the art of navigation, 214; stars and, 118; texts by nakhodas on, 116; treatises on, 100; wayfinding tools over open sea, 200–204, 325nn4–5
Netrani Island, 175, 219
*niga*ʿ (wharves owned by merchants), 26, 28, 33, 275
Norie, John William, 202, 203, 217
Norie's Nautical Tables, 202, 217
Nowruz calendar, 6

Oman, 27, 32, 38, 79, 111, 122, 222; Al-Busaʿidi dynasty and, 83; Al-Qitami's history of, 119, 121; Bahrain occupied by, 251; division of Omani empire, 110; Gujarati merchants in, 225; Imam

of, 120; Kingdom of Hormuz and, 105; maps of coasts, 218; Portuguese conflict with, 120; sultan of, 179, 206
Ottoman Empire, 3, 30, 55, 66, 120; Bani Khaled tribe and, 31; Basra as outlet to Indian Ocean world, 99; dismemberment after First World War, 228; Gulf rulers and, 91; imperial bureaucracy, 56; loss of direct control over Basra, 78–79; Portuguese conflict with, 103, 105; safe-passage documents and, 96, 314n4 (Inscription)
Owen, Captain William Fitzwilliam, 216, 218

Pacioli, Luca, 268
Pahlavi, Reza Shah, 280
Palat, Mohammad Saʿad, 218
pan-Arabism, 224, 235
pearl diving, 22, 45, 81, 248–50, 289; pearl divers' riot at Manama (1926), 263–64; as pillar of Kuwait's port economy, 29
pearl trade, 29, 59, 61–62, 90, 290; Bombay as largest market, 158–59; *chau* manuals, 193; Daly's reform of, 256–61; Japanese cultured pearl, 284–85; *Majlis Al-ʿUrfi* tribunal, 255; oil industry as competitor, 288; pearl market in Bahrain (*Sūq Al-Ṭawāwīsh*), 248–50, 249*fig.*, 253; pearl trade law of Kuwait (1939), 284; *Sālifat Al-Ghawṣ* tribunal, 255–56, 260, 263; weights and measures used in, 194, 195
peasants (*fellāḥs*), 53–54, 64
Pelly, Lewis, 252
Pemba, 219
Peninsular & Oriental Company (P&O), 52
Perim (Mayyun) Island, 212–13
Persia, 2, 5, 82, 113, 153, 160, 248, 276; Arab governors of coastal towns, 79; commodities from, 223, 247; date markets in, 48; merchants from, 60, 153; pearl trade and, 248, 250; political uncertainties of trade in, 55; Qajar dynasty, 56, 57, 122. *See also* Safavid Empire
Persian Gulf, 9, 32, 73, 76*map*; Indian Ocean world and, 4; pearl fishing in, 250; pearl trade around, 29, 158; sidelined in Middle Eastern history, 3; water depths in, 198
Persian language, 11, 51
pirates/piracy, 9, 18, 56, 77, 120; Alexander the Great and, 88; British encounters with, 86–87; British-sponsored peace treaty of 1820 and, 88–92, 95; politics and, 78, 81; Qawasim and, 82–88; raiding, 77–82; suppressed by British Empire, 95
Piri Reis (Ahmad Muhiddin Piri), 120
poetry, 241–43, 246; *madīḥ* poetry, 243; *Nafḥat Al-Yemen* anthology, 243–44; *nahhām* (singer) of dhow crew and, 245; oral recitation of, 244; Persian ʿAjami tradition, 243
Pondicherry, 111
Porbandar (Khormiyan), 69, 123*map*, 140, 219, 276; Al-Khariji's sketch of, 138, 139*fig.*; date market in, 141
"Port-Clearance" documents, 94
Portuguese empire, 3, 7, 12, 55, 101–2, 122, 178; Bahrain occupied by, 251; *cartaz* (safe-conduct pass) issued by, 95, 102–3; Dutch rivalry with, 103–4; Gulf rulers and, 91; Omani conflict with, 120; Ottoman conflict with, 103, 105
Privy Council, 211
Psyche (East India Company brig), 82
Purshottam, Ratansi, 134–35, 318n31

Qais Island, 75, 76*map*, 77, 98
Qajar dynasty, 56, 57, 122
Qānūn Al-Safar fi Al-Kuwayt (Kuwaiti law, 1940), 283–84
Qatar, 37, 118, 248; Bahrain *nakhodas*' relocation to, 262–63; Qatar peninsula, 28, 247
Qawasim tribal confederation, 82–88, 89; British and Omani military campaign against, 85–86, 94; Wahhabis and, 84
qawls (safe-passage documents), 93–94, 95–97
Qeshm Island, 83, 118, 240

Rahmah bin Jaber, 80, 87, 312n13
Rāʿi al-Sālifah (expert in customs of the trade), 284

railroads, 123*map*, 125–26, 127, 130, 138; timber merchants and, 182; Uganda Railway, 230
Rangoon, 154
Ras Al-Hadd, 20*map*, 127n1 (Ch. 9), 220, 222, 230
Ras Al-Khaimah, port of, 86, 88, 90, 99, 180
Red Sea, 31, 90, 118
reefs, 74, 100, 230, 247
Réunion, 111, 208
rice, 42, 127, 133, 135, 249, 261
Rida, Rashid, 174, 227, 228
Ross, Charles, 253
ruznamahs (logbooks), 1–3, 2*fig.*, 4, 5, 19, 137, 232, 233; poetry in, 14; temporal horizons plotted in, 6–7

Sabiliyat, village of, 60
Safavid Empire, 30, 31, 55, 251; decline of, 153; fall of, 82; Portuguese conflict with, 105–6; violence at frontiers of, 78
safe conduct passes, 16
Ṣafḥāt Min Tārīkh Al-Kuwayt [*Pages from Kuwait's History*] (television program), 293–96, 335n6
sailors. *See* mariners/sailors
salaf (loan), 249, 261–63
Salamah Islands, 98, 99
sambuk (type of boat), 47, 63
Saudi Arabia, 281
Saudi–Wahhabi alliance, 84
Savornin Lohman, Alexander de, 109–10
sāyer dates, 49, 50, 67, 128, 151
Sea of Oman, 98–99, 101–3, 109, 121
Selden, John, 104, 109
Seychelles, 154, 208
Shah, Nader, 79
shamāl winds, 291
Sharjah, 37
Shatt Al-'Arab, 18, 47–50, 55, 68, 122, 160; date plantations along, 47–49, 62, 67, 155; landholders in villages along, 60; political geography of, 56
Shi'a Muslims, 83, 84, 253, 260, 282, 332n29
shipbuilders/shipbuilding, 21, 81, 176, 179, 277; cuts of lumber for, 183; growing demand for timber for, 180–81
shipworms, 22
shipwrecks, 62, 177
shipwrights (*ustād*), 23, 182–83, 280, 286, 292
shopkeepers, 23, 305n6
shū'i (type of boat), 63
Singapore, 154, 161
slavery/slave trade, 4, 9, 18, 56, 302; bookkeeping and, 268; debt contracts and, 46; enslaved pearl divers, 248, 259; piracy and, 88, 89; sailors descended from enslaved people, 39–40, 65, 307n50; suppressed by British Empire, 95, 204–8, 205*fig.*, 216
Smith, Adam, 89, 322n11
smuggling, 25, 286–87
Socotra Island, 100
Somalia, 83, 118, 151, 189, 230
Sombart, Werner, 267–68
South Arabia, 3, 100, 118, 155, 162; littorals of, 5; maps of, 220
steamships, 7, 10, 129, 130, 154, 209, 229; industrial capitalism and, 12; in Karachi, 125, 126; letters carried by, 146; requisitioned for British war effort, 286
Straits of Hormuz, 32, 98, 99, 103, 239
Straits of Malacca, 100, 103
Suez Canal, 154
Sufism, 83
sugar, 24, 42, 45, 53, 128, 223, 248; in Bahrain marketplace, 247, 248; exported from Karachi, 186, 276; Javanese, 249; listed on account sheets, 36; provided to mariners on credit, 39; smuggling of, 286; sugar cane, 126; trans-shipments of, 128; varieties of, 52
sukūni (helmsman), 74
Sullivan, George, 206
Sultan bin Saif, 106, 107
Summa de Arithmetica (Pacioli, 1494), 268
Sunni Muslims, 83, 253, 298
supply chains, 53
Sur, nakhodas of, 111–13
Surat, port of, 81, 118, 142; Al-Ibrahim family in, 60–61; decline of, 153; EIC headquarters transferred to Bombay from, 152

Swahili coast, 96, 118, 189, 229, 291, 328n25
Swahili language, 207

tabbāb (boy sailor), 73–74
tamarinds, 186–87, 188, 276
Tanganyika, 111, 230
Tārīkh Al-Kuwayt (Al-Rushaid), 29, 30
Tawfiq, Muhammad, 236
taxes/taxation, 79, 80, 84, 193; avoidance of, 178; export duties, 283; maps and, 216; pearl trade and, 255, 258, 259, 332n29; tax exemptions, 63, 252, 255; tax farmers, 56, 83
Tayseer [*Good Fortune*] (dhow of Al-Failakawi), 280
tea, 22, 38, 150, 223, 249, 254, 277; in Bahrain marketplace, 247, 248, 250; in barter exchanges, 25; exported from Karachi, 186; provided to mariners on credit, 39; in ships' provisions, 24, 36, 65, 73
teak wood, for shipbuilding, 183
telegraph, 7, 10, 50, 52, 130, 145; British network of submarine cables, 308n11; merchants' telegraph addresses, 128–29; speed of communication and, 148–49
textiles, 153–54, 156, 248, 279
Thesiger, Wilfred, 14, 289
Thompson, Perronet, 89
Tigris River, 47, 48, 55
Tikamdas, Gangaram, 258
timber, 46, 182, 185–86; lumber and units of measure, 191, 192*fig.*, 193; shipbuilding and imperial rivalry, 179–80
Tipu Sultan, king of Mysore, 179
tisqām (pearl diving loan), 249
tobacco, 22, 126, 223, 261
Tootie, x
trade manuals, 16
trade routes, 6, 56, 306n18
trading companies, 5, 12
Treaty of Seeb (1920), 120
Treaty of Tordesillas (1494), 103
Trikam Pasa (Gujarati dhow), 69
Trucial Coast, 90, 160, 248
Tulsidas, Hirji, 135

United Arab Emirates, 296
United States, dates sold to, 49, 59, 310n39
Urdu language, 11, 242
ʿUtub tribe, 28, 30, 31, 85, 306n23; Al-Khalifa family and, 251; conflicts with neighboring communities, 79–80

Veravel, 65, 123*map*, 140, 141–42
Villiers, Alan, 35, 43, 101, 232, 245, 316n3; on dates of Shatt Al-ʿArab, 47; on decline of the art of navigation, 214–15; on Indian Ocean trade, 285–86; on Mutrah harbor, 223; on nakhodas' use of Admiralty charts, 218, 219
Vittoria, Francisco de, 315n20
voyaging (*al-safar*), 19, 45, 283

Wahhabis, 31, 84, 86, 87–88
Walji, Haridass, 151
Wamanga (Omani Arabs), 234
Washihiri (Hadhramis), 234
water depths, 176
weather, 6, 47, 53, 64
weight, units of, 191, 195
Welwod, William, 104
Western Ghats, 175, 178, 187
wheat, 27, 29, 127–28, 129, 148, 276; exported from Karachi, 124, 132, 186; grown in Punjab, 126; milled and unmilled, 127–28
women: families of *nakhodas*, 16; "family firms" and, 60–61; mariners' homecoming as time of reprovisioning, 278–79; *Souq Al-Ḥarīm* (Women's Market), 279, 279*fig.*
World War, First, 52, 58–59, 122, 132, 159, 167, 228, 239
World War, Second, 220, 236, 239, 280, 285; economic opportunities opened by, 287; steamships requisitioned for war effort, 286

Yaʿrubi dynasty, 83, 113
Yemen, 37, 38, 151, 189, 243
Yokohama Specie Bank, 158

zāhidi dates, 49, 50, 128
zakat alms, money transfers and, 171
Zamorins, 178
Zand, Karim Khan, 79

Zanzibar, 4, 83, 106, 111, 118, 208, 276; Al-Iṣlāḥ party, 235; British efforts to suppress slave trade and, 204, 207; British officials in, 237; decline of Arab identity in, 235–36; dhow agents in, 233, 328n33; division of Omani empire and, 110; hurricane in, 44; Indian merchants in, 234; maps of, 220; printers in, 10; Rufiji mangrove delta outside, 229, 231, 232; steamship routes to, 154; Sultan of, 95
Zubara, 31, 80, 85

THE CALIFORNIA WORLD HISTORY LIBRARY

Edited by Edmund Burke III, Kenneth Pomeranz, and Patricia Seed

1. *The Unending Frontier: Environmental History of the Early Modern World*, by John F. Richards
2. *Maps of Time: An Introduction to Big History*, by David Christian
3. *The Graves of Tarim: Genealogy and Mobility across the Indian Ocean*, by Engseng Ho
4. *Imperial Connections: India in the Indian Ocean Arena, 1860–1920*, by Thomas R. Metcalf
5. *Many Middle Passages: Forced Migration and the Making of the Modern World*, edited by Emma Christopher, Cassandra Pybus, and Marcus Rediker
6. *Domesticating the World: African Consumerism and the Genealogies of Globalization*, by Jeremy Prestholdt
7. *Servants of the Dynasty: Palace Women in World History*, edited by Anne Walthall
8. *Island World: A History of Hawai'i and the United States*, by Gary Y. Okihiro
9. *The Environment and World History*, edited by Edmund Burke III and Kenneth Pomeranz
10. *Pineapple Culture: A History of the Tropical and Temperate Zones*, by Gary Y. Okihiro
11. *The Pilgrim Art: Cultures of Porcelain in World History*, by Robert Finlay
12. *The Quest for the Lost Nation: Writing History in Germany and Japan in the American Century*, by Sebastian Conrad; translated by Alan Nothnagle
13. *The Eastern Mediterranean and the Making of Global Radicalism, 1860–1914*, by Ilham Khuri-Makdisi
14. *The Other West: Latin America from Invasion to Globalization*, by Marcello Carmagnani
15. *Mediterraneans: North Africa and Europe in an Age of Migration, c. 1800–1900*, by Julia A. Clancy-Smith
16. *History and the Testimony of Language*, by Christopher Ehret

17. *From the Indian Ocean to the Mediterranean: The Global Trade Networks of Armenian Merchants from New Julfa*, by Sebouh David Aslanian
18. *Berenike and the Ancient Maritime Spice Route*, by Steven E. Sidebotham
19. *The Haj to Utopia: The Ghadar Movement and Its Transnational Connections, 1905–1930*, by Maia Ramnath
20. *Sky Blue Stone: The Turquoise Trade in World History*, by Arash Khazeni
21. *Pirates, Merchants, Settlers, and Slaves: Colonial America and the Indo-Atlantic World*, by Kevin P. McDonald
22. *Black London: The Imperial Metropolis and Decolonization in the Twentieth Century*, by Marc Matera
23. *The New World History: A Field Guide for Teachers and Researchers*, edited by Ross E. Dunn, Laura J. Mitchell, and Kerry Ward
24. *Margins of the Market: Trafficking and Capitalism across the Arabian Sea*, by Johan Mathew
25. *A Global History of Gold Rushes*, edited by Benjamin Mountford and Stephen Tuffnell
26. *A Global History of Sexual Science, 1880–1960*, edited by Veronika Fuechtner, Douglas E. Haynes, and Ryan M. Jones
27. *Potosí: The Silver City That Changed the World*, by Kris Lane
28. *A Global History of Runaways*, edited by Marcus Rediker, Titas Chakraborty, and Matthias van Rossum
29. *The City and the Wilderness: Indo-Persian Encounters on the Burmese Frontier*, by Arash Khazeni
30. *The Bloody Flag: Mutiny in the Age of Atlantic Revolution*, by Niklas Frykman
31. *Empire of Convicts: Indian Penal Labor in Colonial Southeast Asia*, by Anand A. Yang
32. *Zanzibar Was a Country: Exile and Citizenship between East Africa and the Gulf*, by Nathaniel Mathews
33. *A Sea of Wealth: The Omani Empire and the Making of an Oceanic Marketplace*, by Nicholas P. Roberts
34. *Monsoon Voyagers: An Indian Ocean History*, by Fahad Ahmad Bishara

www.ingramcontent.com/pod-product-compliance
Lightning Source LLC
Chambersburg PA
CBHW021333230426
43666CB00006B/286
* 9 7 8 0 5 2 0 4 1 5 9 2 8 *